Modeling the Possible

Models are used to explore possibilities across all scientific fields. Climate models simulate the potential future climatic conditions under various emissions scenarios, macroeconomic models investigate the implications of different fiscal and monetary policy initiatives, and infectious disease models study the spread of viral diseases under a range of conditions. Such modeling approaches have not gone ignored by philosophers of science, but they have only recently started to explicitly address modeling the possible. So far, the discussion has been spread across a variety of more or less isolated pockets of debate in the philosophy of science. *Modeling the Possible: Perspectives from Philosophy of Science* draws together these studies, focusing specifically on how various modeling practices probe possibilities and justify claims concerning them.

The volume is divided into three sections, plus an introductory chapter. The introductory chapter provides a state-of-the-art survey of the discussions of modeling possibilities within the philosophy of science, as well as an introduction to the book's main themes and individual papers. The three sections focus on different kinds of possibility concepts, possibility spaces, and how-possibly modeling in practical situations. The chapters contained in this volume address conceptual and theoretical issues while also presenting case studies from various scientific domains: physics, evolutionary and synthetic biology, network science, climate science, economics, and epidemiology.

Essential reading for philosophers of science, epistemologists, and modelers in various scientific disciplines, *Modeling the Possible* is also suitable for anyone interested in model-based scientific inferences, their validity, and the policy conclusions derived from them.

Tarja Knuuttila is a Professor of Philosophy of Science at the University of Vienna, Austria. Knuuttila's research focuses on scientific modeling, as well as artifactuality and interdisciplinarity in science. Her philosophical work is practice-oriented, with an emphasis on synthetic biology, engineering sciences, and economics.

Till Grüne-Yanoff is a Professor of Philosophy at the Royal Institute of Technology, Stockholm, Sweden. His research focuses on scientific modeling, decision theory, and behavioral policy. He often collaborates with psychologists, economists, and engineers on these topics.

Rami Koskinen is a Researcher at the University of Vienna, Austria. Koskinen's research focuses on the general philosophy of science and the philosophy of biology. He has studied especially the epistemology of modal inferences, ranging from scientific modeling to biological engineering.

Ylwa Sjölin Wirling is a Reader in Theoretical Philosophy at the University of Gothenburg, Sweden, and a Pro Futura Scientia Fellow at the Swedish Collegium of Advanced Study. Her research concerns epistemology, philosophy of science, and metaphilosophy. She has worked on epistemological issues pertaining to modality, especially in the contexts of scientific modeling and philosophical inquiry.

Routledge Studies in the Philosophy of Science

The Internet and Philosophy of Science
Edited by Wenceslao J. Gonzalez

New Philosophical Perspectives on Scientific Progress
Edited by Yafeng Shan

Evidence Contestation
Dealing with Dissent in Knowledge Societies
*Edited by Karin Zachmann, Mariacarla Gadebusch Bondio,
Saana Jukola, and Olga Sparschuh*

Conjunctive Explanations
Nature, Epistemology, and Psychology of Explanatory Multiplicity
Edited by Jonah N. Schupbach and David H. Glass

The Aesthetics of Scientific Experiments
Edited by Milena Ivanova and Alice Murphy

Scientific Explanation, Causality, and Agency
A Free Energy Account
Majid D. Beni

Understanding and Conscious Experience
Philosophical and Scientific Perspectives
Edited by Andrei Ionuț Mărășoiu and Mircea Dumitru

Modeling the Possible
Perspectives from Philosophy of Science
*Edited by Tarja Knuuttila, Till Grüne-Yanoff, Rami Koskinen
and Ylwa Sjölin Wirling*

For more information about this series, please visit: https://www.routledge.com/Routledge-Studies-in-the-Philosophy-of-Science/book-series/POS

Modeling the Possible

Perspectives from Philosophy of Science

Edited by Tarja Knuuttila, Till Grüne-Yanoff, Rami Koskinen and Ylwa Sjölin Wirling

Routledge
Taylor & Francis Group
LONDON AND NEW YORK

First published 2025
by Routledge
4 Park Square, Milton Park, Abingdon, Oxon OX14 4RN

and by Routledge
605 Third Avenue, New York, NY 10158

Routledge is an imprint of the Taylor & Francis Group, an informa business

Funded by the European Research Council.

British Library Cataloguing-in-Publication Data
A catalogue record for this book is available from the British Library

ISBN: 978-1-032-37964-7 (hbk)
ISBN: 978-1-032-37968-5 (pbk)
ISBN: 978-1-003-34281-6 (ebk)

DOI: 10.4324/9781003342816

Typeset in Sabon
by KnowledgeWorks Global Ltd.

Contents

Acknowledgments

This project received funding from the European Research Council (ERC) under the European Union's Horizon 2020 research and innovation program (grant agreement no. 818772).

We would like to express our heartfelt gratitude to all of the authors of this collection for their expertise and care in writing the individual chapters. Additionally, we are grateful for their patience throughout the editorial process of the volume. It has been a delightful experience collaborating with each and every one of you.

We extend our immense gratitude to our exceptional helpers at the University of Vienna, Meghan Bohardt, Konstantin Eckl, and Alexander Gschwendtner, as well as Tony Bruce and Adam Johnson from Routledge, for their invaluable assistance during this process.

Tarja Knuuttila, Till Grüne-Yanoff, Rami Koskinen,
and Ylwa Sjölin Wirling
August 2024

Contributors

James DiFrisco, *The Francis Crick Institute, London, UK*

Gregor P. Greslehner, *University of Vienna, Austria*

Till Grüne-Yanoff, *KTH Royal Institute of Technology, Stockholm, Sweden*

Andreas Hüttemann, *University of Cologne, Germany*

Johannes Jaeger, *University of Vienna, Austria*

Tarja Knuuttila, *University of Vienna, Austria*

Andrea Loettgers, *University of Vienna, Austria*

Mary S. Morgan, *The London School of Economics and Political Science, UK*

Alfred Nordmann, *Technical University of Darmstadt, Germany*

Collin Rice, *Colorado State University, US*

Joe Roussos, *Institute for Futures Studies, Stockholm, Sweden*

Ylwa Sjölin Wirling, *University of Gothenburg, Sweden*

Mauricio Suárez, *Complutense University of Madrid, Spain*

Paul Teller, *University of California, Davis, US*

Philippe Verreault-Julien, *Eindhoven University of Technology, Netherlands*

Marcel Weber, *University of Geneva, Switzerland*

Introduction

Tarja Knuuttila, Till Grüne-Yanoff,
Ylwa Sjölin Wirling, and Rami Koskinen

0.1   Modeling the Possible in the Philosophy of Science

Models are used to investigate possibilities in all scientific domains. Climate models simulate potential future climatic conditions under various emissions scenarios. Macroeconomic models examine the possible effects of policy measures on different sectors of the economy. Infectious disease modeling tracks the spread of viral diseases under a range of assumptions about behavioral, medical, and population factors. Model organisms are used to explore the possible effects of new types of medications. Engineers and scientists create synthetic models to research and develop new types of materials and biological structures. It should be obvious that one of the primary objectives of contemporary modeling practice is the study of possibilities, yet the philosophy of science has only recently begun addressing modeling the possible. How can this be the case? At least three distinct philosophical currents have led to the neglect of the modal side of modeling: the dominant representational approach to models, the philosophical emphasis on explanation, and, in some cases, general skepticism about modalities.

The view of models as representations has long been dominant in mainstream philosophy of science. This viewpoint combines the ideas that models are representations and that they provide knowledge by representing certain real-world (social or natural) target systems. The unit of analysis has been that of one model and one target system, with the majority of the analytic work focused on accounting for the representational link between the two. As a result, various explanations of model-based representation have been proposed, including structuralist (e.g., Bueno and Colyvan 2011; Pincock 2012), similarity-based (e.g., Giere 2004; Weisberg 2013), and deflationary accounts (Suárez 2004). The specifics of these approaches do not need to concern us here. What appears crucial from a modal standpoint is what the target systems, according to representational accounts, are supposed to be. This question has received comparatively little attention, though, in the already vast amount of studies on modeling. It is as if it were taken for granted that the ultimate goal of modeling was to represent actual systems, and thus targets would not provide a very interesting subject for philosophical inquiry.

DOI: 10.4324/9781003342816-1

One of the exceptions is Weisberg (2013), who examines the construction of target systems in terms of subjecting an actual phenomenon to abstraction and idealization processes, picking just the features of the phenomenon that the theorist wishes to explore (see also Elliott-Graves 2020). While Weisberg also considers hypothetical and impossible targets, the emphasis is on target-directed modeling that addresses spatio-temporally identifiable actual target systems.

The focus on representation also permeates the discussions of idealization and abstraction: idealizations are expected to distort aspects of real-world target systems, while abstract representations omit some of their features (e.g., Thomson-Jones 2005, Godfrey-Smith 2009). Consequently, the conventional ways of thinking about idealization and abstraction begin with actual real-world systems, and often do not extend to non-actualized possible systems. How is one supposed to think of omitting some features of non-actualized objects or processes, or to distort non-actual systems? More often than not, this question is not even raised—it is rather the distortion of the features of actual systems that are thought to deliver modal information. Batterman and Rice (2014), for example, have argued that the renormalization group strategy renders models caricatures of actual systems that can nevertheless supply an explanatory modal structure.

The philosophical emphasis on explanation has also tended to downplay the modal significance of modeling. In the explanation literature, the explanandum has typically been taken as an actual phenomenon or event to be explained, while the explanans, depending on the account in question, either deduces it from the relevant natural laws and initial conditions (Hempel and Oppenheim 1948) or specifies a causal process or mechanism that lead up to, or produce, the explanandum. Moreover, a successful explanation is ordinarily assumed to be truthful and as complete as feasible (Bokulich 2017). Model-based explanations, in contrast, do not usually qualify as true and ideally complete explanations. As a result, while the causal-mechanistic tradition takes into account how-possibly models (Craver 2006, 2007), they are primarily seen as intermediary stages toward how-actually explanations, instead of considering that how-possibly models might explore non-actualized possibilities (Koskinen 2017).

In addition to focusing on describing and explaining actual processes, entities, and events, or presuming that traditional accounts can account for the modal side of modeling, philosophers of science have occasionally expressed skepticism about modality. For example, Norton (2021) finds the present literature on modality unsatisfying when it comes to understanding events that we cannot experience or that may never occur. While claims about such events abound in science, Norton thinks that speaking of them as possibilities is only defensible if they are allowed by the evidence. He suggests that philosophers should let go of possible worlds, though he also acknowledges that there can be factual non-actualized possibilities. Norton is certainly not alone in his suspicion of objective modalities; within analytic philosophy, for instance, Quine was famous for being skeptical of the intelligibility of modal speech and modal logic—especially ones that quantify over merely possible entities or properties (e.g., Quine 1953).

During the previous decade, the situation has shifted considerably. Several philosophers of science have invoked modalities, particularly in their work on modeling. To begin, idealizations have been seen as providing modal information rather than isolating actual difference makers or causal elements (or making the model tractable). de Donato Rodríguez and Arroyo Santos (2012) and de Donato Rodríguez and Zamorra Bonilla (2009) analyze idealizations as counterfactual, subjunctive, or hypothetical conditionals, where the antecedent expresses ideal conditions in which the idealization holds, and that can exhibit different degrees of contingency. Rice (2018, 2021) argues in his account of holistic idealization that idealizations are ubiquitous misrepresentations of the features, processes, and entities of the model's target system. They allow scientists to obtain modal information on which features of a system are relevant and which in turn irrelevant for explaining some widespread patterns of behavior. More generally, Sjölin Wirling and Grüne-Yanoff (forthcoming) argue that models addressing counterfactual objective possibilities might not be undermined by idealization in the same way as models that target actual systems. In a similar vein, Tan (2019) has approached idealizations in scientific models as counterpossible counterfactuals. Ruyant (2021) presents an anti-realist interpretation of science that is modal in character, based not on laws but rather on manipulations and observations within certain realms of experience, subject to natural restrictions. According to him, scientific models provide norms for their use in specific contexts and seek empirical success in all relevant and accessible scenarios.

While models have traditionally been studied from a representational perspective, there are alternative approaches to modeling that more readily account for the modal uses of modeling. Morrison and Morgan's perspective on models as mediating instruments views models as tools that are, by their construction, partially independent of theory and data and, hence, able to mediate between them (Morrison and Morgan 1999). Morrison and Morgan's primary focus is on how scientists learn by constructing and manipulating models. Their account, which emphasizes the epistemic enabling of the world in a model (Morgan 2012) and the partial independence of models from any target systems, is similar to indirect representation as portrayed by Weisberg (2007) and Godfrey-Smith (2006). However, Weisberg and Godfrey-Smith regard models as abstract or fictional entities, and their approaches are more reliant on representation than Morrison and Morgan's more practice-oriented account, which underscores how scientists gain understanding by working with models.

Building on Morrison and Morgan's insight on learning from models as well as philosophical discussions on artifacts and multimodality, Knuuttila has argued for viewing models as epistemic artifacts (e.g., Knuuttila 2011, 2021). According to her, the artifactual approach can better accommodate the modal dimension of modeling than the representational approach, which focuses on a relationship between a model and some determinable target. The philosophical gist of the artifactual approach is to focus on the theoretical or empirical question or task that the model, as a specifically constructed artifact, is designed to address. The questions posed to the model do not need to concern actual phenomena, entities, or processes but can also address possible and impossible ones. Scientific models are rendered by symbolic, semiotic, computational and material resources, and constrained in such a way that they facilitate solving some question at hand. Typically, such questions concern more generic phenomena. Knuuttila, like Le Bihan (2017), argues that models bring modal understanding by facilitating the study of different dependency patterns that might generate various types of phenomena. Such phenomena can be actual or non-actual, robust, necessary, or impossible.

The explorative nature of models has been articulated by Gelfert (2016), whose discussion of models as proof-of-principle demonstrations does not just address actual target systems. Massimi (2019) discusses further functions of exploratory modeling, arguing among other things that perspectival hypothetical models can give scientists knowledge of objective possibilities by way of what she calls "physical conceivability", i.e., a form of imagining possibilities that are constrained or "driven" by physical laws. The notion of conceivability has long been seen as crucial for modal reasoning by epistemologists of modality. Drawing from modal epistemology, Sjölin Wirling (2022) suggests that scientific models can function as "epistemic counterparts" that enable us to learn what is *de re* possible for an actual entity—by investigating what happens in a model that is relevantly similar to the entity in question.

Other modal notions besides possibility and necessity that are especially relevant to science include classic modal notions like contingency. For example, debates in evolutionary biology are often couched in terms of contingency (e.g., whether evolutionary history would come out the same or radically different if we "replayed the tape of life") rather than directly in terms of possibilities (e.g., Gould 1989). In scientific modeling practices, there are also more technical concepts that are not explicitly modal, but whose underpinnings can be identified to be modal. Evolvability (see Weber, this volume) provides one example of a notion that contains an implicit reference to modality (i.e., a disposition to change or evolve). The closely related notion of biological robustness has also been argued to be best understood as an expanded modal notion (Koskinen 2023). While the notion itself does not contain any explicit or implicit reference to classical modal notions, biological robustness can be given a modal reading similar to that of evolvability that essentially relies on quantification over possible states of affairs (Wagner 2014). Alternativity (see Morgan, this volume) is another concept that often comes with a strong modal coupling, while probabilistic notions, like chance (Suárez, this volume) and randomness, provide another very important family of modal-adjacent concepts for philosophical scrutiny.

Given the emerging discussion of modalities in modeling within the philosophy of science, it is still difficult to provide even a patchy map, let alone a thorough description of the different ways in which possibilities, necessities, and other related modal notions have been addressed within the philosophy of science. In the rest of this introduction, we will review the proposals and analyses provided by the contributions of this volume, according to the kinds of possibility concepts they put forth (section 0.2), the various ways they discuss possibility spaces (section 0.3), and how they address how-possibly models, or plausible and prospective scenarios (section 0.4). Section 0.5 broadens the scope to include broader philosophical implications of modeling the possible, followed by brief descriptions of each article in its own terms.

0.2 Possibility Concepts for Scientific Use

Upon closer inspection, the concept of possibility used in science and, in particular, in scientific modeling fractures into many different notions. Perhaps the most widely noted distinction is that between epistemic and objective possibility. According to standard definitions of epistemic possibility, p is epistemically possible when p is *compatible with* (Vetter 2015, 215) or *not ruled out by* (Chalmers 2011, 61) the relevant body of knowledge, or when *not-p is not part of that body of knowledge* (Weatherson and Egan 2011). In contrast, objective possibilities express something about the world—not only how it is, but how it could be. This includes counterfactual possibilities that differ from what is true in the actual world, acknowledging that the world could have been different from how it actually is, and that there is more than one way the world can be in the future, even if there is just one way it will be (Deutsch 1990; Williamson 2016).

Both possibility notions are extensively used by scientists, even if the distinction between epistemic and objective possibility is not always made explicit. Nevertheless, it is relevant for the analysis and justification of many scientific practices, including various modeling activities. Sjölin Wirling and Grüne-Yanoff (forthcoming) argue that heeding the distinction helps, for example, determining when how-possibly explanations are *sui generis*, which models can be used to reject necessity claims, or whether constraint-based reasoning can provide explanations. In their contribution to this volume, Sjölin Wirling and Grüne-Yanoff highlight that when seeking to justify possibility claims based on models, one must first determine the kind of possibility claim one seeks to justify, as different strategies may be called for in cases concerning different modalities.

Joe Roussos' contribution in this volume also draws on the distinction between epistemic and objective possibility when discussing climate modeling and conclusions about possibilities therefrom. He notes that scientists with a "possibilist" view seem to interpret both the individual models in the CMIP (Coupled Model Intercomparison Project) model ensembles, and climate storylines, as epistemic possibilities. In contrast, some philosophers defending such a possibilist view (e.g., Katzav 2014) appear to base their defense on an objective possibility notion. It seems that there is plenty more work to be done in this domain in order to resolve such apparent disagreements.

When seen from a vantage point of scientific practice, the epistemic/objective possibility distinction seems to yield an asymmetry: while objective possibilities often are seen as useful, epistemic possibilities are not—they do not yield *sui generis* how-possibly explanations, do not support understanding, or are not "real possibilities" in some relevant sense. Indeed, skeptics like Norton (2021) suggest disregarding such possibilities altogether: they are "dependent on our beliefs and knowledge and thus fail as a conception of possibility in the world" (Norton 2021, 22) and "by deriving possibility from ignorance, [they] allow the assertion of possibilities where prudence would dictate silence" (Norton 2021, 23). Both of these criticisms derive from the standard accounts of epistemic possibility, which define it as consistency with current knowledge. This makes epistemic possibility obviously dependent on knowledge, and it also opens it to the charge of *justification from ignorance*: one can be perfectly justified in claiming that p is epistemically possible even though scientific knowledge relevant to p is very scarce—it is just a matter of judging the relation between p and a body of knowledge, whatever it contains. In this volume, Roussos and Verreault-Julien address this issue.

Interestingly, according to Verreault-Julien, it might be possible to retain a concept of epistemic possibility that is knowledge- or belief-dependent, yet avoiding the justification of ignorance charge. Such accounts have been proposed in the literature, suggesting that p is epistemically possible just in case the truth of p is in some sense *supported by* a given body of epistemically prioritized (e.g., known, believed, justified) propositions (compare Przyjemski 2017). If suggestions like these are accepted, then the epistemic possibility concept might fracture further into a number of alternatives.

Such a subdivision is already in common use for objective possibilities. Metaphysical, logical, nomological, and contingent possibility are just some examples of such differentiations of objective possibility often found in the literature; nomological possibility is often further differentiated along disciplinary lines, including physical, biological, and economic possibility. It is often an open question just how these notions are best understood in a way that is relevant to scientific practice. For instance, in this volume, both Marcel Weber, and Tarja Knuuttila and Andrea Loettgers, discuss how we are to understand biological possibility, given that there are few if any biological laws. Moreover, these differentiations are often also relevant for understanding science. For example, a long-standing controversy about the relevance of the results of theoretical modeling in economics can be explained as a disagreement about whether certain economic models allow concluding mathematical or economic possibilities (Grüne-Yanoff and Verreault-Julien 2021).

As noted above, there are plenty of skeptics regarding objective possibilities and their relevance to science, especially in the strongly empiricist camp. In order to assuage at least some of the worries raised by empiricist criticism, friends of possibility modeling should consider the following two questions: first, can all scientifically relevant possibility claims be extracted out of the actual, and if so, in what way? And second, is there a way to characterize (and perchance constrain) the set of scientifically relevant possibility claims? Both of these questions are taken up in several contributions to this volume.

Mauricio Suárez argues in his contribution that it is propensities that ground objective possibilities. He considers propensities as abstract place holders for the concrete properties of complex dynamical systems that generate objective possibilities (and chances), at whatever level of description. The objective possibilities of interest exist in virtue of very special dynamical processes acting in highly constrained and specific scenarios. Because the modalities here are relative to specific setups, the resulting kind of possibility is more strictly and narrowly defined than mere logical, metaphysical, or even physical possibility—yet because they are grounded in the concrete properties of actual systems, a theoretically transparent connection is established between the actual and the possible, thus allowing the extraction of the latter out of the former.

Suárez describes the resulting concept as "local and actual possibilities" (15), by which he refers to those objective possibilities that are relevant to scientists. Similar intuitions have been voiced in climate science, where authors like Katzav promote a notion of "real possibility" (see Roussos' chapter). Paul Teller in his contribution seeks to make this intuition explicit. For him, an actual possibility is a state of the world that is not ruled out by what has occurred in the actual world up until a certain time. He contrasts this notion with the possible worlds approach to modality, dominant until recently in the philosophical literature (see Vetter 2011 for an overview of the changing landscape). Teller argues that it is actual possibility, not possibility in the sense that there is a possible world in which the relevant claim is true, which is of relevance to scientists. Interestingly, he also contrasts actual possibility with both metaphysical and nomological possibility concepts.

Collin Rice seeks in his chapter to counter such identification or constraining of possibility concepts relevant for scientists. Specifically, he examines three ways of determining which possibilities contribute to scientific understanding: only those sufficiently close to the actual world, only those relevant to intervention, or only those of interest to scientists. He finds each of them wanting, promoting instead his all-possibilities-count view: using models (or theories, or anything else) to learn about what would occur in a possible system always deepens our scientific understanding of real phenomena. Consequently, while there might be many different kinds of possibilities, none can be safely excluded from being relevant for science.

0.3 Possibility Spaces

Possibility spaces offer one way of conceptualizing and exploring possibilities. They can be logical, mathematical (e.g., probability spaces), scientific (e.g., phase and state spaces), or practical (e.g., design spaces in engineering).

The space of logical possibilities provides the most general space of the possible, including all those possibilities that are logically consistent. Such spaces have been offered as semantics for modal concepts. The possibility spaces relevant to science are more restricted, being limited, for example, by scientific laws or the history of a system. Possibilities, both actualized and unactualized, are contained within a possibility space that can be thought of as the collection of all possibilities that a system can realize based on our current scientific understanding and assumptions about its operation. Possibility spaces in science are typically constructed and studied using formal methods, the creation of possibility spaces being an essential part of modeling practice. The dimensions of the possibility space are determined by the dependencies, variables, and parameters specified in the construction of a model system. Scientific possibility spaces can be large and multidimensional, especially when simulating a complex system like an organism, an environment, or a network of social connections. Scientists are frequently only able to explore portions of such vast spaces.

Given that the building and analysis of possibility spaces in science are inextricably linked to modeling practices, what modal status should such possibility spaces have? Philosophers' answers vary. On one end of the spectrum are philosophers like Williamson (2018), who regard phase spaces in physics as "abstract mathematical representations of the possible states of a physical system" (189) that permit the "exploration of objective modal space" (197). On the other end, Wilson thinks that much of the discussion about possible worlds and possibility spaces is inflated and that the emphasis should be on localized possibility spaces that are more akin to control and design spaces in engineering that facilitate "locating the proper solution to a particularized problem" (Wilson 2018, 296).

Another issue is that different scientific fields have diverse approaches to articulating and grounding possibility spaces, both theoretically and experimentally. Physicists can design possibility spaces using laws, principles, and well-established formal methods. The phase and state spaces investigated in physics are the spaces of all conceivable states of a physical system. Other fields, like biology, lack comprehensive laws and principles for anchoring possibility spaces and giving the possibilities under consideration a more objective status. The contributions of the volume addressing possibility spaces approach them from the perspectives of different disciplines, and distinct modeling methods. While Andreas Hüttemann considers physical spaces, grounding his account of objective possibilities on physical laws and invariances, the next two contributions discuss biological possibilities. Marcel Weber addresses evolvability in evolutionary biology as an accessibility in a modal genotype-phenotype space, while Tarja Knuuttila and Andrea Loettgers consider how the combinatorial biological possibility spaces are structured. James DiFrisco, Johannes Jaeger and Andrea Loettgers extend their discussion also to social systems, highlighting the challenges of studying complex, self-organizing systems, whose possibility spaces are difficult to capture by modeling.

Hüttemann's approach to possibility spaces comes closest to that of Williamson (2016, 2018). Both of them take nomic necessity or possibility to be a species of objective possibility. Though there also are other kinds of objective possibilities, the authors seem to agree that the objective possibilities relevant to science are likely grounded in laws of nature. Hüttemann links laws to invariances in nature, arguing that the very notion of law presupposes that something changes while something else, expressed by the law (equation), remains invariant. Invariances and the laws based on them should be understood modally; they do not concern only the actual behavior of systems, but also their possible behaviors, and behaviors that are obtained by some sort of necessity. Coming closer to modeling, Hüttemann distinguishes between system-external and system-internal generalizations. The latter constrain the behavior of a system, establishing a possibility space. Hüttemann seeks to show that state spaces studied by models can be understood modally, and considered as objective if grounded on invariances that can be established empirically.

Most modeling taking place in other sciences than physics does not trade in scientific laws. How then to understand their modal character? Within biological sciences, biological possibilities have been grounded on the genomic level. The materiality of genes and proteins and scientists' ability to sequence, analyze, and manipulate genomes computationally and experimentally have provided a basis for thinking about biological possibility as accessibility within a space of all possible genomes. Within philosophical discussion and evolutionary biology, Dennett (1995) and Wagner (e.g., Wagner 2014) have referred to the "Library of Mendel" as such a space of biological possibilities. It is important to note that within this approach, biological possibility is always relative to a given genome and the space of possibilities is hyper-astronomical in size, consisting mostly of nonviable genomes. Moreover, the very idea of defining a complete space of biological possibilities has been criticized, for instance, by Maclaurin and Sterelny (2008), though this does not preclude possibility spaces, like morphospaces in evolutionary biology that specify an appropriate range of possibilities, given specific explanatory objectives (Weber this volume). The idea of a single universal DNA-based library of possibilities has also been challenged on empirical grounds by recent synthetic biology work on alternative genetic systems (e.g., Hoshika et al. 2019; for philosophical discussion, see Koskinen 2019). For example, recent results with genetic systems that incorporate unnatural bases to form both 6 and 8-letter genetic alphabets suggest that it might be possible to achieve biological functionality with chemical systems that diverge considerably from the properties of natural DNA and RNA.

Dennett's and Wagner's formulation of biological possibilities in terms of genotype spaces serves as a starting point for the contributions of both Weber, and Knuuttila and Loettgers in this volume. Weber argues that evolvability is essentially a *modal* concept that can be studied in terms of accessibility in a genotype-phenotype map space. The *genotype-phenotype* or *GP-map* specifies for any given type of organism what phenotypes can be produced from what genotypes. Weber argues that GP-maps "possibilify" the process of random genetic change in showing how evolvability is possible. Within a GP-map, those genotypes that are adjacent to the initial genotype are considered accessible (for accessibility metrics, see Huber 2017). Weber studies small RNAs that provide a basic model system that, because of their unique characteristics, enables the building of a GP-map. Given enough time, a nearly infinite population size, and/or a sufficiently high mutation rate, any RNA form might evolve from another. However, in the actual world, some transitions are inaccessible because of their unlikeliness. Weber argues that judgments of (im)possibility are influenced by both the explanatory objectives of the biological model in question and the evolutionary conditions of actual populations. Thus, even though Dennett's Library of Mendel appears to be an objective, almost endless space of possible genomes, the GP-maps are limited by their applicability and relevance.

One crucial feature of the accessibility in the GP map is that what is biologically possible is constrained by the need to preserve the life of some organisms under the genetic changes considered: one cannot transition in the genotype space from one viable form to another via a lethal form. In contrast, synthetic biologists have argued, as Knuuttila and Loettgers point out, that the engineering approach is not similarly constrained, and can thus explore those areas of possibility space that are not available for evolution (Elowitz and Lim 2010). Instead of concentrating on genotype spaces, Knuuttila and Loettgers study synthetic biology through Armstrong's (1997) combinatorial theory of possibility. It is based on combinations of states of affairs that consist of particulars, and universals that are instantiated by particulars. The combinations can be complex, involving structural universals in which any number of universals might be in various types of relationships to one another. The upshot is that possibility spaces can be highly structured. According to the reconstruction of Knuuttila and Loettgers, Armstrong's theory addresses two important sides of the synthetic biology program. First, the combinatorial theory of possibility fits synthetic biology's goal of (re)constructing biological entities and functions by combining standardized biological parts. Second, in building synthetic constructs, synthetic biology probes various kinds of biological possibilities. Knuuttila and Loettgers suggest that the design principles used to build synthetic constructs could be understood as structural universals in Armstrong's terms. According to Armstrong, structural universals as such are mere abstractions in need of instantiation by actual particulars. One way of understanding the more basic science work within synthetic biology is to view it as an exploration of whether the various kinds of hypothetical design principles, often transferred from engineering to biology, are realizable in living biological organisms.

While biological systems are inherently dynamic, it is not clear that Armstrong's logic-based combinatorial theory could address the dynamics of, e.g., synthetic genetic networks that model interactions `between genes and proteins. But the difficulty of dealing with dynamic systems does not just arise for metaphysical theories such as Armstrong's but also for many scientific modeling methods. In discussing network models, DiFrisco, Jaeger, and Loettgers argue that in general, they are not able to take into account the time-dependence of system behaviors and properties and the relationship between network structure and dynamics (DiFrisco and Jaeger 2019). DiFrisco, Jaeger, and Loettgers make a clear distinction between network models and the systems they are used to study. They point out that in the case of networks, the possibility spaces of the real-world systems are of a higher dimension and complexity than the possibility spaces of the models used to study them. Moreover, the behavior and structure of natural and social systems are intrinsically time-dependent, while it is difficult to model such time-dependency.

Network models, according to DiFrisco, Jaeger, and Loettgers, define and generate abstract spaces of possibilities that should not be taken as spaces of possibilities that existed prior to modeling activities. Instead, they depend on the specific formalisms, variables, and parameters used. The model construction frequently employs transdisciplinary *model templates* such as small-world or scale-free networks (Knuuttila and Loettgers 2016, 2023). Consequently, the dimensions of resulting possibility spaces are governed not only by the (typically very few) characteristics of natural or social systems, but also by the properties of the model templates used and their practical aspects, such as mathematical tractability. Including dynamics would expand the space of possibilities spawned by current network models, addressing some of their shortcomings, yet current dynamical models have their own set of limitations.

0.4   Exploring How-Possibly in Practical Contexts

How-possibly models, as their name suggests, are directed toward possibilities; instead of revealing how things are in actuality, they depict how things possibly could be. Traditionally, in the philosophy of science, how-possibly models have been mostly invoked in the context of scientific explanation. Thus, in the majority of cases in the philosophy of science literature, they have not been considered as having any independent epistemic value that is distinct from explanation. In the recently popular mechanistic approach, for example, how-possibly models are identified with how-possibly explanations (HPEs). These have been conceived of as incomplete explanations that function only as heuristic devices towards more evidentially complete how-actually explanations (e.g., Craver 2006; Craver and Darden 2013). As examples of mechanistic how-possibly explanations, Craver and Darden mention neuroscientific hypotheses that account for the surface characteristics of a phenomenon (e.g., neural excitation), but lack the evidential confirmation for all the actual parts and processes that are responsible (e.g., the action-potential mechanism). Similar attitudes can be found elsewhere throughout the philosophy of biology, where how-possibly explanations are typically identified with speculative evolutionary scenarios in contexts where decisive historical evidence is hard to come by (Brandon 1990; Resnik 1991; Reydon 2012).

The original notion of how-possibly explanation goes back to William Dray's (1957) account of specific explanations in the historical sciences, the purpose of which is to control for epistemic surprise in the face of unexpected events. For example, we know for a fact that the Roman Empire fell (and thus already know it was possible for Rome to fall). However, we might still be wondering how on earth this was possible to begin with, i.e., what kinds of circumstances could have led to this outcome. We can now construct Dray-type HPEs that make it less surprising that the Roman Empire fell. In effect, these amount to various historical scenarios that fulfill certain disciplinary standards of plausibility, but do not necessarily have full evidential support due to lack of source material. These HPEs are modal in that (a) they show how or why something else was possible to begin with and (b) they might remain merely possible scenarios themselves. Thus, Dray is not first and foremost interested in reaching a detailed how-actually account of a historical chain of events. Yet, his account is still clearly explanatory. The notion of a how-possibly explanation lends itself to a variety of research strategies or heuristics. These are not so much concerned with the different types of modalities that an HPE is supposed to track, but rather the kinds of questions it is used to answer.

In recent years, however, the tight linkage between how-possibly models and the actualist-representationalist ideal of explanation has started to dissolve and some philosophers of science now treat how-possibly models as more *sui generis* epistemic tools. One of the reasons for this development lies in the recognition that there is a multitude of uses of how-possibly models that typically involve very different contrast classes. Recent literature has stressed how scientific modeling can be used to track very different types of modalities depending on whether the modelers are interested in epistemic or objective possibilities. Another important distinction concerns the potential contrast classes for the notion of "how-possibly." While earlier post-positivistic philosophy of science tended to regard the gold standard of explanation in providing a "why-necessarily" account of any pending scientific question, more recent approaches have contrasted "how-possibly" explanation with "how-actually" accounts. Disregarding the fact that a complete "why-necessarily" explanation for a phenomenon might be difficult to come by in practice, in theory it renders any further inquiry into "how-possibly" as scientifically redundant (Rosenberg 2006). After all, if we know why something had to happen as it did, out of necessity, there is hardly any reason left to inquire how it could have happened. Interestingly, however, this is not so in the case of the contrast class of "how-actually." If researchers have shown how a scientific phenomenon of interest actually works, there are important scenarios where it is still sensible to ask if another possible way to produce the said phenomenon could exist (Koskinen 2017). This is especially true in contexts where the subject matter is deeply contingent, like the biological or social sciences.

The chapters by Mary S. Morgan, Joe Roussos, and Gregor P. Greslehner all have in common that they investigate complex modeling practices in areas of science whose subject matter is deeply influenced by historically contingent dynamics. The kinds of economic, climate, and epidemiological models considered in these chapters combine counterfactual scenario-probing and "what-if" or alternative world exploration with astute actual world predictions that have to take into account not only factual aspects of the world, but also different societal stakeholders in order to best inform political decision-making. While these chapters have important lessons for the epistemology of modality, they start from a more practice-oriented approach, mining scientific modeling practices in localized cases for philosophical insight.

The practices that are studied in these articles also take a more exploratory stance towards different kinds of possibilities where it is not always clear what the exact relationship between individual modal results is. For example, beyond some very general features of global circulation, it is not at all clear if different types of climate models can be realistically seen as occupying a single well-defined space of possibilities. Indeed, as Roussos argues, it seems that different types of climate models are used to track entirely different types of modalities (i.e., objective, epistemic) depending on the goal of research. In his chapter on epidemiological modeling, Greslehner emphasizes the important dual role of modal models: while many aim to predict and characterize possibilities that are likely to be realized, some are decidedly focusing on unlikely (but potentially dangerous) what-if scenarios. This type of tandem strategy of how-possibly modeling is also common in climate science, where stakeholders such as governments want to be prepared for a variety of outcomes.

Morgan likewise focuses on the practical side of modal modeling and uses examples from economic history to paint a broader philosophical picture. In her chapter, Morgan outlines different formulas for creating and appraising alternative model worlds in economics: using theoretical reasoning as well as data and empirical evidence to assess the plausibility of an alternative world, and ultimately combining (or "triangulating") these two methods. To function properly and avoid being reduced to a mere game that scientists play to entertain themselves, she concludes that the exploratory strategy of spinning alternative worlds requires "*both* imagination *and* background knowledge *and that they must work together to be effective.*" Mere imagination alone can easily lead to impossible world accounts, while sticking to known facts and established theory does not enable us to access new knowledge the way reasoning with alternative model worlds does.

Verreault-Julien, on the other hand, identifies a particular kind of epistemic how-possibly explanations—exemplifying them with deep neural network models (DNNs)—which draw on models that are opaque in the sense that scientists do not understand the process(es) they involve and thus not how they yield the results that they do. Verreault-Julien argues that such models lack explanatory value: they purportedly identify epistemic possibilities, in the standard sense that they identify systems not contradicting current knowledge, but scientists do not understand how these models work, thus undermining the claim that these models have been checked against current knowledge. Verreault-Julien suggests resolving this issue by making DNNs functionally transparent and by providing evidence for the epistemic possibility claim. This solution suggests that one is not bound by the standard definition, according to which only the absence of contradiction constitutes evidence for epistemic possibility.

Finally, in his account of "prospective modeling," Alfred Nordmann takes the emphasis on the practical to its zenith. One of his key insights is how our understanding of what is possible is, in an important sense, dependent on our ability to materially act in the world, and especially to manipulate it for our purposes. Prospective models, as Nordmann calls them, are specifically designed to link our understanding of possibilities not with how the world might be, but rather how we can achieve something—how something that is possible can be realized. This goes well with findings from cognitive psychology, which inform us that the human recognition of a possibility *as* a possibility to begin with is correlated with how controllable the target is perceived to be (Byrne 2005).

0.5 Broader Philosophical Implications

As is evident from the opening section above, the matters raised by possibility modeling connect with several debates in the philosophy of science, in particular concerning scientific representation, explanation, and idealization. But they also connect with a host of other philosophical issues. In this section, we will briefly review a few of those.

One set of debates with clear connections to the current volume is that in *modal epistemology*. Modal epistemology is the philosophical inquiry into how we can come to have knowledge of what is possible (and impossible). This has been an increasingly lively field of research since the early 1990s. In the last 20 years, the field has seen a turn towards an interest in primarily *empirical* methods and a more *pluralist* (or "non-uniformist," as it is also known) approach. That marks a change from the previously dominant focus on conceivability, perceived as an *a priori* method, as the one method for finding out modal truths of significance for various debates in metaphysics. As part of this turn, several authors have suggested that much of our knowledge of what is possible is likely to come from, or be significantly informed by, science (e.g., Bueno and Shalkowski 2014; Fischer 2017; Hirvonen, Koskinen and Pättiniemi 2021; Mallozzi 2021; Nolan 2017; Williamson 2016). But just how science is supposed to be producing that modal knowledge has not yet been explored in any great detail. Inquiry into when and how science models the possible obviously contributes to filling this lacuna. Indeed, given how much of modern science involves modeling of various sorts, such investigations are likely to be central to doing so. In this volume, Sjölin Wirling and Grüne-Yanoff's contribution explicitly highlights several points of contact between the literatures on modeling and on modal knowledge.

Imagination is a theme that occurs regularly both in the philosophical literature on scientific modeling, and in modal epistemology. Many modal epistemologists have taken imagination to be crucially involved in how we come to know modal truths (e.g., Dohrn 2021; Kung 2010; Nichols 2006; Williamson 2007; Yablo 1993), and many philosophers of science—especially so-called *fictionalists* (e.g., Frigg 2010; Frigg and Nguyen 2016; Levy 2015; Toon 2012; Levy and Godfrey-Smith 2019; Salis 2021)—think of imagination as central to scientific reasoning. The phenomenon of possibility modeling obviously lies at the intersection of these two debates, so there is every reason to think that closer attention to how the possible is modeled can shed light on issues of interest to both of these discussions.

Scientific explanation and scientific understanding are other topics for which the current volume has implications. The literature on the relation between models and explanations is lively, and more specifically several authors have suggested that a capacity for correct counterfactual reasoning, and/or a grasp of a certain possibility space pertaining to some *explanandum* phenomenon, is centrally involved in explaining and understanding (e.g., Le Bihan 2017; Rice 2021; Saatsi 2019; Woodward 2003). Discussions about the conditions under which models can be taken to support modal claims, such as that undertaken by Sjölin Wirling and Grüne-Yanoff in their contribution, are of clear relevance to this type of view. Furthermore, Collin Rice's and Philippe Verreault-Julien's respective contributions both take off from an apparent challenge to the idea that a grasp of possibilities is important to scientific explanation or scientific understanding, namely that it would seem that not all possibilities are equal in this respect. Rice and Verreault-Julien tackle this challenge in different ways. Verreault-Julien discusses how epistemic how-possibly explanations based on epistemically opaque models may yet have explanatory value. Rice, focusing instead on objective possibilities, argues that contrary to appearance, all such possibilities *can* contribute to increased scientific understanding.

Finally, getting to grips with modeling of the possible can also have implications for our understanding of how the *impossible* figures in modeling. The philosophy of science literature has seen a recent surge of interest in the role of counterpossibles and counterlegals in modeling and in scientific reasoning more generally. Counterpossibles are counterfactual conditionals with a metaphysically impossible antecedent (counterlegals have a nomologically impossible antecedent—these may or may not coincide, depending on whether one takes the metaphysical possibilities to outstrip the physical possibilities). On the standard Lewis-Stalnaker semantics for counterfactuals, all counterpossibles are vacuously true. But this is regarded as problematic by an increasing number of philosophers, not least because counterpossibles apparently figure so frequently in scientific reasoning, and, in particular, in model-based reasoning (Iranzo-Ribera 2022; Jenny 2018; McLoone 2020; McLoone, Grützner, and Stuart 2023; Tan 2019; Wilson 2021; Dohrn forthcoming). This problem points to the need for new semantics for counterpossibles, but it also raises a number of interesting questions concerning how scientists think about the relevance of considering impossible scenarios in their work.

0.6 Summary of Chapters

Ylwa Sjölin Wirling and Till Grüne-Yanoff's contribution, "Through the Prism of Modal Epistemology: Perspectives on Modal Modeling," explores one of the many connections between possibility modeling and other philosophical debates: the epistemology of modality. This chapter also acts as a gateway to the rest of the volume, as it introduces some central conceptual and epistemological issues associated with modeling the possible and describes some known accounts of such practices. They start by noting that learning what is possible from modeling raises the question of what the conditions are under which we really can trust that a model reveals a genuine possibility. Then, they move on to review some of the existing attempts that philosophers of scientific modeling have made to answer this question. In doing so, they use work in the epistemology of modality to highlight, on the one hand, the justificatory strategies implied in these accounts of modal modeling, and, on the other hand, prospects and challenges in the further development.

Collin Rice's "Which Possibilities Contribute to Scientific Understanding?" takes off from the idea that knowing what is possible can be important to scientific explanation and/or scientific understanding. An apparent challenge for this idea, whether one has in mind epistemic or objective possibilities, is that not all possibilities are equal in this respect. Scientists take some possibilities more seriously than others, and some possibilities appear too far away from the actual world, or simply irrelevant on some other grounds. In his chapter, Rice meets this challenge head-on by defending the view that despite such appearances, *all* possibilities can contribute to improved scientific understanding. Along the way, he considers three attempts to capture the difference between possibilities that do contribute to understanding and possibilities that do not: only possibilities sufficiently close to the actual world matter, only possibilities relevant to intervention and manipulation matter, and only possibilities of interest to scientists matter. Rice finds all of these insufficient, and in their place, advances the All Possibilities Count view.

Paul Teller's chapter, "Actual Possibility," exemplifies an attempt to develop objective possibility concepts of relevance to science. Teller contrasts his view with the possible worlds approach to modality—where, roughly, it is possible that p just in case there is a possible world in which p is true—which has been much discussed in the literature on modal metaphysics. He offers the notion of something being *actually possible*. While there are forerunners in the literature who have outlined similar notions, Teller's contribution offers more detail and develops the concept in a novel way. The result is a time-indexed notion of possibility, where what is possible in the relevant sense depends on what has occurred in the actual world up until a certain time. Roughly, on this view, a statement made at some time t, presenting some event or state of affairs concerning a time t' later than t, presents an actual possibility just in case the event or state of affairs that will or will not occur at t' is not ruled out by what has happened up to t.

Mauricio Suárez's "The Possibilities in Propensities: Emergence and Modality" focuses on the relationship between possibility and probability, and the different kinds of probability spaces that underpin scientific modeling practices. Suárez's project is an answer to Humphrey's Paradox, which effectively leads to a dilemma where a propensity theorist either must give up some of Kolmogorov's classic axiomatization of probability calculus or deny the so-called identity thesis between probabilities and propensities. In his chapter, Suárez opts for the latter strategy. More specifically, he proposes a new indexed approach to a central problem in the interpretation of probability, reminiscent of a typed solution. Under this picture, propensities do give rise to (or "ground") probabilities, but they cannot figure in the resulting probability functions themselves, as they belong to a different type. In short, propensities are any system properties that enable or generate objective chances. Moreover, Suárez argues that there is no privileged domain or level of description for propensities, but that they can rather take the form of emergent system properties. His main motivation for this stance does not stem from metaphysical considerations. Instead, it is guided by paying heed to scientific modeling practices concerning objective chance.

Andreas Hüttemann's "Invariance, Modality, and Modelling" focuses on the relation between scientific modeling and *objective* or *de re* modal features of systems. He discusses how scientific modeling gives knowledge about *possible* states of systems, addressing also the ways in which the behavior of target systems is *constrained*. Hüttemann argues that the concept of invariance is particularly helpful for exploring this relationship. Hüttemann sets out by analyzing the concept of invariance in terms of scientific laws, which he approaches as nomologically necessary generalizations. As examples of laws, he discusses Galileo's law of free-falling bodies and Hooke's law of elasticity, which he distinguishes from accidental generalizations, noting though that laws and accidental generalizations should be understood as endpoints of a spectrum of invariances. Hüttemann argues that invariances are empirically accessible as independence claims, in the same way as other (in)dependence claims in science are. Furthermore, he suggests that (some) modal features of the behavior of systems can be understood in terms of invariance relations. Hüttemann concludes by examining the links between empirically accessible modal features of the behavior of systems and some aspects of our modeling practices such as abstractions and idealizations.

Marcel Weber points out in his "Modeling the Biologically Possible: Evolvability as a Modal Concept" that biological modalities have not as yet received much attention from philosophers. Yet, it is widely agreed that there are biological constraints on physically possible states of affairs, such that not everything that is physically possible is also biologically possible, even if everything that is biologically possible is also physically possible. Furthermore, biologists use concepts that appear to be modal in nature, such as the concept of *evolvability* in evolutionary developmental biology, or "evo-devo." Weber focuses on what kind of modality underlies the concept of evolvability. This concept seeks to capture the capacity of an organism or a lineage to sustain genetic changes that enable it to evolve or to evolve adaptively. Weber construes evolvability as a kind of accessibility in a modal space that presents the problem of how to specify this modal space and the relevant accessibility relation. While there may not be a general way of defining such a relation, there exist model systems for which it is possible, e.g., evolving small RNAs. The modal space, and similarity metric turn out to be quite distinct from those constructed by philosophers like Daniel Dennett and David Lewis. Even though the small RNAs examined by Weber are quite special, attending to the way in which biological possibilities are modeled in this case harbors some general lessons about biological modalities, in particular their dependence on the explanatory goals of the models modeling modalities.

Tarja Knuuttila and Andrea Loettgers' "The Combinatorial Possibilities of Synthetic Biology" studies the modeling practice within synthetic biology that makes creative use of various kinds of models: mathematical models and their simulations, synthetic models and even electronic versions of synthetic models. They focus on how synthetic modeling, and synthetic biology more generally, address possible biology in exploring the ways in which new, useful functions can be achieved by recombining and rewiring well-characterized biological components, thus building novel biological parts and systems. Given the combinatorial nature of the modeling practice of synthetic biology, Knuuttila and Loettgers examine the construction of synthetic genetic circuits by applying David Armstrong's combinatorial theory of possibility (Armstrong 1986, 1997). Armstrong's theory is based on combinations of states of affairs that consist of a particular having a property, or a relation between particulars. Knuuttila and Loettgers argue that apart from the idea of combination, two other features of Armstrong's theory are crucial for synthetic biology: structural universals and the requirement of their instantiation. One of synthetic biology's primary goals is to understand biological organization, natural or otherwise. Such a search for general design principles of life can be cast in terms of Armstrong's structural universals that synthetic biology probes by attempting to realize/instantiate them in actual biological organisms.

In their "Beyond Networks: Explaining the Dynamics in the Natural and Social Sciences" James DiFrisco, Johannes Jaeger, and Andrea Loettgers study network models that have spread from their origin in graph theory and condensed-matter physics across the life and social sciences. During this process of "model template transfer," models developed in one scientific context became adapted to new domains of application, while still retaining many of their original characteristics and limitations. One of these limitations is an excessive focus on static structure rather than dynamics. Even when network dynamics *are* considered, the focus mainly lies on linear analysis around steady states, and the structure of a model is usually treated as time-invariant. DiFrisco, Jaeger, and Loettgers argue that such disregard of dynamics severely limits the applicability of the network approach to systems whose properties depend on transient behavior and self-organizing, time-variant structure. Such processes are ubiquitous in important classes of complex systems, such as living organisms, ecosystems, neural and cognitive systems, social networks, and the economy. The space of possibilities of these natural and social systems is of a much higher dimension and complexity than that of the models used to study them. DiFrisco, Jaeger, and Loettgers examine several case studies to illustrate particular idealizations, and their limitations and consequences that in turn suggest ways to go beyond the inevitable constraints of the present modeling practices.

Philippe Verreault-Julien's contribution, "Three Strategies for Salvaging Explanatory Value in Deep Neural Network Modeling," like Rice's paper, starts from the apparent problem of sorting the possibilities which have scientific relevance and explanatory value from those that do not. Verreault-Julien focuses in particular on epistemic how-possibly explanations that receive their support from DNNs. These models are opaque in the sense that scientists do not understand the process(es) they involve and thus not how they yield the results that they do. One approach to the problem of saying which (epistemic) possibilities have explanatory value in science is to say that only possibilities which are sufficiently *supported* by what is known—in the sense that what is known somehow indicates, though of course inconclusively, that the explanation is correct—are worthwhile. Explanations that are possible only in the sense of not being strictly inconsistent with what is known can be set aside. Verreault-Julien notes that on this view, possibilities based on opaque models such as DNNs appear to lack explanatory value: since scientists do not understand how these models work, it would seem that DNNs might not constitute evidence for the explanations they suggest. In response to this problem, Verreault-Julien proposes three ways in which DNNs can be considered to give a positive indication of the truth of an explanation despite their opacity: by being functionally transparent, by giving evidence for the objective possibility of the explanation, and by being pursuit-worthy.

Joe Roussos, in his chapter "Modelling Climate Possibilities," examines the pressing case of climate modeling and argues that many of its central practices contain an important modal dimension. He focuses especially on two connected subjects. The first is the connection between climate change and extreme weather events (EWEs) using climate models. Two prominent strategies for assessing EWEs are discussed and analyzed: the risk approach and the storyline approach. While both aim at estimating the role climate change has on the prevalence of EWE events, their modal characters seem to differ in important ways, with the first being primarily aimed at objective counterfactual possibilities, while the role of the latter is to give specific kinds of epistemic how-possibly explanations. Interpreting climate model ensembles is the second focus of the chapter. Here, Roussos examines especially the contrast between probabilistic and non-probabilistic approaches. The chapter looks at the types of modal statements made by scientists and how they are supported in each instance. It concludes with a consideration of how modal modeling figures specifically in informing policy decisions. Despite some claims to the contrary, Roussos argues that the modal dimension of many climate modeling results does not lessen their relevance for decision-making. To the contrary, they are often about either real possibilities whose relevance for future climate conditions cannot be denied, or they are given a modal reading by design for ease-of-use decision-making purposes.

Alfred Nordmann's "Prospective Modeling" contrasts *What might be the case* and *What can be done*. Nordmann calls the modeling of what can be done "prospective modeling" and argues that it involves the demonstration of technical power. In his view, prospective modeling takes a special place among the different ways of modeling the possible. Instead of representing what might be the case, the technical achievement of building a prospective model opens up a field of action. In Nordmann's terms, "prospective" is not opposed to "retrospective" and does not refer to "future prospects." Demonstrating that something can be done and therefore done again, prospective models exhibit possibility in the sense of *potentia*. While many debates about modeling move in the sphere of the veridical and concern ways in which models represent objects of inquiry, prospective models afford working knowledge and mimetic practice. Beginning with examples from art, architecture, and archaeology, Nordmann moves on to show instances of prospective modeling in various sciences. Though prospective models are only one way of modeling the possible, they prove to be ubiquitous in scientific and other practices: whenever something becomes salient for having been made, it can become productive as the prospective model that affords specific reenactments.

Mary S. Morgan's "Alternative Worlds: Reasonable Worlds? Plausible Worlds?" seeks to identify criteria by which scientists distinguish between possible and impossible worlds. Expanding on her "models as mediators" account, she argues that scientists employ models both to explore their theories and their relevance to the world jointly, with the goal to judge the quality of models in terms of the alternative accounts of the world that they offer: are those worlds plausible and reasonable, or impossible or nonsensical? In particular, Morgan investigates two dimensions of judgment of these alternative worlds found through modeling. The first, reasonability, focuses on "judging the paths of model reasoning on the way," seeing how the elements of the model knit together along a possible path to an outcome. The second, plausibility, judges the congruence of both the pathways implied by model reasoning and their outcomes with what is already known about those aspects of the world. This might involve synthesizing something that already exists in the world, or alternatively designing a possible object and then to "engineer" that something into an actual world object. Plausibility judgment then is driven by the confidence in both data and model being more carefully aligned to the question.

In his chapter, "Are Pandemics a Necessary Evil? The Role of Epidemiological and Immunological Models in Understanding and Preventing Diseases," Gregor P. Greslehner investigates the modal elements of pandemics in the aftermath of COVID-19: what is to be expected in the future in relation to the individual and population level of pandemic occurrences. By addressing the central question of whether pandemics are a "necessary evil," we can delve into hitherto undiscovered facets of the philosophy of epidemiology. The idea that infections inevitably and naturally become "milder" or less virulent, is especially critically assessed in the chapter. Greslehner suggests that epidemiological modeling provides a fruitful case to widen our understanding of the role of how-possibly modeling in the sciences. Somewhat similar to climate modeling, epidemiological modeling practices need to strike the right balance between robust, accurate prediction of likely events and the exploration of rare, but potentially deadly, unfolding of "what if" scenarios.

References

Armstrong, David M. 1986. "The Nature of Possibility." *Canadian Journal of Philosophy* 16(4): 575–94.

Armstrong, David M. 1997. *A World of States of Affairs*. Cambridge: Cambridge University Press.

Batterman, Robert W, and Collin C Rice. 2014. "Minimal Model Explanations." *Philosophy of Science* 81(3): 349–76.

Brandon, R. N. 1990. *Adaptation and Environment*. Princeton: Princeton University Press.

Bueno, Otávio, and Mark Colyvan. 2011. "An Inferential Conception of the Application of Mathematics." *Noûs* 45(2): 345–74.

Bueno, Otávio, and Scott A Shalkowski. 2014. "Modalism and Theoretical Virtues: Toward an Epistemology of Modality." *Philosophical Studies* 172(3): 671–89.

Bokulich, Alisa. 2017. "Models and Explanation." In *Springer Handbook of Model-Based Science*, edited by Lorenzo Magnani, and Tommaso Bertolotti, 103–18. Cham: Springer International Publishing.

Byrne, R. M. J. 2005. *The Rational Imagination: How People Create Alternatives to Reality*. Cambridge, MA: The MIT Press.

Chalmers, David. 2011. "The Nature of Epistemic Space." In *Epistemic Modality*, edited by Andy Egan, and Brian Weatherson, 60–107. New York: Oxford University Press.

Craver, Carl F. 2007. *Explaining the Brain: Mechanisms and the Mosaic Unity of Neuroscience*. Oxford: Oxford University Press.

Craver, Carl F. 2006. "When Mechanistic Models Explain." *Synthese* 153(3): 355–76.

Craver, Carl F, and Lindley Darden. 2013. *In Search of Mechanisms Discoveries Across the Life Sciences*. Chicago: The University of Chicago Press.

de Donato Rodríguez, Xavier, and Alfonso Arroyo Santos. 2012. "The Structure of Idealization in Biological Theories: The Case of the Wright-Fisher Model." *Journal for General Philosophy of Science* 43(1): 11–27.

de Donato Rodríguez, Xavier, and Jesús Zamora Bonilla. 2009. "Credibility, Idealisation, and Model Building: An Inferential Approach." *Erkenntnis* 70(1): 101–18.

Dennett, Daniel C. 1995. *Darwin's Dangerous Idea: Evolution and the Meanings of Life*. New York: Simon & Schuster.

Deutsch, Harry. 1990. "Real Possibility." *Noûs* 24: 751–55.

DiFrisco, James, and Johannes Jaeger. 2019. "Beyond Networks: Mechanism and Process in Evo-Devo." *Biology and Philosophy* 34(6): 54.

Dohrn, Daniel. forthcoming. "The Science of Counterpossibles vs. the Counterpossibles of Science." *British Journal for the Philosophy of Science*.

Dohrn, Daniel. 2021. "A Humean Modal Epistemology." *Synthese* 199(1): 1701–725.

Dray, William. 1957. *Laws and Explanation in History*. Oxford: Clarendon Press.

Elliott-Graves, Alkistis. 2020. "What Is a Target System?" *Biology and Philosophy* 35(2): 1–22.

Elowitz, Michael, and Wendell A Lim. 2010. "Build Life to Understand It." *Nature* 468(7326): 889–90.

Fischer, Bob. 2017. *Modal Justification via Theories*. Cham: Springer International.

Frigg, Roman. 2010. "Models and Fiction." *Synthese* 172(2): 251–68.

Frigg, Roman, and James Nguyen. 2016. "The Fiction View of Models Reloaded." *The Monist* 99(3): 225–42.

Gelfert, Axel. 2016. *How to Do Science with Models: A Philosophical Primer*. Cham: Springer International.

Giere, Ronald N. 2004. "How Models Are Used to Represent Reality." *Philosophy of Science* 71(5): 742–52. https://doi.org/10.1086/425063.

Godfrey-Smith, Peter. 2006. "The Strategy of Model-Based Science." *Biology and Philosophy* 21(5): 725–40.

Godfrey-Smith, Peter. 2009. "Abstractions, Idealizations, and Evolutionary Biology." In *Mapping the Future of Biology: Evolving Concepts and Theories*, edited by Anouk Barberousse, Michel Morange, and Thomas Pradeu, 47–56. Boston Studies in the Philosophy of Science. Dordrecht: Springer Netherlands.

Gould, Stephen J. 1989. *Wonderful Life: The Burgess Shale and the Nature of History*. New York: W. W. Norton.

Grüne-Yanoff, Till, and Philippe Verreault-Julien. 2021. "How-Possibly Explanations in Economics: Anything Goes?" *Journal of Economic Methodology* 28(1): 114–23.

Hempel, Carl G, and Paul Oppenheim. 1948. "Studies in the Logic of Explanation." *Philosophy of Science* 15(2): 135–75.

Hirvonen, Ilmari, Rami Koskinen, and Ilkka Pättiniemi. 2021. "Modal Inferences in Science: A Tale of Two Epistemologies." *Synthese* 199(5–6): 13823–43.

Hoshika, Shuichi, Nicole A Leal, Myong-Jung Kim, Myong-Sang Kim, Nilesh B Karalkar, Hyo-Joong Kim, and Alison M Bates et al. 2019. "Hachimoji DNA and RNA: A Genetic System with Eight Building Blocks" *Science* 363 (6429): 884–7.

Huber, Max. 2017. *Biological Modalities*. Ph.D. thesis, University of Geneva. Retrieved from https://archive-ouverte.unige.ch/unige:93135

Iranzo-Ribera, Noelia. 2022. "Scientific Counterfactuals as Make-Believe." *Synthese* 200 (6): 1–27.

Jenny, Matthias. 2018. "Counterpossibles in Science: The Case of Relative Computability." *Noûs* 52(3): 530–60.

Jones, Martin R. 2005. "Idealization and Abstraction: A Framework." *Poznan Studies in the Philosophy of the Sciences and the Humanities* 86(1): 173–218.

Katzav, Joel. 2014. "The Epistemology of Climate Models and Some of Its Implications for Climate Science and the Philosophy of Science." *Studies in History and Philosophy of Modern Physics* 46: 228–38.

Knuuttila, Tarja. 2011. "Modelling and Representing: An Artefactual Approach to Model-Based Representation." *Studies in History and Philosophy of Science* 42(2): 262–71.

Knuuttila, Tarja. 2021. "Epistemic Artifacts and the Modal Dimension of Modeling." *European Journal for Philosophy of Science* 11(3): 1–18.

Knuuttila, Tarja, and Andrea Loettgers. 2016. "Model Templates within and between Disciplines: From Magnets to Gases—and Socio-Economic Systems." *European Journal for Philosophy of Science* 6(3): 377–400.

Knuuttila, Tarja, and Andrea Loettgers. 2023. "Model Templates: Transdisciplinary Application and Entanglement." *Synthese* 201(6): 1–23.

Koskinen, Rami. 2017. "Synthetic Biology and the Search for Alternative Genetic Systems: Taking How-Possibly Models Seriously." *European Journal for Philosophy of Science* 7(3): 493–506.

Koskinen, Rami. 2019. "Multiple Realizability and Biological Modality." *Philosophy of Science* 86(5): 1123–33.

Koskinen, Rami. 2023. "Kinds of Modalities and Modeling Practices." *Synthese* 201(6): 1–16.

Kung, Peter. 2010. "Imagining as a Guide to Possibility." *Philosophy and Phenomenological Research* 81(3): 620–63.

Le Bihan, Soazig. 2017. "Enlightening Falsehoods: A Modal View of Scientific Understanding." In *Explaining Understanding: New Perspectives from Epistemology and Philosophy of Science*, edited by Stephen Grimm, Christoph Baumberger and Sabine Ammon, 111–36. New York: Routledge.

Levy, Arnon. 2015. "Modeling without Models." *Philosophical Studies* 172(3): 781–98.

Levy, Arnon, and Peter Godfrey-Smith eds. 2019. *The Scientific Imagination*. New York: Oxford University Press.

Maclaurin, James, and Kim Sterelny. 2008. *What Is Biodiversity?* Chicago: The University of Chicago Press.

Mallozzi, Antonella. 2021. "Putting Modal Metaphysics First." *Synthese* 198(Suppl 8): 1937–56.

Massimi, Michela. 2019. "Two Kinds of Exploratory Models." *Philosophy of Science* 86(5): 869–881.

McLoone, Brian. 2020. "Calculus and Counterpossibles in Science." *Synthese* 198(12): 12153–74.

McLoone, Brian, Cassandra Grützner, and Michael T Stuart. 2023. "Counterpossibles in Science: An Experimental Study." *Synthese* 201(1): 1–20.

Morgan, Mary S. 2012. *The World in the Model: How Economists Work and Think*. Cambridge: Cambridge University Press.

Morrison, Margaret, and Mary S Morgan. 1999. "Models as Mediating Instruments." In *Models as Mediators: Perspectives on Natural and Social Science*, edited by Mary S. Morgan and Margaret Morrison. Cambridge: Cambridge University Press.

Nichols, Shaun. 2006. *The Architecture of the Imagination: New Essays on Pretence, Possibility, and Fiction*. Oxford: Oxford University Press.

Nolan, Daniel. 2017. "Naturalised Modal Epistemology." In *Modal Epistemology After Rationalism*, edited by Bob Fischer, and Felipe Leon, 7–28. Cham: Springer International.

Norton, John D. 2021. "How to Make Possibility Safe for Empiricists." In *Rethinking the Concept of Laws of Nature: Natural Order in the Light of Contemporary Science*, edited by Yemima Ben-Menahem, 129–159. Cham: Springer.

Pincock, Christopher. 2012. *Mathematics and Scientific Representation*. Oxford Studies in the Philosophy of Science. Oxford: Oxford University Press.

Quine, Willard Van Orman. 1953. *From a Logical Point of View*. Cambridge: Harvard University Press.

Przyjemski, Katrina. 2017. "Strong Epistemic Possibility and Evidentiality." *Topoi* 36: 183–95.

Resnik, David B. 1991. "How-Possibly Explanations in Biology." *Acta Biotheoretica* 39: 141–49.

Reydon, Thomas A. C. 2012. "How-Possibly Explanations as Genuine Explanations and Helpful Heuristics: A Comment on Forber." *Studies in History and Philosophy of Biological and Biomedical Sciences* 43: 302–10.

Rice, Collin. 2018. "Idealized Models, Holistic Distortions, and Universality." *Synthese* 195(6): 2795–819.

Rice, Collin. 2021. *Leveraging Distortions: Explanation, Idealization, and Universality in Science*. Cambridge, MA: The MIT Press.

Rosenberg, Alexander. 2006. *Darwinian Reductionism: Or, How to Stop Worrying and Love Molecular Biology*. Chicago: The University of Chicago Press.

Ruyant, Quentin. 2021. *Modal Empiricism: Interpreting Science without Scientific Realism*. Cham: Springer International Publishing.

Saatsi, Juha. 2019. "Realism and Explanatory Perspectives." In *Understanding Perspectivism: Scientific Challenges and Methodological Prospects*, edited by Michela Massimi, and C.D. McCoy, 65–84. New York: Taylor and Francis.

Salis, Fiora. 2021. "The New Fiction View of Models." *British Journal for the Philosophy of Science* 72(3): 717–742.

Sjölin Wirling, Ylwa. 2022. "Extending Similarity-Based Epistemology of Modality with Models." *Ergo* 4: 45.

Sjölin Wirling, Ylwa, and Till Grüne-Yanoff. forthcoming. "Epistemic and Objective Possibility in Science". *The British Journal for the Philosophy of Science*. 10.1086/716925.

Suárez, Mauricio. 2004. "An Inferential Conception of Scientific Representation." *Philosophy of Science* 71(5): 767–79.

Tan, Peter. 2019. "Counterpossible Non-Vacuity in Scientific Practice." *Journal of Philosophy* 116(1): 32–60.

Toon, Adam. 2012. *Models as Make-Believe: Imagination, Fiction, and Scientific Representation*. London: Palgrave-Macmillan.

Vetter, Barbara. 2011. "Recent Work: Modality without Possible Worlds." *Analysis* 71(4): 742–54.

Vetter, Barbara. 2015. *Potentiality: From Dispositions to Modality*. Oxford: Oxford University Press.

Wagner, Andreas. 2014. *Arrival of the Fittest: Solving Evolution's Greatest Puzzle*. Penguin.

Weatherson, Brian, and Andy Egan. 2011. "Epistemic Modals and Epistemic Modality." In *Epistemic Modality*, edited by A. Egan, and B. Weatherson, 1–18. New York: Oxford University Press.

Weisberg, Michael. 2007. "Three Kinds of Idealization." *Journal of Philosophy* 104(12): 639–59.

Weisberg, Michael. 2013. *Simulation and Similarity: Using Models to Understand the World*. Oxford: Oxford University Press.

Williamson, Timothy. 2007. *The Philosophy of Philosophy*. Malden: Blackwell Publishing.

Williamson, Timothy. 2016. "Modal Science." *Canadian Journal of Philosophy* 46(4–5): 453–92.

Williamson, Timothy. 2018. "Spaces of Possibility." *Royal Institute of Philosophy Supplements* 82(July): 189–204.

Wilson, Alastair. 2021. "Counterpossible Reasoning in Physics." *Philosophy of Science* 88(5): 1113–24.

Wilson, Mark. 2018. *Physics Avoidance: And Other Essays in Conceptual Strategy*. Illustrated edition. Oxford: Oxford University Press.

Woodward, James. 2003. *Making Things Happen*. New York: Oxford University Press.

Yablo, Stephen. 1993. "Is Conceivability a Guide to Possibility?" *Philosophy and Phenomenological Research* 53(1): 1–42.

Possibility Concepts for Scientific Use

1 Through the Prism of Modal Epistemology

Perspectives on Modal Modeling

Ylwa Sjölin Wirling and Till Grüne-Yanoff

1.1 Introduction

Philosophers of science have recently begun to investigate modeling practices from a modal perspective, for at least two reasons. First, many scientists explicitly describe their modeling results in modal terms. For example, Maynard Smith and Price claim that their "main reason for using [the Hawk-Dove game model] was to test whether it is *possible* even in theory for individual selection to account for 'limited war' behaviour" (Maynard Smith and Price 1973, 15, our emphasis). Second, even where modelers do not use explicit modal language, philosophers of science have sometimes offered a modal interpretation, in order to address philosophical issues that have been left unanswered by prior accounts of scientific modeling or to explain a number of putatively neglected aspects of certain modeling practices.

The notion of modal modeling raises what we elsewhere have dubbed "*the epistemic question for modal modeling*" (Sjölin Wirling and Grüne-Yanoff 2021): if scientific modeling practices deal with modals, or if they are philosophically reconstructed to trade in modal claims, then what are the conditions for their success, and in virtue of what can models perform this function (under said conditions)? While philosophers of science recently have shown an increased interest in this question, answers remain fragmented and largely tied to the documentation and analysis of specific practices or reconstructions thereof.

At the same time, the philosophical field of *modal epistemology* is devoted to how one can come by knowledge of modal truths. That is, modal epistemologists have long considered a more general version of the epistemic question for modal modeling. Despite this common interest, there has so far been very little interaction between modal epistemology and the philosophy of scientific modeling. We think this is a missed opportunity for both sides. Hence, the aim of this chapter is to open these two fields up to one another and showcase how more exchange between them may be fruitful. We will do so by analyzing three of the existing attempts at an answer to the epistemic question for modal modeling—Sugden's (2000) credibility account,

DOI: 10.4324/9781003342816-3

Massimi's (2019) physical conceivability account, and Rice's (2018, 2019) universality account—by considering them through the prism of work in the epistemology of modality.

The prism analogy is helpful in spelling out what we think this analysis achieves. Modal epistemologists have not (yet) systematically investigated scientific modeling as a source of modal justification.[1] However, modal epistemologists often focus on less complex cases and more mundane modal claims than philosophy of science does. Accounts also tend to concentrate on one single source of modal justification. We can expect the cases at issue with modal modeling to typically be more complex and perhaps mix several potential sources of modal justification. But just as a prism is useful in the analysis of light by breaking it up into its constituent colors, modal epistemology can help us identify the relevant potential sources of modal justification involved in modal modeling practices. Identifying a convergence between modal justifications in modeling practices and those discussed in modal epistemology—as indeed we do—is an interesting finding in and of itself. Moreover, modal epistemology has identified challenges for various justificatory strategies; it will be useful to consider the form these may take in the context of modal modeling, and whether the accounts in philosophy of science might offer some novel answers to them that modal epistemologists can learn from. Of course, there is a particular way in which prisms distort and reverse images. But as with binoculars, their distorting effect might prove important to getting a good view of the landscape.[2]

1.2   Modal Modeling in Science

With "modal modeling" we mean modeling practices that aim at delivering *modal* information. This is in contrast to modeling that aims at delivering information about what actually is, was, or will be the case. In practice, there is often no sharp separation between these two aims. Scientists are typically interested in acquiring modal information insofar as it contributes to our understanding of the actual world. Nevertheless, one can *conceptually* distinguish modeling practices by their immediate results, even if those results are then employed to achieve some further objective. So, by way of this preliminary definition, modal modeling practices are identified by aiming at delivering modal information as their immediate results.[3] Under the "modal modeling" flag, we also include practices of modelers whose intentions are not explicitly directed at modal information, but of which philosophers of science nevertheless offer a modal interpretation, in order to address philosophical issues or conceptual challenges that are left unanswered when standard modeling accounts are applied to the practice in question.

The clearest examples of modal modeling are perhaps those connected with *how-possibly explanations* (HPEs). There is no general consensus on how to characterize HPE practices, but most philosophers of science seem to agree at least that (i) HPEs involve modal claims, and (ii) models play a crucial role in supporting HPEs (see, e.g., Bokulich 2014; Grüne-Yanoff 2009, 2013; Reutlinger, Hangleiter, and Hartmann 2018; Rohwer and Rice 2013; Verreault-Julien 2017, 2019; Weisberg 2013, chap. 7; Ylikoski and Aydinonat 2014). These are clear cases of modal modeling: scientists draw (or should be interpreted as drawing) the conclusion that such-and-such is possible on the basis of doing something with a model. There are several different kinds of modeling practices used to support HPEs and different purposes for which the resulting HPEs are employed in a broader scientific context that have been highlighted by philosophers of science. Whether properly distinct from HPE practices or not, further examples of purported modal modeling include toy, exploratory, or hypothetical modeling practices that support possibility claims that serve, for example, to refute necessity or impossibility claims (Grüne-Yanoff 2009), as proofs of principle (Gelfert 2018), or otherwise help delineate the space of possibility regarding a phenomenon (Massimi 2019). Some support "explanations in search of observations" (Sugden 2011), offering representations of possible properties of possible explananda for the purpose of understanding such phenomena in case such explananda should become actual. Disciplines that study unactualized possibilities include synthetic biology, where models are used to represent minimal cells and alternative genetic systems, even though such targets might turn out to be only partially realizable (Knuuttila and Koskinen 2020).

Philosophers of science have also argued that certain modeling practices that are not overtly modal should be re-interpreted in modal terms in order to account properly for the epistemic contribution of the models in question (e.g., Rice 2018; Verreault-Julien 2017; Ylikoski and Aydinonat 2014). This is typically because the models in question do not satisfy standard criteria of representational accuracy. If those models cannot be shown to represent actual targets accurately, these authors ask, what kind of epistemic functions can such models then play? Their answer: these models can contribute to a deeper scientific understanding of the studied phenomena by enabling users to draw correct *counterfactual inferences* about actual target systems, thus providing a core ingredient for successful scientific explanation.

As preliminarily defined above, modal modeling is not limited to models that support possibility claims or counterfactual claims. Nguyen (2020) contests the idea that toy models like Schelling's (1971) checkerboard model support "mere" possibility claims. Nevertheless, Nguyen's interpretation of these modeling practices is also modal in nature. He suggests that facts about the toy models should be translated into claims about the real world that (a) are less specific than the model facts (e.g., to qualitative trends from real values); and (b) ascribe a *capacity* or *susceptibility* to the target. For example, in Akerlof's (1970) "market for lemons" model, asymmetric information prevents car trades from occurring, despite the fact that at any given price sellers are willing to sell their car and buyers willing to buy it. When properly interpreted, the claim supported by the model is something like an asymmetric information state in this (particular, real world) market increases the market's susceptibility to fail to reach Pareto-efficient equilibrium. This is a supposedly true claim about an actual target. But it is a modal claim in the sense that it ascribes a *de re* modal property—a susceptibility, which is a form of disposition—to an actual target.

It is now evident that the preliminary definition of modal modeling as modeling aiming to deliver modal information has a rather wide scope. To draw conclusions about dispositional properties and (to some extent) counterfactual conditionals on the basis of modeling is commonplace, as are phrasings of model results in terms of epistemic possibility (what *might possibly* be true for all we know). Considering these many facets, some might even wonder whether there is any truly *non*-modal modeling. In any case, many of these practices might not be thought to be modal in a sense that raises any new or interesting questions. While this issue of properly delineating modal modeling as a distinctive phenomenon is important and deserves further attention, we will not take on that task here. We think it is sufficiently clear that there are modeling practices—including HPE modeling and certain exploratory or hypothetical modeling—that *are* modal in an interesting sense, which *do* raise distinctive and interesting epistemic questions. These are our concerns in this chapter.

1.3 The Epistemic Question for Modal Modeling

The epistemic question for modal modeling (Sjölin Wirling and Grüne-Yanoff 2021) is a two-part question that asks, first:

1 *under what conditions* can models deliver justified modal conclusions?
   and second:
2 *why*, or *in virtue of what*, do those conditions make models good guides to modal truth?

Much of what modal epistemologists do is to present answers to an analogue of this two-part epistemic question for modal modeling. First, some putative route to modal knowledge is suggested. It is then specified under what conditions the said route *is* indeed a reliable way to reach modal knowledge. A crucial part of motivating *that* the route in question is indeed justificatory is to explain *why* said route is supposedly reliable. To illustrate, some authors suggest that intuition provides justification for modal claims (e.g., Bealer 2002; BonJour 1997). This raises the question of under what conditions, and in virtue of what, intuition can play this cognitive role. Spelling that out is a crucial part of bolstering the claim that intuitions can justify modal claims. Now, philosophers of science, as we saw in the previous section, have highlighted that scientists often appeal to models when making modal claims.[4] This suggestion prompts us to think of how the analogous questions can be answered with respect to *models* as a route to modal knowledge.

The epistemic question for modal modeling emerges as pressing under the assumption that with modal claims, just as with non-modal claims, there is generally a fact of the matter as to whether they are true or false. Exactly in virtue of what modal claims are true or false is a question that keeps exercising philosophers who do metaphysics of modality and on which we will not take a stand here.[5] The important point is that if the truth values of modal claims are not a matter of opinion, then one can be wrong when making them. Insofar as in science one strives to make (at least approximately) true claims and refer to models in the course of doing so, this places certain constraints on those models: one needs to specify the conditions under which they are *reliable* scientific tools for making (approximately) true claims. Modal re-interpretations of models are sometimes motivated by the fact that the models in question do *not* meet standard accuracy conditions in terms of, for example, model-target similarity, or isolation of relevant causal factors, and so, are *not* reliable means for finding out what is actually the case. But modal modelers are not off the hook: models, if they are to contribute to science by teaching us about *modal* truths in some form or other, must reasonably be taken to satisfy some modal counterpart of such requirements. These requirements may be different from the standard ones, but *some* conditions under which the result of a modeling exercise speaks to the truth of the modal claim must arguably be specified.[6]

Because clearly, not just *any* model will do this. Differently put, even if it were true that all models support, for example, *some* possibility claim, not just any model will provide one with justification for the modal claims that are relevant to the inquiry one is currently engaged in. For instance, economists claim that the invisible hand hypothesis describes a possible scenario, with reference to the Arrow-Debreu model (Arrow and Debreu 1954). That is, the fact that an invisible hand scenario has been successfully modeled provides reasons for the claim that such a scenario is possible (Verreault-Julien 2017). Assuming there is a fact of the matter regarding the possibility (or not) of what the invisible hand hypothesis describes and that economists are justified when they are claiming, with reference to the Arrow-Debreu model, that it *is* possible, there must be something about the Arrow-Debreu model in virtue of which it *does* support this possibility. What is that something? More generally, what must modeling practices be like in order to provide us with justification for modal claims?

It might be objected that many scientific practices have various non-epistemic aims, for example, obtaining a certain chemical synthesis or designing and building a particular structure. When modal modeling occurs as a part of such practices, can the epistemic question be ignored? Maybe so. But we think this does relatively little to undermine the importance of the epistemic question for modal modeling. For one thing, to the extent that true modal claims are required for reaching these non-epistemic goals, the epistemic question will arise even when the ultimate goal is non-epistemic. In any case, several authors in the literature take models to provide modal knowledge—broadly construed, including grasping of modal information that some take to be necessary for understanding or explanatory knowledge—and thus serve an epistemic function.

The second part of the epistemic question—*why* or *in virtue of what* the identified conditions make models good guides to modal truth—can only be responded to once we have some idea of what those conditions are. In one sense, it is an explanatory question, seeking an answer to why the specified conditions are such that they make the models in question able to provide the relevant modal justification. This may seem superfluous: while it might be interesting to know the answer, as long as we know what characterizes a reliable modal modeling practice, that is all the philosopher really needs to provide. But this is a bit too quick. The explanation for why a proposed set of properties or conditions makes modal models justificatory plays an important role in justifying the philosophical claim that these really *do* make modal models justificatory. Especially since it is sometimes difficult to test the reliability of a strategy for modal justification, as the target claims are often about counterfactuals or nonactual possibilities. Insofar as philosophy of science has a role in evaluating actual scientific practices and not just describing them, the story of *why* we should think that a given practice is a good means to achieving the relevant aim remains important.

1.4 Themes from the Epistemology of Modality

We will presently introduce three key *themes* from the modal epistemology literature: imagination, background theory, and similarity judgments. These themes are non-exclusive strategies for understanding modal justification: as we will see later, philosophers of science sometimes have appealed to more than one theme simultaneously in order to account for modal modeling. They are broadly conceived and encompass many, although not every, main strand in the epistemology of modality literature.[7]

Many philosophers have assumed that the *imagination*—or the ability to conceive—is centrally involved in the way we form modal beliefs. This idea is both historically salient—going back to the writings of Descartes and Hume (Gendler and Hawthorne 2002)—and widely thought to be true to the phenomenology of (much) actual modalizing: what we often do when we consider whether something would be possible is to try and imagine it. Several modal epistemologies that rely centrally on imagination suggest that it primarily supports claims about what is possible: if one can conceive of, or imagine, a scenario in which p is the case, one is justified in holding that p is possible (see, e.g., Kung 2010; Yablo 1993). Other accounts, most notably Williamson (2007), connect imagination with the evaluation of counterfactuals, that is, claims of the form "If a were the case, then b would be the case." According to to Williamson, knowledge of possibility is downstream from counterfactual knowledge, and imagination is importantly involved in acquiring the latter. For instance, what would happen to a rock sliding down a slope if the bush that actually stopped it halfway down had not been there? For an answer, one supposes the antecedent—that there is no bush on the slope—and develops the scenario from there on, in the imagination—for example, the rock sliding down the slope, past the place where in reality there is a bush, further down the slope, ending up in the lake below. One thereby comes to know that if the bush had not been there, the rock would have ended up in the lake. Possibility knowledge is a result of such counterfactual development (in imagination): "we assert $\Diamond A$ when our counterfactual development of A does not robustly yield a contradiction" (Williamson 2007, 163). In this case, we can conclude that it is possible that the rock could have ended up in the lake.

But recent work in modal epistemology also makes abundantly clear that imagination as such is far too liberal to be a reliable guide to modal truth—for at least two reasons. For one, the above examples seem to imagine logical possibilities, while most scientists are interested in physical or biological possibility—and it remains an open question whether such possibilities can be reliably identified with imagination. For another, more generally, people can quite easily imagine the impossible.[8] So, any modal epistemology that assigns a central role to the imagination must specify circumstances (modes, topics, or external constraints) under which the imagination *is* a reliable guide to (relevant) modal truth and explain why imagination can (under those circumstances) be trusted as a guide to modal knowledge.

This challenge brings us to our second theme: *background knowledge*. This theme has been common in the literature responding to the challenge from unchecked imagination. That is, many suggest imagination must be constrained by some appropriate background knowledge that prevents us from imagining the impossible or from judging that what we have imagined is possible when in fact it is not. But background knowledge also features in accounts that make no mention of the imagination, so the two themes are independent. That background knowledge would play a role appears, in some sense, obvious. However, it is arguably instructive to consider the *kinds* of background knowledge that modal epistemologists have suggested to be relevant for modal justification. Three stand out as especially popular.

First, it is natural to think that *knowledge of laws* plays an important role in modal epistemology (e.g., Kment 2021; Wilson 2020). For instance, physical possibility is naturally defined (at least partly) in terms of compatibility with the actual laws of nature. It is thus plausible that knowledge of the laws of nature is very helpful for drawing justified conclusions about what is physically possible. Some have correspondingly suggested that metaphysical possibility is (partly) determined by the "laws of metaphysics," and accordingly, knowledge of metaphysical modal truth requires knowledge of metaphysical laws.[9]

Next, some have argued that possibility knowledge and counterfactual knowledge require access to essentialist or *constitutive knowledge*—that is, knowledge of what is constitutive of being a certain (kind of) object or property. Roca-Royes (2011a, 2011b), Tahko (2012), and Vaidya and Wallner (2018) all argue that imagination-based modal epistemologies presuppose that the epistemic subject has access to constitutive knowledge. For some modal epistemologies explicitly based on constitutive knowledge, see, for example, Lowe (2012), Mallozzi (2018), and Jago (2018). Both in the case of law-based and constitutive knowledge-based epistemologies, a major question concerns whether epistemic subjects can be assumed to have access to such (also modal) knowledge and, if so, how *that*, in turn, is acquired.

Finally, some have argued that one needs to have a *theory* concerning the relevant phenomena or entities that one is seeking modal knowledge of. For instance, Fischer (2017) proposes that a modal claim is justified if its truth is implied by a scientific theory that is itself justified. For instance, I am justified in claiming that it is possible for cells to evolve on the basis of something other than RNA/DNA, just in case I am justified in accepting a scientific theory that implies that it is possible for cells to evolve on the basis of something other than RNA/DNA, and I base my claim on that theory. This makes modal knowledge downstream from scientific knowledge more general. Similarly, Bueno and Shalkowski (2014) argue that we arrive at modal knowledge by investigating the relevant properties and objects in question by *both* scientific and common-sense means. Through common-sense observations, we learn that everyday objects have certain properties only contingently because we have observed them to lose these properties under changing conditions. In scientific practices, we deepen such observations by systematically varying interventions on and background conditions of these objects. One is justified in concluding that p is possible, on this view, on the basis that nothing in the body of relevant theoretical knowledge suggests that p is not possible.

The third and final theme is that of *similarity*. A recently influential idea in modal epistemology is that *similarity-judgements* can justify possibility claims (Dohrn 2019; Roca-Royes 2017). For instance, one can draw justified conclusions about what is possible for some individual entity e on the basis of knowledge of what is actually the case with other, distinct entities that are relevantly similar to e. Background knowledge plays a part here too, but the similarity-account originates in an explicit effort to present an epistemology of possibility, which does *not* presuppose that the epistemic subject has access to constitutive knowledge or a full-blown, justified theory of the relevant entities. Nor does it rely on imagination itself as conducive to modal truth, so it warrants separate mention. All that is required for knowledge of nonactual possibility, in this case, is knowledge of actual *token* events involving actual entities and the ability to reliably judge that certain entities (the targets of the prospective possibility judgments) are relevantly similar to certain other entities.

While such accounts might remind readers of similarity-based accounts of representational quality in the philosophy of science (e.g., Giere 1988; Weisberg 2013), it is worthwhile stressing some differences here. Most importantly, perhaps, many of these accounts do not treat modal inferences at all. In the few that do, the role of similarity for modal modeling often remains vague. Weisberg (2013, chap. 7), for example, analyzes modal modeling practices as either involving generalized phenomena or non-existent phenomena as targets or as not having targets at all. In neither case does he give an explicit account of how similarity judgments between models and targets help justify the modal claims—and the characterization of the targets as either "generalized" or "non-existent" indicates that any such account would not be trivial.

One more finding from the general literature on modality worth additionally mentioning is that determining the relevant type of modal claim is highly important for epistemological issues. Modality comes in varieties, and of particular importance is the distinction between epistemic and objective modality. An epistemic modal claim is relative to a body of epistemically privileged (e.g., known, justified, or evidenced) propositions—it expresses something about what might be the case given one's epistemic situation. In contrast, an objective modal claim expresses something about the world, independently of our epistemic situation. Objective and epistemic modal claims require quite different answers to the epistemological questions (Sjölin Wirling and Grüne-Yanoff, forthcoming). While modal claims of both kinds are arguably among the targets of modal models, we focus here—like much of the modal epistemology literature—on objective modality. But objective modality too comes in different flavors. What modality is relevant varies with context and depends partly on our interests (but once the relevant sense *is* determined, whether or not some *p* indeed is possible is arguably not up to us). Modal epistemologists have mainly been concerned with metaphysical modality; that is, what is necessary in the strongest objective sense and what is possible in the least restricted objective sense. Another familiar form of modality is *natural* or *physical* modality.[10] But more restricted forms of modality of interest to scientists include, for example, biological and economic possibility. Such modal claims arguably differ in their truth conditions and hence, unsurprisingly, likely in the ways in which we can come to know them.

1.5   Modal Modeling through the Prism of Modal Epistemology

We will now use these themes from modal epistemology as a prism through which we examine three answers to the epistemic question for modal modeling. It will emerge that all three accounts of modal modeling make use of one or several of the justificatory strategies familiar from the modal epistemology literature. This convergence is in itself a striking and interesting finding. We will in turn use this insight to draw on moves and issues arising in modal epistemology to highlight useful directions for future work both on modal modeling and on the epistemology of modality more generally.

1.5.1   Credible Worlds

Sugden (2000) suggests that we can learn from some models—in particular, toy models in economics—when these models describe *credible worlds*. More exactly, economic toy models are formal structures interpreted by their users as imaginary worlds or scenarios. If the model world is credible, that is a reason to think that the model result (or some equivalent thereof) is possible.[11] That idea has since been taken on board by a number of other authors. For instance, Grüne-Yanoff (2009, 95) writes that "The credibility of a minimal model establishes that it depicts a possible world"; Mäki (2009, 39–40) makes a similar point; and Fumagalli (2016, 437), who is critical of the idea that toy models afford knowledge of actuality, still concedes that "Considerations of models that are credible (...) may enable scientific modelers to acquire epistemically informative insights about the *possible worlds* posited by these models."[12]

The notion of "credibility" in Sugden's original paper is suggestive but in dire need of unpacking. Sugden writes that credibility in models is "rather like credibility in 'realistic' novels" (Sugden 2000, 25), and Grüne-Yanoff (2009) picks up on this analogy with fiction, in an attempt to elaborate on what it means for a model to be credible. When we read a novel, we imagine a fictional world, proceeding from the fictional text but going beyond it by adding detail, drawing out implications, and filling in gaps. It is this imagined world that we assess when we consider whether the novel presented a credible story or not. Analogously, scientists employing a model imagine a model world, proceeding from but going beyond the model description (cf. Frigg and Nguyen 2016). This imagined model world is then assessed for credibility, which, in turn, is offered as evidence of possibility.[13]

Now, this appeal to imagined worlds suggests an affinity with imagination-based modal epistemologies. Indeed, the credibility account faces a challenge analogous to that of reining in the imagination. That we can easily imagine the impossible suggests that the fact that I can imagine a scenario in which p is not very reliable evidence that p is possible. But similarly, it would seem that impossible scenarios can also be credible—at least insofar as we are to take the analogy with credibility in fiction seriously. Grüne-Yanoff stresses that many features of the imagined model/fictional world can deviate from what the actual world is like, yet the imagined world can be judged credible. But a credible model/fictional world, imagined on the basis of the model description/fictional text, must be sufficiently detailed and free of incoherent or contradictory assumptions and implications. So, what matters is, on the one hand, internal coherence. On the other hand, the *development* in a credible, imagined world must be judged plausible *conditional on* the information provided about, for example, preferences and environment. But clearly, fictional scenarios can fulfill these conditions—and so be credible—without being possible (Nolan 2021). Presumably, this is true also of model scenarios.

So clearly, a credibility judgment is not always good evidence of possibility. Does the credibility account have resources to address this issue? Grüne-Yanoff writes that credibility judgments are "driven by empathy, understanding, and intuition" (Grüne-Yanoff 2009, 94–95). This last part indicates that only the credibility assessment of a *competent user* of the model will do, that is, of someone with the right background knowledge. The idea, then, is that only credibility judgments made by competent users can serve as evidence of possibility (though of course not infallible evidence—and it can certainly be undermined by disagreement among competent users, as witnessed in often intense discussions among modelers about the credibility of certain assumptions).

This suggests that the justificatory power of credible scenarios lies not in the fact that they can be generated or imagined, but with the competent assessment of these scenarios in light of appropriate background knowledge. Once unpacked, the credibility account of modal modeling does not identify a new source of evidence for possibility—the credibility of model scenarios. Instead, it closely parallels imagination-involving modal epistemologies that rely ultimately on the appropriate background knowledge. This, in turn, helps us see the way forward in developing the credibility account. The following two questions must be answered: first, what background knowledge is required for reliable credibility judgments, and second, is it likely that users of the model typically possess it?

Presumably, competence with credibility judgments is relative to discipline; for example, different background knowledge is required to assess an economic model and a biological model for credibility, respectively. Can the kinds of background knowledge that modal epistemologists have taken to inform possibility judgments point the way? Although Sugden at one point suggests that credibility requires compatibility with the "general laws governing events in the real world" (Sugden 2000, 25), knowledge of laws is unlikely to be helpful here, since economics is not taken to be a field where there are many, if any, laws; and the laws of nature found in other disciplines are arguably not relevant to the kind of possibility at issue in economics. The idea that *theory* informs credibility judgments seems initially more promising but faces the challenge that toy models are often used in contexts where theory is scarce or put into question (Reutlinger et al. 2018; Sjölin Wirling 2021). What about constitutive knowledge? Mäki suggests that credibility requires compatibility with what he calls a Way the World Works—*"www"*—constraint (Mäki 2009, 39). It is not clear exactly what that involves, but some phrasings suggest an appeal to constitutive knowledge. For instance, he writes that models that violate the www-constraint are to be rejected because they violate *the very nature of the (kind of) system* (e.g., a market) it sets out to describe, or the nature of the sorts of things that populate the system (Mäki 2009, 40), and so does not even represent a *possible* version of such a system (Mäki 2001, 383; 385).[14] Modal epistemologies based on constitutive knowledge have enjoyed a revival recently, so perhaps this option for developing the credibility account is worth exploring.

1.5.2 *Physical Conceivability*

Michela Massimi (2019) suggests that certain exploratory modeling practices deliver possibility knowledge, insofar as they involve what Massimi calls "physical conceiving." The key idea is that if a scientist can physically conceive of p, she is justified in believing that p is possible, and certain forms of exploratory modeling involve this particular form of imagining. Physical conceiving is a form of imagining, but the question of how the imagination is to be constrained in order to avoid reaching into impossibility is conveniently built into Massimi's preliminary definition of physical conceivability:

> p is physically conceivable for an epistemic subject S (or an epistemic community C) if S's (or C's) imagining that p not only complies with the state of knowledge and conceptual resources of S (or C) but is also consistent with the laws of nature known by S (or C).
>
> (Massimi 2019, 872, our emphasis)

In other words, only successful attempts to conceive of p where the subject holds fixed knowledge of the laws of nature provide justification for the belief that p is possible.[15]

According to Massimi, the hypothetical modeling of SUSY (super symmetrical) particles in physics is an example of modeling that provides possibility knowledge because it involves the right kind of constrained conceiving. Scientists have not been able to confirm whether there actually are any SUSYs in nature, but that is ultimately what they want to do. To this end, they investigate the different ways in which it is *possible* that a SUSY particle exists, by using a modeling technique—the pMSSM-19—which produces different "model points" (roughly: fictional model systems), each of which portrays SUSY particles as having mutually inconsistent properties and value assignments (e.g., a given mass value, a given decay mode, etc.) for 19 parameters, *and are consistent with certain nomological constraints* (e.g., R parity conservation, consistent electroweak symmetry breaking). For a given model point produced by pMSSM-19, it is concluded that a hypothetical target, corresponding to the model point, is objectively (physically) possible. That is, it is possible that a particle with such-and-such properties exists. Again, we can observe a convergence with the sorts of justificatory routes that have been proposed in modal epistemology: in this case, imagination is constrained by nomological knowledge. Massimi calls this law-bounded (LB) physical conceivability. This holds some promise as a method for justifying claims of physical possibility—provided that the modeler possesses the relevant knowledge of laws (and this is successfully implemented in the pMSSM-19).

Interestingly, Massimi distinguishes LB-conceivability from *law-driven* (LD) physical conceivability. LD-conceivability of some p is also supposed to provide justification for the belief that p is possible, but according to Massimi, the interplay between nomological background knowledge and imagination is different from that underlying LB-conceivability. She claims that knowledge of laws sustains ("drives") analogical reasoning with models from other fields, guiding the construction of a model indicating what is causally possible for the target system(s) of interest. Massimi's example here is Maxwell's construction of the molecular vortex model to derive the equations describing how electric and magnetic fields are generated by charges, currents, and changes in the fields. The molecular vortex was imagined in analogy with better-understood systems in other fields, specifically hydrodynamics. Maxwell drew on Faraday's law of electromagnetic induction but also on Helmholtz's equations for fluid dynamics in imagining the system from which he could infer what possibly caused electromagnetic induction. It might be tempting to write off the modeling here as a "mere" creative exercise—part of the context of discovery, as it were—that turned out to be very fruitful in the sense of leading Maxwell to formulate a successful theory of actual phenomena. But Massimi clearly states that Maxwell's ether model "delivered *modal knowledge* about what is causally possible in the phenomenon of electromagnetic induction" (Massimi 2019, 872, our emphasis).[16] Since modal knowledge requires both that the possibility claim(s) in question are modal truths and that there is *justification* for taking them to be so, the LD-conceivability is plausibly read as an attempted answer to the epistemic question.

This specifically law-driven analogical reasoning seems interestingly different from the routes to modal knowledge that have been discussed in the modal epistemology literature to date. But at the same time, it is less than clear just how the justificatory route goes and how, if at all, it can be generalized beyond this particular case. Let's see whether the prism of modal epistemology can nevertheless be helpful in sketching an elucidation here.

The analogical reasoning component might be taken to suggest that what we have is some form of similarity-based reasoning, which, as we saw, is among the proposed strategies for modal justification. However, as we discussed in the previous section, similarity notions familiar from philosophy of science discussions of modeling are not directly applicable here—this is not about the similarity of *representans* to *representandum*, but of possible and possibly nonactual *representanda* to actual instances of another domain. In short, the idea would be that with modeling involving LD-conceivability, justification for the modal claims comes from similarity judgments between some source domain (e.g., hydrodynamics) and the target domain (e.g., electromagnetism). In Roca-Royes' similarity-based modal epistemology, what underwrites reasoning from the fact that *a* is F to the conclusion that *b* could possibility be F is the fact that *a* and *b* are similar in the sense that they share some relevant feature(s), and/or are tokens of the same type. While this perspective on similarity for modal purposes seems novel for philosophy of science, it will require substantial refinements before it can be fruitfully applied. Questions to be answered are: what are the objects of the similarity comparison—token objects, or properties of objects?[17] What role do laws of nature play in this—do they act as constraints, or are they subject to the similarity comparisons themselves? Whether this is indeed a workable way to flesh out the LD-conceivability is an open question—a question, however, that modal epistemology has helped raise.

1.5.3 *Universality*

Collin Rice has argued that minimal models enable scientists to draw true counterfactual conclusions about targets.[18] He also attempts to address the epistemic question that arises from this modal modeling claim:

> [I]t seems somewhat mysterious how holistically distorted models can provide *true* counterfactual information about their target systems (...) I will try to offer one possible solution to this problem by appealing to *universality*.
>
> (Rice 2018, 2812)

Suppose one knew that a given model system, despite being highly dissimilar to the target system of interest, was disposed to behave in largely the same way as the target system. In that case, you could justifiably use the model in order to learn about what would happen to the target system under such-and-such circumstances. The idea is that minimal modeling allows us to *delineate universality classes.* Universality, Rice writes, is a "convenient feature of our universe," which in its most general form, is just "the fact that (perhaps extremely) different physical systems will display similar macrobehaviors that are largely independent of the details of their physical components" (Rice 2018, 2812). Whether some systems are in the same universality class is an empirical question—something to be *discovered* (Rice 2018, 2813; Rice 2019, 200).

This discovery is what justifies scientists in using a certain model system to elicit modal information relevant to explaining the target phenomenon. Once we know that some idealized model and target system(s) are in the same universality class, we can draw counterfactual conclusions about the target based on what we learn from the model. Examples of this include the Lattice Gas Automaton (LGA) model vs. real fluid flow, Fisher's linear substitution cost model vs. sex ratios in various animal populations (both in Batterman and Rice 2014), and an optimal foraging model in an infinite population vs. Eider duck foraging behavior (Rice 2018).

The universality account, we think, employs a justificatory strategy that has affinities with that of similarity-based modal epistemologies. This may, at first, seem to go against what Rice claims. Part of what motivates the universality account of minimal modeling is, according to Rice, that the model and target need not be similar in the sense of sharing features or causal mechanisms, in order for us to use the model to learn counterfactual truths about the target. However, what underwrites reasoning from model to modal conclusion about target is the similarity of token behavior, not the similarity of the reasons or causes of that behavior—to be in the same universality class is to exhibit the same macrobehavior. It is similarity in quite a different sense, but a form of similarity all the same. Of course, the idea is that we don't have the similarity knowledge *prior* to the modeling but rather acquire it through the modeling. This is surely significant, but, again, *insofar as we are to draw modal conclusions* about actual targets based on the models, knowledge of this similarity in macrobehavior is apparently a precondition and so, highly relevant to the justification of the modal claims in question.

A looming question for similarity-based modal epistemologies—and one that also arises for standard accounts of scientific modeling that appeal to similarity between model and target features—is that of just what a *relevant* similarity is. As the conceptual problems of a binary similarity relation are well known, *what* the similarity comparison should include is typically made dependent on the purpose of the modeling exercise (Giere 1988). Indeed, Rice notes that "the universality class required to justify a particular instance of idealized modeling will depend on the details of the modeling context; e.g., the target explanandum" (Rice 2018, 2816). That is, *what* behavioral similarity is needed depends on the counterfactual information one is after. But this admission casts into bold relief the fact that the universality account, just like similarity-based modal epistemology, faces the non-trivial question of how we come by the knowledge, or ability to reliably judge, what similarities (or dissimilarities) are relevant to a prospective possibility. It seems intuitively obvious that not just any accidentally identified behavioral similarity between model and target would license modal inference. This question seems, at least initially, somewhat more complicated for the universality account compared to, for example, Roca-Royes' similarity-based modal epistemology. There, the assumption is that individuals with the same properties have the same causal profile, and so coming to know that things are similar justifies concluding that they are disposed to behave similarly (although the question of *relevance* remains). But on the universality account, one needs to establish a similarity of behavior *independent of* the instantiation of certain properties or sameness of type.

However, unlike the discussion so far in modal epistemology, the universality account interestingly offers a strategy to meet these two challenges. The general idea is that the larger minimal modeling practice will help reveal that a model system and a target are in the same universality class. First, as we already discussed, one establishes *behavioral* similarity across different relevant systems. Second, one investigates how counterfactual changes in the features of interest impact this behavior. The resulting counterfactual relations point to dependencies in the behavioral outcomes of the respective systems that therefore should be taken as the relevant similarities. Clearly, these counterfactuals are modal information—thus making the modeling practices here a kind of modal modeling, albeit in a different way than the accounts discussed so far. Third, by performing transformations on a set of models and showing that their behavior converges despite these changes, one justifies inferences to *counterfactual independence*—that is, information about what changes could happen to the system and not make a difference to the observed behaviors. These inferences establish not just *that* the behaviors are similar but also why they are despite all the differences between the systems.

This is significant, and perhaps modal epistemologists can learn from this example how the questions raised by similarity accounts may be addressed. Yet, it is important to note that this procedure seems more plausible in some of the example cases than in others, and hence it is unclear how far, if at all, it generalizes. The case of the LGA model stands out because of the many constraining assumptions it is based on. It relies on the renormalization group strategy, illustrated at the hand of the Kadanoff block spin transformation. Here, the procedure plausibly identifies a possibility space, determined by assumptions of what entities systems consist of and how they interact. It also offers a plausible rescaling procedure and an easy way to infer the macrobehavior of these scaled-up entities. However, in the other minimal modeling cases discussed, the robustness analyses supporting the delineation of a universality class are nowhere near as constrained or thorough. For instance, in the Eider foraging case, no attempt was made to systematically describe the possibility space. Nor was there any attempt to re-scale the models and compare their predictions. Rice only compares three model predictions with actual Eider behavior, concluding that one fits the data better than the other two. It is far from clear that mere comparative similarity of model prediction and empirical data licenses counterfactual conclusions.

In sum, answering the epistemic question for (some cases of) modal modeling in terms of universality classes relies on a strategy that has affinities with similarity-based modal epistemology. Seeing this also enables us to highlight two challenges for such modal epistemologies that carry over to the universality account: first, it is a highly non-trivial matter to establish that there *is* relevant similarity of behavior, given the counterfactual conditionals one is interested in with respect to a target. Second, the universality account inherits the difficult question of relevance, in particular, regarding which behavioral similarities are relevant to justify the counterfactual conclusions of interest. While similarity-based modal epistemology does not provide solutions, seeing these practices through the modal-epistemology prism helped identify what epistemic issues they face and to what extent their strategies can solve them.

1.6 Conclusions

Philosophers of science who suggest that scientists legitimately draw modal conclusions based on certain modeling practices face the epistemic question of modal modeling: under what conditions and in virtue of what do models provide reasons for accepting certain modal claims as true? In this paper, we looked at some answers to this question through the prism of modal epistemology. Our goal was to identify common themes regarding the justification of modal claims. By way of demonstrating this, we introduced three dominant themes from modal epistemology that, although not specifically addressing the role of modeling, identify three general justificatory strategies for modal claims that are applicable to scientific modeling. We then argued that these themes are very much present also in three attempts to answer the epistemic question for modal modeling in the philosophy of science. Seeing just what the underlying justificatory strategies are will be important when one wants to assess the proposed accounts in terms of, for example, viability and how far, if at all, they generalize beyond the particular cases discussed.

Our second goal was to articulate challenges for the philosophy of modal modeling and see to what extent current accounts have the resources to meet them. Scrutinizing the modal modeling accounts through the prism of modal epistemology, we argued that they face (versions of) many of the same challenges that are familiar from modal epistemology more generally. This includes the relationship between background knowledge and imagination and the importance of addressing the question of whether, and if so how, the relevant epistemic subjects possess the knowledge in question. It also includes a model for specifically *modal* inference based on similarity judgments, accompanied by a number of thorny questions concerning how the relevant similarity relations are to be articulated and how epistemic subjects can come to know about them.

Once we see how modal modeling accounts are subject to analogues of challenges facing modal epistemologies, we can use these insights to think about further questions that they raise for the philosophy of modal modeling specifically. Here we will just mention two. First, while it may seem trivial that modal justification often relies on having appropriate background knowledge, an extra wrinkle is added when we consider the contexts in which modal modeling is often said to occur. Practices like HPE and exploratory modeling are considered important means for advancing research in areas where there is a *lack* of confirmed theory or shared background knowledge, or as ways to *challenge* the background assumptions in a field. This highlights the delicate task of striking a balance with respect to the constraining background knowledge required. Second, modal epistemologists disagree on whether the imagination *as such* can provide modal justification or whether it is merely an important cognitive tool for exploration of modal implications or consequences of one's background knowledge. If the latter, questions about background knowledge become pertinent, as already noted. If the former, the question is what mode or kind of imagination *can* do justificatory work. This is a question that has been discussed in modal epistemology, but also, we think in the philosophy of science[19]—although the two literatures remain somewhat disconnected. Here, we see fruitful opportunities for collaboration, especially between modal epistemologists and philosophers of science within the broadly "fictionalist" camp, who take great interest in imagination and its role in how scientific modeling (modal and non-modal) supports, for example, explanation.

These are, in our view, just a few examples of how exchanges between work on modal modeling and modal epistemology more generally can prove fruitful. While we have focused here on how modal epistemology can inform philosophy of science, we think there are good reasons to suspect that there can be very useful influences going also in the other direction—indeed, some of the discussion above already indicates this.

Acknowledgments

Thanks to Tarja Knuuttila, Michela Massimi, Collin Rice, and the audience at *Modeling the Possible* online workshop series at the University of Vienna for their helpful comments. We also want to acknowledge funding from the Swedish Research Council, grant nos. 2018-01353 and 2019-00635.

Notes

1 Sjölin Wirling (2022) is an exception. The potential relevance of models to modal justification in science has also been noted by, for example, Nolan (2017).

2 The prism analogy also indicates the direction of our discussion: we look at how philosophers of science have discussed modal modeling practices by taking up concepts, arguments, and challenges from modal epistemology. We do not deny that the inverse perspective might be fruitful as well, but we reserve such treatment for a later paper.

3 Note that this doesn't preclude that these very practices may *also* directly deliver information about what is actually the case. But we focus here on the *modal* information they are said to deliver.

4 This is *not* to suggest that modeling is the only, or the main, way for scientists to come by modal knowledge—clearly, there needs to be other means. The modal modeling claim is simply that modeling is *one* way for scientists to arrive at justified modal claims.

5 It is also a question that might demand a different answer depending on the kind of modality that is at issue (Sjölin Wirling and Grüne-Yanoff forthcoming).

6 Of course, some pluralism is to be expected here given the diversity of modal models. See also endnote 5.

7 Notably, the themes cut across the distinction between rationalist and non-rationalist modal epistemologies, in the sense that both rationalists and non-rationalists have used them when constructing their accounts of modal knowledge—but they differ on whether they take, for example, the relevant strategy to be *a priori* or not. We recognize that philosophers of science will presumably be interested exclusively in the non-rationalist versions of these modal epistemologies, but insofar as the central themes go, rationalist modal epistemologies may have insights to offer, too.

8 An initial reaction to this problem was to say that we cannot really imagine impossibilities—at most, it *seems to us* that we imagine something impossible, while we are in fact imagining a possible situation that we mistake for and misdescribe as the impossible one (Kripke 1980). Kung (2016) provides a convincing case against such error-theoretic approaches.

9 Of course, a law-based strategy has limited relevance for scientific fields such as biology or economics where there are few or no laws.

10 Philosophers disagree on whether natural and metaphysical modality come apart. So-called necessitarians think that they do not.

11 Sugden's own view is that we can *also* draw general, inductive conclusions about how things are in the actual world, on the basis of these credible models. We set the latter part of his view aside.

12 The toy model scenarios *themselves* are often physically impossible since they involve impossible assumptions (e.g., infinite population size in the Hawk-Dove model, and limitless supplies of food for prey in the Lotka-Volterra model). This indicates that when a certain model world is credible, this means at most that some part of it is possible or that some de-idealized description of it is possible.

13 For a more detailed version of this analysis, see Sjölin Wirling (2021).

14 See also Gelfert's comment that minimal models "shine a spotlight on the essential character of a phenomenon" (Gelfert 2019, 10–11).
15 As we argue in Sjölin Wirling and Grüne-Yanoff (forthcoming), we think Massimi's account is regrettably unclear on what the relevant notion of possibility, to which physical conceivability is supposedly a guide, is supposed to be. Insofar as it is supposed to be a guide to objective possibility, we think the definition of physical conceivability needs to be amended. We suppress these concerns here, however.
16 Compare Hon and Goldstein (2021, 326) who also conclude that Maxwell's use of modeling "concerns physically possible causes and is no longer based solely on analogy."
17 Hesse (1963) is an early proponent of understanding similarity in this way. One of her examples is that sound and light are similar in this sense because echoes are similar to reflection, loudness to brightness, pitch to color, and so on. See Khosrowi (2020) for a more recent appeal to similarity as the similarity between properties and Sjölin Wirling (forthcoming) for another way in which this may be relevant to modal modeling.
18 This originally builds partly on joint work with Robert Batterman (see Batterman and Rice 2014).
19 See, for example, Salis (2020).

References

Akerlof, George A. 1970. "The Market for 'Lemons': Quality Uncertainty and the Market Mechanism." *The Quarterly Journal of Economics* 84(3): 488–500.
Arrow, Kenneth J, and Gerard Debreu. 1954. "Existence of an Equilibrium for a Competitive Economy." *Econometrica* 22: 265–90.
Batterman, Robert. W, and Collin Rice. 2014. "Minimal Model Explanations." *Philosophy of Science* 81(3): 349–76.
Bealer, George. 2002. "Modal Epistemology and the Rationalist Renaissance." In *Conceivability and Possibility*, edited by T. S. Gendler, and J. Hawthorne, 71–125. Oxford: Clarendon Press.
Bokulich, Alisa. 2014. "How the Tiger Bush Got Its Stipes: 'How Possibly' vs. 'How Actually' Model Explanation." *The Monist* 97(3): 321–38.
BonJour, Laurence. 1997. *In Defense of Pure Reason: A Rationalist Account of A Priori Justification*. Cambridge: Cambridge University Press.
Bueno, Otávio, and Scott A Shalkowski. 2014. "Modalism and Theoretical Virtues: Toward an Epistemology of Modality." *Philosophical Studies* 172(3): 671–89.
Dohrn, Daniel. 2019. "Modal Epistemology Made Concrete." *Philosophical Studies* 176(9): 2455–75.
Fischer, Bob. 2017. *Modal Justification via Theories*. Cham: Springer International.
Frigg, Roman, and James Nguyen. 2016. "The Fiction View of Models Reloaded." *The Monist* 99(3): 225–42.
Fumagalli, Roberto. 2016. "Why We Cannot Learn from Minimal Models." *Erkenntnis* 81(3): 433–55.
Gelfert, Axel. 2018. "Models in Search of Targets: Exploratory Modeling and the Case of Turing Patterns." In *Philosophy of Science. European Studies in Philosophy of Science*, vol 9., edited by A. Christian, D. Hommen, N. Retzlaff, and G. Schurz, 245–69. Cham: Springer International Publishing.
Gelfert, Axel. 2019. "Probing Possibilities: Toy Models, Minimal Models, and Exploratory Models." In *Model-Based Reasoning in Science and Technology*, edited by M. Fontaine, C. Barés-Gómez, F. Salguero-Lamillar, L. Magnani, and Á. Nepomuceno-Fernández, 3–19. Cham: Springer International.

Gendler, Tamar, and John Hawthorne. 2002. *Conceivability and Possibility*. Oxford: Clarendon Press.

Giere, Roland N. 1988. *Explaining Science: A Cognitive Approach*. Chicago: University of Chicago Press.

Grüne-Yanoff, Till. 2009. "Learning from Minimal Economic Models." *Erkenntnis* 70(1): 81–99.

Grüne-Yanoff, Till. 2013. "Appraising Models Nonrepresentationally." *Philosophy of Science* 80(5): 850–61.

Hesse, Mary B. 1963. *Models and Analogies in Science*. Notre Dame, IN: Notre Dame University Press.

Hon, Giora, and Bernard R Goldstein. 2021. "Maxwell's Role in Turning the Concept of Model Into the Methodology of Modeling." *Studies in History and Philosophy of Science* 88: 321–33.

Jago, Mark. 2018. "Knowing How Things Might Have Been." *Synthese* 198: 1981–99.

Khosrowi, Donal. 2020. "Getting Serious about Shared Features." *British Journal for the Philosophy of Science* 71(2): 523–46.

Kment, Boris. 2021. "Essence and Modal Knowledge." *Synthese* 198(S8): S1957–79.

Knuuttila, Tarja, and Rami Koskinen. 2020. "Synthetic Fictions: Turning Imagined Biological Systems into Concrete Ones." *Synthese* 198: 8233–50.

Kripke, Saul. 1980. *Naming and Necessity*. Oxford: Blackwell.

Kung, Peter. 2010. "Imagining as a Guide to Possibility." *Philosophy and Phenomenological Research* 81(3): 620–63.

Kung, Peter. 2016. "You Really Do Imagine It: Against Error Theories of Imagination" *Noûs* 50(1): 90–120.

Lowe, E. J. 2012. "What Is the Source of Our Knowledge of Modal Truths?" *Mind* 121: 915–50.

Mallozzi, Antonella. 2018. "Putting Modal Metaphysics First." *Synthese* 198(S8): 1937–56.

Mäki, Uskali. 2001. "The Way the World Works (www): Towards an Ontology of Theory Choice." In *The Economic World View: Studies in the Ontology of Economics*, edited by Uskali Mäki, 369–89. Cambridge: Cambridge University Press.

Mäki, Uskali. 2009. "MISSing the World. Models as Isolations and Credible Surrogate Systems." *Erkenntnis* 70(1): 29–43.

Massimi, Michela. 2019. "Two Kinds of Exploratory Models." *Philosophy of Science* 86(5): 869–81.

Maynard Smith, J., and George R Price. 1973. "The Logic of Animal Conflict." *Nature* 246: 15–18.

Nguyen, James. 2020. "It's Not a Game: Accurate Representation with Toy Models." *British Journal for the Philosophy of Science* 71(3): 1013–41.

Nolan, Daniel. 2017. "Naturalised Modal Epistemology." In *Modal Epistemology After Rationalism*, edited by B. Fischer, and F. Leon, 7–28. Cham: Springer International.

Nolan, Daniel. 2021. "Impossible Fictions Part I: Lessons for Fiction." *Philosophy Compass* 16: e12723.

Reutlinger, Alexander, Dominik Hangleiter, and Stephan Hartmann. 2018. "Understanding (with) Toy Models." *British Journal for the Philosophy of Science* 69(4): 1069–99.

Rice, Collin. 2018. "Idealized Models, Holistic Distortions, and Universality." *Synthese* 195(6): 2795–819.

Rice, Collin. 2019. "Models Don't Decompose That Way: A Holistic View of Idealized Models." *British Journal for the Philosophy of Science* 70(1): 179–208.

Roca-Royes, Sonia. 2011a. "Conceivability and De Re Modal Knowledge." *Noûs* 45(1): 22–49.

Roca-Royes, Sonia. 2011b. "Modal Knowledge and Counterfactual Knowledge." *Logique Et Analyse* 54(216): 537–52.
Roca-Royes, Sonia. 2017. "Similarity and Possibility: An Epistemology of De Re Possibility for Concrete Entities." In *Modal Epistemology After Rationalism*, edited by B. Fischer, and F. Leon, 221–245. Cham: Springer International.
Rohwer, Yasha, and Collin Rice. 2013. "Hypothetical Pattern Idealization and Explanatory Models." *Philosophy of Science* 80(3): 334–55.
Salis, Fiora. 2020. "Learning through the Scientific Imagination." *Argumenta* 6(1): 65–80.
Schelling, Thomas C. 1971. "Dynamic Models of Segregation." *Journal of Mathematical Sociology* 1(2): 143–86.
Sjölin Wirling, Ylwa. 2021. "Is Credibility a Guide to Possibility? A Challenge for Toy Models in Science." *Analysis* 81(3): 470–8.
Sjölin Wirling, Ylwa. 2022. "Extending Similarity-Based Epistemology of Modality with Models." *Ergo: An Open Access Journal of Philosophy* 4: 45.
Sjölin Wirling, Ylwa, & Grüne-Yanoff, Till. Forthcoming. "Epistemic and Objective Possibility in Science." *British Journal for the Philosophy of Science*
Sjölin Wirling, Ylwa, and Till Grüne-Yanoff. 2021. "The Epistemology of Modal Modeling." *Philosophy Compass* 16(10): e12775.
Sugden, Robert. 2011. "Explanations in Search of Observations." *Biology and Philosophy* 26: 717–36.
Sugden, Robert. 2000. "Credible Worlds: The Status of Theoretical Models in Economics." *Journal of Economic Methodology* 7(1): 1–31.
Tahko, Tuomas E. 2012. "Counterfactuals and Modal Epistemology." *Grazer Philosophische Studien* 86(1): 93–115.
Vaidya, Anand J, and Michael Wallner. 2018. "The Epistemology of Modality and the Problem of Modal Epistemic Friction." *Synthese* 198(S8): 1909–35.
Verreault-Julien, Philippe. 2019. "How Could Models Possibly Provide How-Possibly Explanations?" *Studies in History and Philosophy of Science Part A* 73: 22–33.
Verreault-Julien, Philippe. 2017. "Non-Causal Understanding with Economic Models: The Case of General Equilibrium." *Journal of Economic Methodology* 24(3): 297–317.
Weisberg, Michael. 2013. *Simulation and Similarity: Using Models to Understand the World*. Oxford: Oxford University Press.
Williamson, Timothy. 2007. *The Philosophy of Philosophy*. Malden: Blackwell.
Wilson, Alastair. 2020. *The Nature of Contingency*. Oxford: Oxford University Press.
Yablo, Stephen. 1993. "Is Conceivability a Guide to Possibility?" *Philosophy and Phenomenological Research* 53(1): 1–42.
Ylikoski, Petri, and Emrah N Aydinonat. 2014. "Understanding with Theoretical Models." *Journal of Economic Methodology* 21(1): 19–36.

2 Which Possibilities Contribute to Scientific Understanding?

Collin Rice

2.1 Introduction

Both philosophers of science and epistemologists have recently focused attention on how scientific models are used to provide understanding (de Regt 2017; Elgin 2017; Gelfert 2019; Khalifa 2017; Potochnik 2017; Rice 2019; Saatsi 2019; Sjölin Wirling and Grüne-Yanoff 2021; Strevens 2013).[1] Several of these accounts have suggested that scientific models provide understanding by allowing scientists to answer what-if-things-had-been-different questions via the investigation of counterfactuals or possibilities (Le Bihan 2017; Rice 2019, 2021; Saatsi 2019). In other words, scientific models enable understanding by providing modal information about various possible states of their target system(s)—or of other possible systems—and how the phenomenon of interest would (or would not) change in those counterfactual situations. However, a crucial question for this prominent approach to understanding is, *which* possibilities scientists ought to investigate to generate (or deepen) scientific understanding? The prominent views of understanding found in the literature seem to suggest three possible answers: (1) accuracy/ closeness with respect to the actual world (Strevens 2013; Trout 2007), (2) the counterfactual situations relevant for evaluating the outcomes of interventions (Douglas 2009; Potochnik 2017; Woodward 2003), or (3) the possibilities of interest to the scientists using the model or theory (Elgin 2017; Potochnik 2017; Saatsi 2019).

In this chapter, I use examples from scientific practice to argue that each of these proposals fails to capture the range of ways idealized models enable understanding in science. Accounts that require accuracy (or closeness) with respect to the actual world miss the understanding enabled by scientific models that describe possibilities in which difference-making factors of the real system are highly idealized. Interventionist accounts miss the understanding enabled by investigating distant possibilities that do not provide information about the results of manipulating real systems. Accounts focused on the interests of scientists miss the understanding provided by learning about possibilities that are of interest to non-scientists (but not scientists).

DOI: 10.4324/9781003342816-4

In contrast with these three approaches, I will argue that *learning about what would occur in a possible system, or a possible state of a real system, always contributes to scientific understanding of real phenomena*—although sometimes in ways that are not currently of interest to anyone. More specifically, I will contend that grasping more possibilities, more relationships between those possibilities, or answers to more what-if-things-had-been-different questions always deepens one's scientific understanding of a phenomenon to some degree.[2]

It is worth noting from the outset that I think what makes these examples instances of *scientific* understanding is only that the understanding produced comes from investigating scientific models (or theories), methods, or evidence. However, since I do not think there is a hard/clear demarcation between scientific and non-scientific ways of understanding, the arguments presented here can easily be expanded to address questions about which possibilities enable understanding more generally.[3]

Moreover, since I maintain that understanding *always comes in degrees* rather than being a threshold phenomenon, I will argue that grasping these possibilities provides genuine understanding (and deepens it) without requiring that the agent first meet some kind of *minimal* understanding threshold. While my claim that grasping possibilities deepens understanding is compatible with stipulating some further minimal requirements for understanding (e.g., as Khalifa 2017 does), I think the cases and arguments show that grasping these possibilities provides (and deepens to some degree) understanding independent of such requirements.

The account of understanding defended here is still interested in grasping truths, but it includes many modal truths about possibilities that go well beyond grasping truths about the actual world.[4] I will also defend the claim that grasping these non-actual possibilities generates scientific understanding even if one fails to grasp what is true of the actual world—i.e., grasping the actual world is not necessary for scientific understanding. While some proponents of modal accounts of understanding seem to hint at this particular view (e.g., Le Bihan 2017; Verreault-Julien 2019; Rice 2021), it has not been independently argued for. My aim here is to independently motivate what I will call the All-Possibilities-Count (APC) view and explore some of its implications for philosophical theorizing about the nature of (scientific) understanding.

In section 2.2, I argue against accounts that appeal to accuracy or closeness with respect to the actual world to determine whether a scientific model allows for genuine understanding. Then, in section 2.3, I critique accounts that appeal to considerations of intervention/manipulation to determine which possibilities are relevant for scientific understanding. Next, in section 2.4, I argue that we should also reject accounts that appeal to the interests of scientists to determine which possibilities enable scientific understanding. In response to these cases, section 2.5 proposes the APC view. Section 2.6 responds to some possible objections, and section 2.7 concludes.

2.2 Closeness to the Actual World Improves Scientific Understanding

In order to get a handle on the nature of scientific understanding, several philosophers have appealed to the understanding provided by scientific *explanations* (de Regt 2017; Khalifa 2017; Strevens 2008, 2013; Trout 2007). While it is widely agreed that explanations are a primary way that science generates understanding, some philosophers have gone even further in arguing that scientific understanding (or, at least, understanding why) is *only* produced by grasping an explanation. Moreover, several of these accounts claim that scientific understanding is only produced by grasping a *correct* or *true* explanation. For example, J. D. Trout argues that "scientific understanding is the state produced, and only produced by grasping a *true* explanation (Trout 2007, 585–6, my emphasis). Similarly, Michael Strevens contends that "An individual has scientific understanding of a phenomenon just in case they grasp a *correct* scientific explanation of that phenomenon" (Strevens 2013, 510, my emphasis). Along similar lines, Kareem Khalifa argues that "S has minimal understanding of why p if and only if, for some q, S believes that q explains why p, and q explains why p *is approximately true*." (Khalifa 2017, 126, my emphasis). Indeed, ever since Hempel argued that the premises of a deductive-nomological explanation needed to be true (to make the argument sound), many philosophers have held that scientific explanations, and the understanding derived from them, must be based on (approximately) true (or correct) descriptions of the actual world.

One interpretation of this kind of view is that scientific understanding is only generated when the agent grasps information about the actual world. Indeed, Strevens first tells us that part of grasping an explanation involves grasping "that the states of affairs represented by the propositions [in the explanans] in fact obtain" (Strevens 2013, 511). Moreover, he notes that "In my view, an explanation is correct only if its constitutive propositions are true" (Strevens 2013, 512). Thus, it seems that Strevens's account of scientific understanding implies that grasping states of affairs in the actual world matters most for determining whether someone has scientific understanding. Or, at least, grasping what is true of the actual world seems to be a necessary condition for having any scientific understanding of the phenomenon. The argument for this conclusion seems to be something like the following:

P1. Grasping a correct explanation is required for scientific understanding. (Strevens's simple view of understanding).
P2. Correct explanations must describe features of the actual world that are responsible for (e.g., difference-making causes of) the explanandum.
C. Therefore, grasping what is true of the actual world is necessary for scientific understanding.

A major challenge for this kind of view is that many models that enable scientific understanding are *highly idealized* and describe possibilities that are quite different from the actual world. Of course, both Strevens and Khalifa explicitly recognize that many of the explanations provided by scientists employ idealizations in a variety of ways (Khalifa 2017; Strevens 2013). Strevens responds by arguing that the *explanatory content* of these idealized models is nonetheless true because the idealizations "in fact make true claims about difference-making" (Strevens 2013, 512). Specifically, in Strevens's view, the idealizations involved in scientific explanations claim, correctly, that the idealized features made no difference to the explanandum (Strevens 2008, 2013).

While this enables Strevens's view to capture some additional cases, this kind of view fails to capture the sundry idealized models that provide scientific understanding by directly distorting difference-making features (Elgin 2017; Potochnik 2017; Rice 2018, 2021). The explanatory content of these models cannot be rendered true (of the actual world) by arguing that the distorted features do not make a difference to the explanandum. As a quick example, optimality models are routinely used to generate understanding of biological traits. However, despite the scientists' assumption that the processes involved in the selection of the trait are the key difference-making features, these adaptationist models often directly distort those processes by describing selection as taking place in an infinitely large population, with random mating, no intergenerational overlap, a perfect correlation between some resource and fitness, fixed environment, etc., (Potochnik 2007; Rice 2013, 2021). In other words, these models generate understanding of biological traits by describing a counterfactual situation that is drastically different from the actual systems in which these traits evolved—even if we restrict our focus to difference-making features. As a result, these models' idealizations cannot be (re)interpreted as making true claims about the lack of difference-making in the actual world. Nonetheless, grasping what would occur in the possible states of real systems described by various optimality models has clearly advanced biologists' understanding of various adaptive traits.

In response to these kinds of cases, an alternative version of the above approach might appeal to *closeness* to the actual world to determine the degree to which an individual understands the phenomenon. This would allow for some departures from the actual world, while still maintaining an important role for correctness in determining the *degree* to which one understands. Indeed, Strevens tells us that the correctness of an explanation comes in degrees and that this directly relates to the degree to which the individual who grasps that explanation understands. Similarly, Khalifa's view requires understanding to be provided via explanations that only need to be *approximately* true and his view allows for a plethora of ways that this empirical requirement might be satisfied (Khalifa 2017, Ch. 6). Thus, according to an alternative interpretation of these views, the closer to the actual world (i.e., the more correct or the more empirically confirmed) the model is, the greater the degree of understanding the model will provide. I do not know of any philosophers that have explicitly endorsed this kind of "closeness" view, but it is certainly implied by several accounts that suggest that more accuracy with respect to the actual causes or mechanisms that produced the phenomenon will provide a better explanation and, thereby, a deeper understanding of the phenomenon (Craver 2007; Strevens 2008). This seems to imply that, given two models for the same real-world phenomenon, if model M_1 represents a counterfactual situation that is closer to the actual world than model M_2—at least with respect to the difference-making features—then *ceteris paribus* grasping the explanation provided by model M_1 will produce a greater degree of understanding than grasping the explanation provided by M_2. This kind of view allows models that distort difference-making features to provide *some* understanding of the phenomenon—although a greater degree of understanding would be produced by a model that more accurately represented the system's difference-making features.

While this alteration helps us see how models that misrepresent difference-making features can improve scientific understanding, several cases from scientific practice make it clear that closeness to the actual world does not always improve the degree to which one understands. In particular, scientific models that describe systems that depart from the actual world in rather drastic ways can sometimes improve our understanding of a phenomenon better than models that are closer to the actual world. For example, the Hardy–Weinberg (HW) equilibrium model describes a counterfactual situation in which the population is infinitely large, organisms mate randomly, and there is no selection, migration, or mutation. These idealizations entail that the model drastically distorts many of the features that make a difference to the real-world phenomena the HW-model is applied to (Levy 2011; Morrison 2015; Stoneking 2017). However, despite these rather drastic departures from the actual world, the HW-model shows us that in such a counterfactual situation, if we have a pair of alleles, A_1 and A_2, at a particular locus and in the initial population the ratio of A_1 to A_2 is p to q, the distribution for all succeeding generations will be:

$$p^2 A_1 A_1 + 2pq A_1 A_2 + q^2 A_2 A_2$$

regardless of the distribution of genotypes in the initial generation. Although this situation is rather distant from what occurs in any actual biological population, "The Hardy-Weinberg law enables us to understand fundamental features of heredity and variation" (Morrison 2009, 133). The question, of course, is what does this understanding consist of? Elsewhere, I have suggested that one way the HW-model provides understanding is by enabling scientists to answer a range of what-if-things-had-been-different questions concerning a rather distant counterfactual situation (Rice 2019, 2021). Specifically, the model shows scientists what would occur in a system in which the features idealized by the model were radically different from how they are in the real-world case(s). Moreover, that distant possibility is precisely the counterfactual situation of interest to biologists seeking to (artificially) isolate the heredity processes involved in the transmission of variation across generations from other evolutionary factors that make a difference in every real-world population.

If this is correct, then we might ask whether a model that more accurately represented (or approximated) real biological populations would enable a greater degree of understanding of the phenomenon of interest. Here, I argue, the answer is no. The primary reason is that, without those specific idealizations, the mathematical framework used in deriving these results within the HW-model would no longer be applicable (Morrison 2015; Rice 2021). Thus, removing (or relaxing) those idealizations would result in the model no longer being able to display the key features of heredity and variation that the biologists who use the model aim to understand. Put differently, the drastic idealizations employed by the HW-model are what enable it to answer precisely those what-if-things-had-been-different questions that are of interest to the biologists. Thus, while a more accurate model may be able to provide alternative pieces of understanding (e.g., about what would happen under actual circumstances), the what-if-things-had-been-different questions of interest in this case are best answered, I argue, by exploring the particular counterfactual situations described by the HW-model. Moreover, a model without those idealizations would not allow for the derivations of stability of variation across generations and so would not enable the desired understanding. As a result, the understanding provided by the HW-model would not be provided by a model that more accurately described the difference-making features of real biological populations. I conclude that, in certain scientific contexts, grasping what would occur in a more distant counterfactual situation can promote scientific understanding to a greater degree than grasping a possibility that is closer to the actual system.

2.3 Interventionist Counterfactuals, Causal Patterns, and Scientific Understanding

One way to try to capture the above kinds of cases is to suggest that these more distant possibilities enable understanding because they inform us about the outcomes of interventions. In other words, one might take these cases to show that scientific models describe possibilities that improve our understanding by allowing us to investigate worlds whose consequences we could bring about via some (hypothetical) manipulation of the features of real systems. Since (multiple) interventions can bring about a drastically different system, models that investigate distant possibilities can yield understanding by telling us about what would happen as a result of (perhaps multiple) interventions. Although this view hasn't been explicitly argued for in the literature, as Philippe Verreualt-Julien notes, "The epistemology of understanding has traditionally been related, if not reduced, to the epistemology of causal explanations" (Verreualt-Julien 2019, 17).

For example, James Woodward seems to hint at something like this view at various points throughout his work on causal explanation. Although Woodward never explicitly presents a theory of scientific understanding, he states that "explanations provide understanding by exhibiting a pattern of counterfactual dependence between explanans and explanandum—a pattern of counterfactual dependence of the special sort associated with relationships that are potentially exploitable for purposes of manipulation and control." (Woodward 2003, 13). This seems to suggest that most scientific understanding (or at least the kind provided by explanations) is produced via information about counterfactuals that describe the outcomes of possible interventions. Indeed, Woodward also tells us that one of his goals is to capture "how important our ability to intervene and manipulate nature is in the development of a scientific understanding of nature" (Woodward 2003, 12). As a result, one might think it is learning about various counterfactual situations involving the outcomes of interventions that are the key to scientific understanding.

More recently, Angela Potochnik (2017) has argued that all scientific understanding is achieved through the grasping of *causal patterns* (Potochnik 2017, 91). Moreover, she argues that "these patterns qualify as causal on Woodward's manipulability approach to causation" (Potochnik 2017, 95). However, since we can improve our grasp of these causal patterns via idealized representations that depart significantly from the truth, Potochnik argues that it is information about these patterns of manipulability—rather than accuracy or truth—that promotes understanding. In short, scientific understanding is improved by investigating possible (idealized) systems that enable us to learn about causal patterns of manipulability that are embodied in real phenomena. Thus, like Woodward's account, Potochnik's discussion of scientific understanding focuses on the counterfactual situations in which manipulations change some of the causal variables of the real system.

There are two kinds of cases that raise problems for this sort of manipulationist approach to analyzing scientific understanding. The first set of cases involves the use of idealized models to investigate counterfactual dependence relationships used to provide *noncausal* explanations (Ariew, Rice, and Rohwer 2015; Bokulich 2018; Chirimuuta 2018; Lange 2013; Rice 2013, 2021). Because they are not causes of the explanandum, the features changed in these counterfactual situations (typically) do not represent the outcomes of possible interventions or manipulations. For example, many idealized models in science enable understanding by citing counterfactual dependencies between the explanandum and the system's statistical properties, mathematical necessities, topological features, or fundamental tradeoffs (Ariew et al. 2015; Huneman 2010; Lange 2013; Rice 2013; Walsh 2015). These explanations typically work by showing how changes to these noncausal features (e.g., the dimensionality of space-time) would change the occurrence of the explanandum, that is, these models explain by providing counterfactual information about ways the real system might have been different that would have resulted in the explanandum failing to occur. Therefore, if all explanations (at least have the potential to) generate understanding, and some scientific explanations are noncausal, then some scientific understanding is produced without telling us about the outcomes of causal interventions.

Another set of cases involves the use of idealized models to investigate possibilities that are so distant that they are unable to tell us about the outcomes of interventions on real systems. This can happen in at least two different ways. First, as Woodward's account stresses, in order to use an idealized model to learn about the outcomes of an intervention we need to assume that the causal relationships we are investigating can be manipulated independently such that we can 'hold fixed' the rest of the causal processes in the system. The problem is that many idealized models change so many features of real systems simultaneously that they do not enable us to know what would happen if those idealized features were changed individually. As a quick example, the Prisoner's Dilemma certainly helps us understand the evolution of altruism. However, the model does not represent a counterfactual situation that could be brought about via a set of specific (and surgical) interventions on real evolving populations (Rohwer and Rice 2016). Indeed, the model idealizes (i.e., changes) several causal and noncausal features of real populations that cannot be manipulated independently.

A related kind of issue arises when the possible system (or counterfactual situation) of interest is simply too distant from the real world to be generated by (any number of) hypothetical interventions in the real-world. For example, some possibilities that are of interest to scientists involve rather distant ways that the universe might have been had the conditions of the Big Bang been very different from what they actually were. These models change multiple fundamental constants of the universe to show the importance of those constants' particular values for enabling the formation of planets, the evolution of life on Earth, etc. Consequently, cosmological modelers are often interested in exploring a range of possibilities that could not possibly be produced by human intervention (Rice 2021). In short, many of the possibilities investigated by scientists that promote their understanding of the phenomenon are so distant from the actual world that no amount of human intervention could bring them about.

2.4 Scientists' Interests Determine the Possibilities that Promote Scientific Understanding

The cases described in the last section show that scientists are sometimes interested in understanding what occurs in very distant counterfactual situations that go beyond what could possibly be produced via interventions on real-world systems. Put differently, the possibilities of interest to scientists are not restricted to the outcomes of interventions, but instead include a much wider range of possibilities that may be quite unlike our actual world.

Recognizing this, another way philosophers have tried to specify which possibilities improve understanding is to appeal to the possibilities of interest to scientists.

For example, although her account focuses on causal manipulability patterns, Potochnik's view also requires that, for a model to provide understanding, it must depict the causal pattern *of interest to the scientists*. As Potochnik puts it, "which causal pattern is focal, and thus which causal pattern provides understanding, depends on the researcher's specific interests" (Potochnik 2017, 102). Conversely, "Phenomena embody lots of causal patterns; grasping any old causal pattern embodied by some phenomenon won't lead to an understanding of that phenomenon. The grasped causal pattern must relate in the right way to the inquiry for it to produce understanding" (Potochnik 2017, 116). This certainly seems to suggest that the only causal patterns that provide understanding are those that scientists are interested in. Along similar lines, several other philosophers have appealed to the interests of scientists to delimit a set of what-if-things-had-been-different questions of interest for providing an explanation or producing understanding (Elgin 2017; Saatsi 2019; Woodward 2003). According to these views, the possibilities that produce scientific understanding are those that are of interest to scientists within a particular context or research program.

I will now raise some challenges to this kind of view. First, of course, scientific interests change over time. Thus, just because scientists aren't currently interested in some set of possibilities doesn't mean they won't be later. Although both Potochnik and Saatsi discuss how the possibilities of interest to scientists change over time, I argue that whether grasping those possibilities enables us to better understand the phenomenon should not depend on whether those alternative possibilities ever become of interest to scientists. Such a criterion seems rather arbitrary given the kinds of things that influence what scientists study, for example, personal, public, or political interest in certain outcomes (Kourany 2021). What is more, if we consider an individual who has an extensive grasp of what would occur in a range of possibilities it seems odd to withhold our attribution of scientific understanding to them until those possibilities become of interest to practicing scientists. Rather, we should say that *they understood it before it was of interest to scientists*. For example, if someone in the 1800s had been able to correctly answer a range of what-if-things-had-been-different questions regarding the possible outcomes of climate change, we should say that they understood something about climate change even if those possibilities would not become of interest to scientists until many decades later.

More generally, even if scientists are never interested in a set of possibilities, learning about those possibilities can still enable us to answer a range of what-if-things-had-been-different questions and, thereby, deepen our understanding of the phenomenon. Indeed, non-scientists can deepen their understanding of a phenomenon by exploring possibilities or counterfactuals of interest to them—even if the community of scientists does not see those possibilities as relevant or interesting. For example, exploring possible scenarios concerning various outcomes of climate change for particular communities (or localities) improves our understanding of climate change whether or not those particular possibilities are ever explicitly of interest to any scientist. If those possible outcomes are of interest to those living in those areas, or those with particular historical ties to the land, then they are relevant possibilities for deepening human understanding of the phenomenon. In short, I see no reason to privilege the possibilities of interest to scientists over those that are of interest to non-scientists.

A possible reply here would be to suggest that, although exploring possibilities of interest to non-scientists might improve understanding in a more general sense, only those possibilities that are of interest to scientists will be able to generate *scientific* understanding. However, while the interests of scientists are certainly crucial for determining how science is practiced and what understanding the scientific community aims to generate, I think it goes too far to suggest that what is of interest to scientists should be used to *demarcate* what counts as accomplishing the epistemic achievement of scientific understanding. Just as there is lots of knowledge that is not of interest to us, there can be lots of scientific understanding that is not of interest to scientists. Just as I can meet the conditions for knowing things that you are not interested in; people can scientifically understand things that are not of interest to scientists.[5]

Going further, I suggest that grasping not just the possibilities themselves, but also *the relationships between them* can deepen understanding (Le Bihan 2017). Grasping relationships between multiple possibilities improves our 'modal sense' of the space of possibilities and our ability to navigate that space effectively for our epistemic (and non-epistemic) purposes. This again shows how grasping possibilities that are not of interest to scientists, but might be systematically related to those that are, can deepen our understanding of a phenomenon. Consider two examples. Example one: a student deepens their understanding of how biological traits evolved by learning about what intelligent design claims and why the evidence favors evolution over intelligent design. This is a common occurrence in many of my philosophy of biology courses. Example two: a person deepens their understanding of planetary motion by learning about how the universe would have been different if planets had circular orbits rather than ellipses. In both these cases, the agent deepens their understanding by grasping what would occur in a nonactual system and relating it to the actual systems that are currently of interest to the scientific community. More generally, human understanding can be deepened by bringing the possibilities of interest to various scientific and non-scientific communities into conversation with one another. As a result, grasping what would occur in distant possibilities (or counterfactual scenarios) that are not the focus of practicing scientists can still enable us to better understand natural phenomena—in a scientific way—because they will improve our grasp of the possibility space into which the actual and counterfactual situations we are interested in are embedded.

2.5   The All-Possibilities-Count View of Scientific Understanding

While the failure of the above proposals does not entail that no restriction on which possibilities contribute to understanding will be successful, I think those failures provide strong reasons for considering the proposal that no restriction on the range of possibilities will be sufficient to capture the plethora of ways that humans (can) scientifically understand our world. As a result, in what follows, I propose that using models (or theories, or anything else) to learn about possibilities (e.g., what would occur in a possible system or possible state of actual systems) *always deepens our scientific understanding of real phenomena*—although sometimes in ways that are not currently of interest to scientists or anyone else.

Let me now lay out the All-Possibilities-Count (APC) view of scientific understanding more explicitly:

> **APC:** An agent or community has scientific understanding of a phenomenon if and only if they grasp what would occur in some possible situation(s) that enables them to correctly answer a range of what-if-things-had-been-different questions about the real-world phenomenon. All possibilities, relationships between those possibilities, and answers to what-if-things-had-been-different questions that are grasped by the agent or community contribute to understanding. The more possibilities that are grasped, the more connections between those possibilities that are grasped, and the more what-if-things-had-been-different questions that the agent/community is able to correctly answer about the real-world phenomenon, the deeper their understanding of the phenomenon.

The first part of this view is just the modal view of understanding that has been defended by various authors in the literature (Le Bihan 2017; Rice 2019, 2021; Saatsi 2019). The second part adds the 'all possibilities count' component of the view. The third part clarifies how grasping possibilities relates to the *depth* of one's understanding. A few other clarifications will help make the above view more precise.

First, rather than restricting the possibilities that can contribute to scientific understanding to what is biologically, physically, or logically possible, the above view places limits on human understanding due to the limits on *what is graspable.*[6] While we should appeal to the cognitive capacities of humans when investigating the limits of human understanding, this should not be about what humans find *interesting*, but instead focus on *what is graspable* by human beings.[7] Moreover, I think the examples considered here show that systems that are biologically or physically impossible (in the sense that they are incompatible with the history of life on Earth or violate the laws of physics) can still deepen our understanding of the actual world. Moreover, it seems that, when it can be done, grasping what is logically impossible can show us how logical constraints play a role in accounting for what occurs in the actual world (more on this below). However, I also assume there will be possible states of a system that, while possible in one of the above senses, are simply not graspable by human beings due to our cognitive limitations. Nonetheless, I maintain that *if they were* grasped, such possible states of the system would deepen our understanding. It is perfectly fine to acknowledge that there can, in principle, be pieces (or degrees) of understanding that go beyond what human beings will ever be able to achieve in practice.

Second, in addition to grasping possibilities, the above view also requires that the agent/community be *correct* about those counterfactual situations. That is, just answering a range of what-if-things-had-been-different questions is insufficient. One's answers to those questions must be (at least approximately) correct concerning what would occur in those counterfactual situations. I do not have a specific account of what makes counterfactuals true to offer here, but I think it is clear that there can be true statements about what would occur in counterfactual situations and that scientists are often interested in discovering those modal facts.

Third, the above view makes it explicit that it isn't just grasping possible states of the system that can further deepen one's understanding, but also grasping relationships *between* those possibilities. I take it that these relationships can take a wide variety of forms, but that together they function to give the agent a "modal sense" of how to navigate a network of possibilities. The biologist learning about intelligent design is a clear example of this.

Finally, it is worth noting that since many of these features come in degrees (e.g., the number of possibilities grasped or the correctness of one's evaluation of those situations), the above view is able to account for the common intuition that understanding comes in degrees and can be deepened along a variety of dimensions. Indeed, the details of the above view not only show how scientific understanding can come in degrees, but they also identify several dimensions along which understanding might be deepened. What is more, the view does *not* require that all possibilities deepen scientific understanding to the same degree. For a variety of reasons, grasping one possibility or another might deepen scientific understanding (particularly in the ways we desire) more than grasping another possibility. Nonetheless, I contend that grasping possibilities always deepens scientific understanding to some degree.[8] In short, while grasping certain possibilities will provide a greater degree of understanding than grasping other possibilities, *acquiring the above kinds of modal information always deepens scientific understanding to some degree.*

Having laid out the various pieces of the APC view of understanding, I turn now to some of its interesting implications that deserve additional philosophical attention going forward. One interesting result is that lots of possibilities that we are capable of grasping can improve our understanding of a phenomenon whether or not we are interested in that phenomenon, or those possible states of the system. For example, after learning that Newtonian mechanics has been rejected, a physics student may not be interested at all in learning about that theory. Nonetheless, I argue that the student's grasping of how the universe would be different if Newtonian mechanics were true deepens their understanding of motion even if they are never interested in that counterfactual situation (or possible world). Similarly, grasping what would be true if intelligent design were the case can deepen biologists' understanding of evolutionary phenomena even if they are not interested in that hypothesis.

In addition, the above view entails that achieving the epistemic aim of scientific understanding is somewhat independent of being able to explain or perform successful interventions. Most importantly, the above view shows how grasping possibilities can provide understanding without having an explanation (Lipton 2009; Rohwer and Rice 2013; Rice 2021; Verreault-Julien 2019). While claiming that grasping possibilities deepens understanding is compatible with views that claim that grasping an explanation is necessary for minimal understanding (e.g., Khalifa 2017), adding this constraint is not required. Specifically, the cases above do not depend on the agent also grasping a correct explanation of the phenomenon in order for their grasping of possibilities to provide understanding. This is important because it highlights the value of acquiring understanding as its own epistemic achievement (and concept) that should not be analyzed solely through appeals to (or relationships with) other concepts. We ought to investigate scientific understanding on its own terms rather than constraining what counts as genuine understanding by connecting it exclusively with explanation or manipulation. The above views have much to tell us about which kinds of things scientists want to understand in different contexts and this will certainly be relevant to telling us which possibilities we need to investigate to acquire the desired understanding. But, they do not directly tell us what scientific understanding is or why grasping possibilities improves our degree of scientific understanding.

Finally, once we recognize that grasping possibilities always improves scientific understanding, we ought to reassess what it means to say that one individual understands a phenomenon *better* than another. At first glance, the above account might be taken to suggest that we should just compare two individuals by counting the number of possibilities and relations that they grasp. We could then say that if individual A grasps more possibilities/relationships than individual B, then A understands the phenomenon better than B. But this won't do for a number of reasons. First, as I noted above, how well an individual grasps a possibility and what would occur in that counterfactual situation comes in degrees. For example, suppose I understand that, if the pressure of a given gas went from 1 atm to 1.25 atm, then the temperature of the gas would rise. However, you can use the ideal gas law to calculate exactly what the resulting temperature would be. Although we grasp the same number of possibilities and relationships, your grasp of those possibilities is superior to mine, which results in you having a deeper understanding of the phenomenon. Second, it seems like we ought to consider the diversity/distance of the possibilities grasped in determining one's depth of understanding. For example, suppose I have a relatively good grasp of how the temperature of a gas would change as a result of the pressure changing from

1 atm to 1.1, 1.2, 1.3, and 1.4 atm. You, however, grasp how the temperature would change in response to changing the pressure to .1, .5, 8, and 10 atm. Although we grasp the same number of possibilities, your set is more diverse, which suggests that you have a deeper understanding of the relationship between pressure and temperature. This relates to old debates in the philosophy of science about the value of unification and how to measure it. Measuring generality with respect to real systems is difficult, but the even wider range of systems that are possible (or possible states of real systems) makes such comparisons even more challenging. Such questions concerning the variety of, and systematic relationships between, the possibilities grasped will require more philosophical attention going forward, and will have serious implications for how epistemologists think about novice versus expert comparisons. More generally, it appears that while all grasping of possibilities might contribute to scientific understanding *tout court*, measurements of depth or breadth of understanding may require us to include how contextual factors (e.g., the range of possibilities of interest) influence comparisons of understanding.

2.6 Objections and Replies

A serious objection to the APC view I proposed above is that there is a strong intuition that models or theories that are in direct contradiction with our best scientific knowledge do not provide genuine understanding. For example, young earth creationists claim to explain the existence of the Grand Canyon due to a great flood that covered the earth, but it seems they do not genuinely understand the phenomenon given that the Canyon was actually carved by the Colorado River over several millions of years. However, according to the APC view, it seems that they are simply exploring a distant possibility and, therefore, are deepening their understanding (assuming they are correct about what would occur in that counterfactual situation). So how can we preserve the intuition that scientific theories like Evolution or Combustion Theory provide genuine understanding, while theories like Creationism and Phlogiston theory do not?

I suggest that these cases are often wrongly interpreted as showing us that theories like Creationism or Phlogiston cannot provide *any* understanding because they provide incorrect information—even about counterfactual situations. In contrast, I think what is required to account for these cases is to show that our best scientific models and theories provide a much *deeper* understanding than that provided by Creationism or Phlogiston theory. However, I maintain that engaging with those possibilities can still improve scientists' understanding of the phenomenon, our currently favored theories, and the understanding they provide.

For example, engaging with alternative theories such as independent creation has certainly deepened our understanding of evolutionary theory and how things would have been different in other possible evolutionary systems (Sober 2000). Similarly, investigating the creationists' story about the Grand

Canyon can correctly tell us that "it is possible that a flood forms the Grand Canyon, but only if the world would have been a very different place" or that "it is impossible to generate it with known initial conditions and a great flood as possible process" (Verreault-Julien 2019, 15). That is, considering these possibilities can show us that a great flood producing the Canyon is impossible in systems with similar initial conditions and laws of geology, but a great flood *could* produce the Canyon in a radically different possible world. Nonetheless, because the creationists' explanation would incorrectly answer many other what-if-things-had-been-different questions that would be correctly answered by the explanation that appeals to principles of geology and the Colorado River, we get a clear sense of how/why our best scientific theory/explanation of the phenomenon deepens our scientific understanding to a much greater degree. For example, the creationists' explanation of the Canyon *incorrectly* suggests that had there been a great flood and the same initial conditions and laws of geology, then the Canyon would have formed (Verreault-Julien 2019, 14). So, although investigating those possibilities can provide some correct counterfactual information about the phenomenon, the creationist explanation itself provides lots of incorrect information about the very possibilities that our best scientific theories would answer correctly. Thus, even if considering the counterfactuals described by creationism deepens our understanding to some degree, we need not grant that such a theory enables anywhere close to the depth of understanding provided by our best scientific theories. In addition, we might consider adding the requirement that in order for the grasping of a possibility to deepen one's understanding, one must have some grasp of whether that possibility is (or could be) actual. This would enable grasping of these counterfactuals while acknowledging that they are nonactual to contribute to understanding, but it would rule out grasping them (and no other possibilities) while believing they are the actual world.

This raises another potential objection. The above suggestion seems to require that one grasp which possibilities are actual and which are non-actual. But this, one might think, just collapses into the view that grasping the actual world is required for scientific understanding. In response, I argue that one need not grasp much of what is true about the actual world to know that a particular possibility is non-actual. For example, scientists may know their idealized models represent systems that are non-actual even if they have rather limited knowledge about what is true of the real system. Knowing that a possibility is non-actual is importantly different from knowing what is actual. As a result, although we may want to require one who understands to have some sense of which of the possibilities are non-actual, this does not require them to grasp what is true of the actual world in some robust sense.

Finally, it is worth asking whether the counterfactual situations that contribute to understanding need to be genuinely possible. For example, many counterfactual conditionals involve impossible antecedents that describe *impossible* situations.[9] Does grasping these impossibilities (or what would occur in these impossible systems) also improve understanding? This is a difficult question that deserves further investigation than I have the space to adequately provide here. In other work, I have tried to address some cases involving mathematically impossible antecedents (Rice 2021, 113–115). My general reply is that in order for us to assess whether the answer to a what-if-things-had-been-different question is *correct*, it needs to be the case that there is a fact of the matter about what would occur in that counterfactual situation. While this may be difficult to determine for counterfactual conditionals with impossible antecedents, it is not always impossible. Therefore, although the above cases focus on using idealized models to explore genuine possibilities, the APC view appears friendly to expanding the view to include *fictional* impossible systems or situations as long as we can work out how to evaluate (or say true things about) those counterfactual situations.

2.7 Conclusion

This chapter has considered several ways of determining which possibilities contribute to scientific understanding. I have argued that grasping what would occur in various possibilities, how those possibilities relate to one another, and being able to correctly answer more what-if-things-had-been-different questions always deepens scientific understanding. Adopting this APC view allows us to embrace the epistemic contributions of science beyond manipulation and control, to value the contributions of non-scientists to scientific understanding, and to emphasize the importance of exploring what is possible for understanding our world.

Notes

1 Thanks to Kareem Khalifa, Soazig Le Bihan, Philippe Verreault-Julien, and Ylwa Sjölin Wirling for helpful comments on earlier versions of this chapter.
2 I take grasping to be the—typically internal—relationship between the agent (or group) who understands and the possibilities, relationships between possibilities, and answers to w-questions that constitute one's understanding.
3 Thanks to Philippe Verreault-Julien for suggesting that I be clearer on this point.
4 While I will sometimes talk of models representing possible systems, what is important is that the models provide access to accurate modal information. Targeting a possible system is one way that scientific models can accomplish this, but it is certainly not the only way.
5 As I mentioned in the introduction, I think a more promising suggestion is that what makes some understanding count as scientific is that it is produced by the investigation/use of scientific models, theories, methods, experiments, or evidence. Yet, I also don't think we need, or should want, a clear demarcation between scientific understanding and other ways of understanding given that I think much of the understanding developed outside of scientific communities uses similar kinds of methods, experimentation, evidence, and modeling.
6 Thanks to Marcel Weber whose comments pressed me to be clearer on this point.
7 Of course, each group or individual will also have its own limits on which possibilities they can grasp. For example, a white male will not be able to grasp what it is like to have lived their life as a black woman.

8 However, I also think the above cases show that sometimes grasping what is true
   of a highly idealized non-actual system will deepen scientific understanding more
   than grasping what is true of the actual system.
9 Thanks to Ylwa Sjölin Wirling for raising this excellent question/objection.

References

Ariew, André, Collin Rice, and Yasha Rohwer. 2015. "Autonomous Statistical Explanations and Natural Selection." *British Journal for the Philosophy of Science* 66(3): 635–58.

Bokulich, Alisa. 2018. "Searching for Non-Causal Explanation in a Sea of Causes." In *Explanation Beyond Causation: Philosophical Perspectives on Non-Causal Explanation*, edited by J. Saatsi, and A. Reutlinger, 141–63. Oxford: Oxford University Press.

Chirimuuta, Mazviita. 2018. "Explanation in Computational Neuroscience: Causal and Non-Causal." *British Journal for the Philosophy of Science* 69: 849–80.

Craver, Carl. 2007. *Explaining the Brain: Mechanisms and the Mosaic Unity of Neuroscience.* Oxford: Oxford University Press.

de Regt, Henk. W. 2017. *Understanding Scientific Understanding.* New York: Oxford University Press.

Douglas, Heather. 2009. *Science, Policy, and the Value-Free Ideal.* Pittsburgh: University of Pittsburgh Press.

Elgin, Catherine. Z. 2017. *True Enough.* Cambridge, MA: MIT Press.

Gelfert, Alex. 2019. "Probing Possibilities: Toy Models, Minimal Models, and Exploratory Models." In *Model-Based Reasoning in Science and Technology*, edited by Ángel Nepomuceno-Fernández, Lorenzo Magnani, Francisco J. Salguero-Lamillar, Cristina Barés-Gómez, and Matthieu Fontaine, 3–19. Berlin: Springer International Publishing.

Huneman, Philippe. 2010. "Topological Explanation and Robustness in the Biological Sciences." *Synthese* 177: 213–45.

Khalifa, Kareem. 2017. *Understanding, Explanation and Scientific Knowledge.* Cambridge: Cambridge University Press.

Kourany, Janet. 2021. "How the Facts Might Give Us Socially Responsible Science." In *The Routledge Handbook of Feminist Philosophy of Science*, edited by Sharon Crasnow, and Kristen Intemann. New York: Routledge.

Lange, Marc. 2013. "Really Statistical Explanations and Genetic Drift." *Philosophy of Science* 80(2): 169–88.

Le Bihan, Soazig. 2017. "Enlightening Falsehoods: A Modal View of Scientific Understanding." In *Explaining Understanding*, edited by S. Grimm, and C. Baumberger, 111–35. New York: Routledge.

Levy, Arnon. 2011. "Makes a Difference. Review of Michael Strevens' Depth." *Biology and Philosophy* 26: 459–67.

Lipton, Peter. 2009. "Understanding Without Explanation." In *Scientific Understanding: Philosophical Perspectives*, edited by H. W. de Regt, S. Leonelli, and K. Eigner, 43–63. Pittsburgh: University of Pittsburgh Press.

Morrison, Margaret. 2009. "Understanding in Physics and Biology." In *Scientific Understanding: Philosophical Perspectives*, edited by H. W. de Regt, S. Leonelli, and K. Eigner, 123–145. Pittsburgh: University of Pittsburgh Press.

Morrison, Margaret. 2015. *Reconstruction Reality: Models, Mathematics, and Simulations.* Oxford: Oxford University Press.

Potochnik, Angela. 2007. "Optimality Modeling and Explanatory Generality." *Philosophy of Science* 74: 680–691.

Potochnik, Angela. 2017. *Idealization and the Aims of Science*. Chicago: University of Chicago Press.

Rice, Collin. 2013. "Moving beyond Causes: Optimality Models and Scientific Explanation." *Noûs* 49(3): 589–615.

Rice, Collin. 2018. "Idealized Models, Holistic Distortions, and Universality." *Synthese* 195: 2795–819.

Rice, Collin. 2019. "Understanding Realism." *Synthese* 198(5): 4097–121.

Rice, Collin. 2021. *Leveraging Distortions: Explanation, Idealization and Universality in Science*. Cambridge, MA: MIT Press.

Rohwer, Yasha, and Collin Rice. 2016. "How Are Models and Explanations Related?." *Erkenntnis* 81(5): 127–1148.

Saatsi, Juha. 2019. "Realism and Explanatory Perspectives." In *Understanding Perspectivism: Scientific Challenges and Methodological Prospects*, edited by Michela Massimi, and C. D. McCoy, 65–84. New York: Taylor and Francis.

Sjölin Wirling, Ylwa, and Till Grüne-Yanoff. 2021. "The Epistemology of Modal Modeling." *Philosophy Compass* 16(10): e12775.

Sober, Elliott. 2000. *The Philosophy of Biology*. 2nd ed. Boulder, CO: Westview Press.

Strevens, Michael. 2008. *Depth: An Account of Scientific Explanation*. Cambridge, MA: Harvard University Press.

Strevens, Michael. 2013. "No Understanding without Explanation." *Studies in History and Philosophy of Science* 44: 510–15.

Stoneking, Mark. 2017. *An Introduction to Molecular Anthropology*. Hoboken, NJ: Wiley-Blackwell.

Trout, J. D. 2007. "The Psychology of Explanation." *Philosophy Compass* 2: 564–96.

Verreault-Julien, Philippe. 2019. "Understanding Does Not Depend on (Causal) Explanation." *European Journal for Philosophy of Science* 9: 17–37.

Walsh, Denis. M. 2015. "Variance, Invariance and Statistical Explanation." *Erkenntnis* 80: 469–489.

Woodward, James. 2003. *Making Things Happen: A Theory of Causal Explanation*. Oxford: Oxford University Press.

3   Actual Possibility

Paul Teller

3.1   Introducing Actual Possibility and Some Preparatory Considerations

I have a map to a treasure at the end of a tunnel in a deep cave. On exploration, we discover a large rock blocking the way. It's possible we can remove this rock. That would be, epistemically possible – removing the rock isn't ruled out by what we know. And we do remove it, only to find the tunnel completely caved in. There is a robust sense of "possible" on which, while earlier it was epistemically possible to get the treasure, it never was really, actually possible because of the immovable barrier we didn't know about.

We often use the word "possible" in something like the sense just illustrated, but there has been very little interpretive examination of the notion. About it, Deutsch writes:

> Real possibility is a future-oriented modal-temporal notion. What is really possible at a time t is what is true at some time t' later than t in some possible world just like the actual world up to and including t.
>
> (Deutsch 1990, 752)

I will add quite a lot of detail to Deutsch's analysis. Rather than "real" I will instead call it "actual possibility" to avoid any implication that other kinds of possibility are not, in their own way, "real" and to emphasize that this notion is "actuality based" in depending not only on general truths but on specific matters of fact as they occur in the actual world. A short summary of the view I will develop: A statement made at t, presenting some event or state of affairs concerning a time t' later than t, presents an actual possibility just in case the event or state of affairs that will or will not occur at t' is not ruled out by what has happened up to t. Much of the exposition in section 3.3 will be to make out the intuitive idea of "not being ruled out by". As my introductory example is at pains to make clear, actual possibility differs from epistemic possibility and care must be taken not to conflate the two.

Before examining actual possibility in section 3.3, section 3.2 will develop the basic tool that I will use, one that also should have broader application.

DOI: 10.4324/9781003342816-5

For at least 50 years analysts have usually worked on alethic possibilities in terms of the conception of possible worlds, what I will call "global possibilities," whether they are taken literally or treated as ersatz possible worlds. I will urge that for an important range of problems we are better off basing our analysis on what I will call "local" possibilities, given with individual statements, such as the possibility that it will rain in Davis on 25 February 2022.[1] While I will offer some broad support for this redirection, importantly, the proof will be in the pudding, seeing how treatment of actual possibilities works most smoothly when approached from the point of view of local possibilities.

To be stressed: my claim is that starting with local rather than with global possibilities has certain advantages when discussing actual possibilities. Though I won't argue it here, I expect that similar considerations apply to work on epistemic possibility. Examining questions in metaphysics may well be another matter, though the advantages of using local rather than global possibilities for issues in metaphysics are certainly worth pursuing.

Section 3.3 will go to work on actual possibility, and section 3.4 will broaden application from statements to other forms of representation.

A word on methodology: I will not offer a traditional philosophical analysis that aims at truth. Instead, I will fashion a piece of conceptual engineering or a Carnapian rational reconstruction of a segment of our thinking about what it is for something (a state of affairs, what is picked out by a statement[2] ...) to count as a possibility. That is, I will offer a model of an important facet of our expectations about the future. It is a virtue of such a model that it does not suffer excess precision, that it leaves out details that would only get in the way of application and understanding, for example, details of how probability is to be understood. The aim of generality also requires omitting details that will vary with application, for example, as I will explain below, details about laws or regularities. The standards of evaluation of such a model will not be truth or falsity but rather the extent to which the model successfully characterizes certain parts of our thinking about possibility and the extent to which, in so doing, the model illuminates what that thinking involves or what it comes to. I will argue for this kind of success for the model by providing illustrative examples that, it will be clear, can easily be multiplied.

A note for the metaphysically squeamish: Some are broadly skeptical of possibility as discussed in analytic philosophy, especially when possibility is characterized in terms of possible worlds. I will show how we can, instead, characterize possibility in terms of individual statements with any specific kind of possibility picked out, as I will explain, by a characterizing kind of constraint. Metaphysical possibilities, if one chooses to consider such, are then given by statements that are consistent with truths of metaphysics operating as the constraints. The metaphysically squeamish can restrict attention to local possibilities given by the constraints that they find intelligible and so endorse the kinds of local possibility determined by these constraints.

The concept that I am examining is not new – as mentioned it has gone under the name, "real possibility". Frantisek Gahér (2003) cites Diogenes Laertius as presenting a conception along the lines of the one I will develop (169–70). Gahér formulates the idea by saying that a "proposition is possible which admits of being true, there being nothing in external circumstances to prevent it being true,…" (183). Gibbs (1970) discusses a variety of notions of possibility that we use in everyday life. The one closest to the conception I am developing: "A thing is naturally possible if and only if nothing that already is or has been actual is incompatible with the actuality of that thing" (340). Katzav (2014) uses the term "real possibility" with apparent intent similar to mine but does not clearly distinguish it from epistemic possibility. Above, I have mentioned Deutsch's "Real possibility" (1990). Broadly, I regard my development as further refinement of what Deutsch has presented, differing in approach by working with local rather than global possibilities.[3]

3.2   From Possible Worlds to Local Possibilities

Today and at least since the 1960s, possible worlds have been the analytic tool of choice when discussing alethic modality. A possible world is supposed to be something that, somehow, covers every way a world could be. Lewis takes these possible worlds to be real and concrete. Here, I will, without argument, reject that speculation.[4] Lewis acknowledges a linguistic, "ersatz," alternative. He supposes some appropriate "world-making language" and then considers maximally consistent sets of sentences expressed in this language.[5] The ersatz possible worlds are these maximally consistent sets of sentences, understood to be abstracta, functioning "to represent the entire concrete [real] world in complete detail, as it is or as it might have been" (Lewis 1986, 137). Note the use of modal language in explaining what ersatz possible worlds are.

Lewis characterizes different kinds of modality in terms of constraints as restrictions on the quantifiers involved in quantifying over all possible worlds (1986, 7). Logical possibility: what is consistent with the constraints of classical logic. Metaphysical possibility: what is consistent with the generalizations of metaphysics. Physical possibility: what is consistent with the laws of physics. And so on.[6] [7] As I will urge that we work, not with possible worlds (global possibilities), but with individual statements (local possibilities), I will appeal to the constraints directly – constraints that individual possibility statements may or may not satisfy.

On the mainstream account of possibility, the relation between local and global possibilities is that a local possibility, given by a statement, S, is the set of all global possibilities – entire possible worlds – that make S true. On this account of possibility, the local possibility that there is a talking donkey is a set of possible worlds – the set of metaphysically consistent sets of sentences that include the sentence, "There is a talking donkey"[8]

Lewis (1986, 142–65) discusses a number of problems with ersatz possible worlds. These can be grouped as follows: First, there are unexplicated appeals to primitive modality. Most importantly, note the use of modal language in explaining what possible worlds are - see quotation from p 137, three paragraphs back. This appeal is compounded with the requirement that the collection of sentences making up an ersatz possible world be consistent (Lewis 1986, 150–2). Can't that be covered by requiring logical consistency among all the world's sentences and constraints? An ersatz possible world will have sentences, for example, "there is a talking donkey." It will also have sentences describing the exact placement of matter. Lewis (1986, 55–56) takes consistency to require getting right just which placements of matter constitute a talking donkey. This is the problem of constitution writ very large: It's not just what we, in the real world, would count as a talking donkey constituted with real matter with real properties, it's what metaphysically would so count, including metaphysically possible constitutions not possible in the real world.

Lewis's (1986, 165) next problem for ersatz possible worlds is that describing many metaphysical possibilities requires terms and concepts we don't, indeed assuredly never will, have. In particular, Lewis takes many possible worlds to have "alien" properties, properties that are not only not instantiated in the actual world, but ones to which we, stuck in the actual world, have no access of any kind.

Yet another difficulty with ersatz possible worlds: All human languages are, probably unavoidably, vague. Lewis comments that "Some further idealization is needed to make sure there are determinate truth values at all: the worldmaking language had better be disambiguated and precise" (Lewis 1986, 142). The problem applies to understanding local possibilities, whether we work with local possibilities directly or as analyzed as sets of complete possible worlds. "There is a talking donkey" is vague. Just what counts as talking? Or, for that matter, a donkey? If we take the possibility that there is a talking donkey to be the collection of possible worlds with a talking donkey, somehow exactly described, just which worlds will count as the ones with a talking donkey?[9]

These four problems, unexplicated appeals to primitive modality, the need for consistency of all kinds, the need for alien terminology, and vagueness will arise generally for any account of ersatz possible worlds in terms of sentences, statements, or propositions.

Shifting from global to local possibilities by itself does nothing to ameliorate these difficulties. But the special case of actual possibility that I will be developing will face a considerably reduced slate of problems.[10] Appeal to unanalyzed primitive modality will be reduced to assuming an open future.[11] By narrowing the scope of inquiry from broadly metaphysical possibilities to the actual possibilities that are accessible to us, the problem of inadequate vocabulary won't arise. This is no "solution" to the problem, broadly put. Rather, I am narrowing the scope of inquiry to a domain where the problem does not occur. By looking at possibilities given narrowly by single statements, we don't have to worry about consistency among statements as one builds up an ersatz possible world. There is also the central requirement of consistency with the constraints and the problem of vagueness. These will get extended discussion in the treatment of actual possibilities.

The shift from global to local possibilities has independent motivation. In the first instance, what makes a theory that appeals to possible worlds attractive is the simple and natural truth conditions it provides for modal statements (the Leibniz biconditionals): what is possible is what is true in some possible world. What is necessary is what is true in all possible worlds. These truth conditions in turn provide a natural and attractive semantic account of the duality between possibility and necessity: what is possible is what is true in some possible world, hence the negation of which isn't necessary. What is necessary is what is true in all possible worlds, and so is something the negation of which isn't possible.

When working with possible worlds taken as concrete, these truth conditions may be substantive. But having forsaken concrete possible worlds in favor of linguistic ersatz possible worlds, the foregoing truth conditions are redundant. "Possibly S" is true just in case S is true in some ersatz possible world that satisfies the constraints that characterize the kind of possibility in question. This last holds just in case S is itself consistent with the relevant constraints. "Necessarily S" is true just in case S is true in all ersatz possible worlds consistent with the constraints. But that holds just in case S is implied by the constraints. So the truth conditions for possibility and necessity are already given by consistency with and being implied by the relevant constraints. The detour through possible worlds is superfluous insofar as truth conditions for modal statements are concerned. The constraints provide the modally relevant truth conditions and can be applied directly to statements giving local possibilities.

Are there considerations that require treatment with global possibilities that can't be done with the local alternative? There is a formal richness in the possible world formulation. It supports supplementation with an accessibility relation that in turn supports the full Kripke possible world semantics. As far as I can see, such formal considerations are rarely, if ever, required for most, if not all issues about specialized kinds of possibility, in particular, actual possibility as I will develop the notion below.

There are two cases for which I find it plausible that they require treatment with entire (ersatz) possible worlds: the conventional approach to analyzing counterfactual conditionals in terms of possible worlds and a nearness or similarity relation, and various possible world accounts of fictional discourse. But it is not clear to me why such analyses could not be carried out by considering only perhaps very long but finite statements. In any case, if there are problems that require treatment with whole possible worlds this can still be done by building whole (ersatz) possible worlds from statements that constitute local possibilities. In other words, I propose to begin analysis with local possibilities and regard whole worlds, if there is need for them, as composed of local possibilities, rather than the conventional opposite direction. Consequently, if whole worlds are required for treatment of counterfactual conditionals, fictional discourse, or other applications they are still available.

Perhaps working with whole possible worlds is sometimes required, but in certain cases doing so can make solution of some problems, if not impossible, at least requiring unnecessarily extreme maneuvers. Consider, for example, the appeal to possibilities in functional, aka modal, approaches to mental contents.[12] In these approaches, one seeks to analyze mental contents in terms of an agent's disposition to behave in possible situations. If these possibilities are understood as sets of possible worlds, as seems to be what is almost always done, there is not yet any appeal to the differences in mode of presentation that is required when it comes to contents as the objects of propositional attitudes. Possible worlds are supposed to be independent of us, in particular independent of the accidents of how we represent them, which is antithetical to distinguishing modes of presentation. At best, the modes of presentation have to be reintroduced as an add-on. For this issue it is best to drop the possible worlds in favor of the notion of local possibilities given by individual statements, with a conception of statement that recognizes the distinctions made with modes of presentation.

3.3   Actual Possibilities

Suppose at noon I shake a die in a cup and put the cup on the table, open end down. Later, having lifted the cup and ascertained the outcome of "6," should we say that at noon the outcome, "5," was then possible? There is a strong inclination to say yes, and a strong inclination to say no. I submit that both are right but with different kinds of possibility. Epistemic possibility: It was consistent with what we knew at noon that the outcome was "5." Actual possibility: *Before* noon, "5" was an objectively actual possibility about something that might occur in the future. But when at noon the die in fact landed on "6," "5" was no longer an actual possibility.

One more example: I run a ski resort and need to know whether I should close the slopes for the day because the heavy snowfall last night might have created an (actual) possibility of an avalanche. Because the case involves a distribution of matter this would more specifically be a physical (actual) possibility. Does the amount and configuration of snow rule out or permit an avalanche? There are two features of this case worth underscoring.

1) A physical possibility of an avalanche would not just be what is permitted by the laws of physics; it would be what is permitted by the laws of physics *and* the distribution of snow on the ground. We might call this a locally situated or contextual kind of physical possibility that turns not just on the laws but also on relevant matters of particular physical fact. Examples one hears sometimes tacitly presuppose such matter of fact, but usually physical possibility is characterized as what is consistent with the laws of physics, period.

2) The phrase used in the paragraph before last, "the heavy snowfall last night *might* have created an (actual) *possibility* of an avalanche", has a nested modality. The "might" is epistemic. But the possibility that might have arisen would be an actual possibility. If it were again epistemic, the modality would be redundant.

To begin a general characterization I take one modal circumstance as primitive: I presuppose an open future (if nothing else, most likely required by quantum mechanics).[13] The leading idea: Local, actual possibilities are given by statements, the truth of which are not ruled out by how things have been up to now. More broadly, the constraint that characterizes local, actual possibility at time t is that the truth of the statement is not ruled out by the state of affairs at and before t. "t" is a parameter ranging over times to which actual possibilities are relative. I will often refer to time t as "the present" as a way of indicating that t is the time to which some possibility for a future occurrence is relative.[14] Since actual possibility is relative to a time, it also encompasses passed possibilities. Since Krakatoa exploded in 1883, it is not now actually possible that it exploded in 1882. But relative to 1881, if circumstances in 1881 did not rule out an explosion in 1882, it was actually possible that Krakatoa would explode in 1882.

Actual possibility fits most naturally with A-theories of time on which, relative to a time t, there is something "not yet real" about what happens at any t' later than t (henceforth, t' > t). But actual possibility will fit with B-theories of time given some ways of understanding B-theories. B-theorists think of themselves as outside of time, looking down, a-temporally, on a whole history of the universe. From the point of view of a B-theorist, looking down on the entire history of the universe, what holds at t holds at t a-temporally. But from the point of view of an agent at t, there is a distinction between what has happened in that agent's past and what does (or might) happen in that agent's future. So, consistent with the basic idea of B-theories, we can take there to be, *at t*, no fact of the matter of what happens at t' > t. That is, for the a-temporally located B- theorist, there is, for all t, all the facts that obtain at t. But I assume that there are no *additional* facts *at t* about what happens at t' > t. In this sense, and absent determinism, in such a formulation of a B-theory, relative to an agent at t, t's future is open.[15]

As for necessity, using the formula that what is necessary is what is implied by the constraints, actual necessities about the future are just those statements that are already determined to be true. This includes natural laws, if there are such. Below I will instead work with reliable regularities. The truths about the past and present are, by being part of them, determined by what has occurred up to now and so all count as actually necessary.

Some readers will be asking, what are these actual possibilities? Remember that Lewis takes his ersatz possibly worlds to be the maximal consistent sets of sentences in a world-making language, understood as abstract representations that "represent the entire concrete [viz, real] world in complete detail, as it is or it might have been" (Lewis 1986, 137). My attitude towards actual possibilities is analogous, but without any broad appeal to modality. A statement, S, or its content, counts as an actual possibility relative to a time t, just in case S concerns some future, relative to t, event or condition that will or will not occur and that this event or condition has not been ruled out by what has occurred up until t. So an actual possibility is a statement or its content (take your pick) viewed in its status as not having been ruled out by prior conditions.

An actual possibility is like a forecast, differing by the cancelation of the implication that what is forecasted is assured or likely to happen. For an actual possibility, it is only required that what is described is still a live option.

Some comments about actual possibility:

a It is natural to hear "it is (presently) possible that P" and "It is (presently) necessary that P" as epistemic possibility and necessity. But, again, what is epistemically possible can fail to be actually possible: what isn't ruled out by what we know may be ruled out by facts we don't know, as in the initial examples.

b As time goes on, new actual possibilities can arise. The drought in California has made devastating fires (actually) possible. Likewise, we can lose possibilities: plentiful rain has eliminated the (actual) possibility of devastating fires.

c Actual possibilities are open-ended. With the exception of mathematics,[16] our statements are almost, if not always, vague. Where being vague is a characteristic of statements, I will say that the corresponding actual possibilities are open-ended. For example, the actual possibility that an asteroid hit the Earth in the 21st century is open-ended insofar as there is no fixed size required to count as an asteroid. The actual possibility that we will find a cure for cancer by the end of the century is open-ended insofar as there are no fixed criteria for how effective a treatment needs to be to count as a cure. What the open-endedness of possibilities comes to will become clearer below when I discuss how to address the vagueness of the statements used to specify actual possibilities.

d Are all actual possibilities ultimately physical possibilities of the locally situated or contextual sort illustrated in the avalanche example? You will conclude, "yes" if and only if you are a physical reductionist. For example, I love boating, but it is not (actually) possible for me to buy a yacht – I don't have nearly enough money. This will count as an example of a physical (im)possibility if and only if the financial is reducible to the physical.

e Can actual possibilities be characterized within a possible world framework? Yes, as was done by Deutsch (1990). We consider a family of possible worlds that are duplicates up to the present time. Or we consider one "branching" world with one trunk, what has happened up to now, and then many futures, all the ones not ruled out by what is in the trunk. But so doing would no more than duplicate the analysis so far given and extended below, dragging in the possible worlds that are superfluous.

There is no need to go up to the metaphysically possible worlds and then back down to a set of these, picked out as the worlds consistent in the relevant way with the local possibility given as a statement. The statement and the constraints are doing all the work.

Next, I need to discuss how in this context I want to understand the notion of constraint. The notion of actual possibility that I have presented is characterized in terms of the constraint of what has been the case up to the present time. A first problem: is there still any room for a notion of (actual) physical, economic, or other sorts of specialized kinds of actual possibility? The subject matter of what was and is the case may be (roughly) partitioned: what is or was physically the case, what is or was economically the case, and so on. Specialized kinds of actual possibility can then be characterized by restricting the constraints of the past and present to the relevant subject matter. For example, colonizing Mars by 2030 is now technologically but not economically (actually) possible.

But *how* should we take what has happened up to now to constrain the future? The natural option is to appeal to the mediation of natural laws. The notion of a natural law is a tangled affair. Do they apply only at some "fundamental level"? In the next section, an example will appeal to legal constraints – how is one to fit legal constraints into our picture? Many such details are not relevant to the uses for which the present model is intended, other details will vary with application, so this network of problems is here appropriately set aside. I will do this by instead appealing to the fact that the world exhibits regularities, understanding "regularity" as appropriate to the problem to which the model is to be applied, for example, when called for by the application, regularities enforced by legislative rather than natural laws.

Should we take what regularities have ruled out to be what they have ruled out with logical or mathematical certainty? Let me cast some doubt with the illustration of sample spaces as used in statistics. A statistician's sample space is the set of *possible* (the term universally used by statisticians) outcomes of a random experiment or other situation. For instance, the sample space for flipping a coin is usually taken to be the two outcomes, heads or tails. Outcomes, such as the coin landing on its edge, that are judged to be too unlikely to need consideration, are not included among the (actual) possibilities. The chance of the coin landing and remaining on its edge is so unlikely that we won't get into trouble by simply ignoring it.

I propose to understand "ruled out" so that such utterly discountable eventualities count as ruled out. What will count as "utterly discountable"? This will vary with context. Contextual considerations, including our current interests and values, determine what standards are appropriate. The world then determines whether or not the appropriate standards are met. So "not ruled out by," equivalently "constrained by," is understood in terms of past and present circumstances that are applied with whatever regularities in fact obtain, and where this application is tempered by also excluding things that are, is so unlikely that they need not be taken into consideration; with the question of just how unlikely fixed by contextually determined standards.[17] The "discountable" is to be understood not as believed to be discountable, but, given the standards in play, discountable in fact.

I will summarize the foregoing by saying that the operative notion of constraint is *what is not ruled out in practice.*

A last, and important, consideration: the proposal makes actual possibility turn on individual events that have occurred in the actual world: what is not ruled out by what has *actually* occurred so far or what was not ruled out by what had *actually* occurred until then. In depending on individual events that have actually occurred I will say that actual possibility is "actuality based". The notion of actual possibility does not include "possibilities" in which everything, or things very broadly, had been different "from the beginning," for example, in which the laws of physics were different; or in which what is actual is otherwise not relevant, allowing, as Lewis does, possibilities like that of a talking donkey. In being actuality-based, actual possibility (and epistemic possibility as well) importantly differs from logical, conceptual, and metaphysical possibility, as well as physical and other more specialized varieties of possibility when these are characterized purely in terms of laws as constraints. An actual possibility is tied to the actual world through the role of what has actually happened up to now as constraints, which I take to include existing regularities. This would appear to make statements of actual possibilities carry a kind of content about the actual world in ways in which most other kinds of modality do not. This is an important reason that I use the term "actual" rather than "real" to characterize possibilities about what might happen in the future. Assuming that knowledge implies truth, epistemic possibility also carries actual world content.[18]

To what extent is my account of actual possibility beset by the difficulties Lewis raised for ersatz possible worlds? The first problem was the unexplicated appeal to primitive modality. Remember that Lewis's ersatz possible worlds are maximally consistent sets of sentences in a world-making language, understood as abstract representations of how a world might be, with the "might" understood as what is allowed by the constraints of metaphysics. Actual possibilities are statements considered in their alethic content, with that alethic character explicated in terms of the constraint of what has happened up to now. No need here, as there was for Lewis, for any completely general appeal to the facts of metaphysics. There was also the problem of metaphysical constitution, what configurations of matter metaphysically could constitute, for example, a talking donkey. Actual possibility will also need to appeal to an account of constitution, for example, what future actually possible configurations of matter are to be counted as a statue of Donald Duck. Here, readers will have to appeal to their solution to the problem of constitution as applied to the real world, which will have whatever metaphysical considerations their solution will require. But having shifted the constraints to ones of the actual world rather than metaphysics broadly this is likely, depending on the solution, to have more modest requirements.

The third problem was the need for a world-making language. Having avoided the ersatz metaphysically possible worlds, and working entirely with local possibilities expressed in our language, we have no need for a world-making language with its exotic vocabulary. We can start with the possibilities that we can give in our language, perhaps somewhat different ones if expressed in English, French, or Chinese. This enables us to consider possibilities to which we have access, the ones that can be stated in a language that we have, leaving it open just what currently unknown possibilities might come to light as we develop new concepts and languages.[19]

I'll turn now to the difficulties of vagueness that, as far as I know, have nowhere been addressed in the modality literature. Problems about vagueness afflict local possibilities every bit as much as global ones. Local possibilities are given with statements, and the best we can do is make these statements with the vague vocabulary we have. What I will do here is sketch an approach to dealing with vagueness for the special case of actual possibility.

Let's explore the problem with an example. Jane and Tom are in a race. We consider three possibilities. Jane wins, Tom wins, or there is a tie. We expect that exactly one of these outcomes will occur. Note the open-endedness of these possibilities. Most cases will be clear-cut. But in some eventualities it won't be clear whether one should say that Jane won, or that Tom won, or that it was a tie. What should be said in such a situation will vary, from case to case. This is a deeply contextual matter. Broadly in such cases, the guiding consideration will be: does the statement expressing the possibility fit the case well enough not to spoil intended applications?

Here is a schematic sketch of how this kind of case can unfold: consider the possibility given by the statement, "Tom will win." Until the race is finished, the vagueness in the statement remains unaddressed. At the end of the race, the judges will, if needed, have to resolve the vagueness, whereby "resolving the vagueness" I mean making a decision as to how the case will be classified. I will regiment this kind of case as follows: Contextual considerations, including the interests and values of the relevant parties, determine what standards for resolving the vagueness are appropriate. With these standards set, it is what happens ("the world") that determines whether the standards are met.

To generalize, the vagueness of actual possibilities is resolved when a decision is required, with various interested parties playing the role of the judges in the foregoing example.

Note that in many cases, there will not be a unique standard that is appropriate. That is, several different options may be maximally, and equally, appropriate. Narrowing among such options can be arbitrary. Also note that all this is still an idealization. My description of what happens in such cases is highly stylized. But I have considerably deidealized the original idealization for ersatz possible worlds of a world-making language with no vagueness.

What if there are no interested parties whose interests and values provide the basis for the standards, for example, for the possibility that tomorrow there will be more black than grey cats in Rome? Black? Grey? In Rome? No one could determine a truth value. No one cares. With no standards that can be set, the possibility just remains open-ended.

Turning to one last issue to be considered here: Broadly, philosophers have distinguished between objective and non-objective possibility. (e.g., Williamson 2017; Wirling and Grüne-Yanoff, to appear) The force of "subjective"/ "objective" here is to say that the modality in question does/does not in some ways depend on us.

Both epistemic and actual possibility count as non-objective in ways that are similar in some ways, but different in others. For epistemic possibility, it is the agent's[20] knowledge base that comprises or determines the constraints, that is, the conditions the satisfaction of which determine whether a statement counts as epistemically possible for the agent. Insofar, the constraints are indexed to agents. For actual possibility, the constraint is not having been ruled out (in practice) by what has occurred until now, and what has occurred until now is independent of us.[21] But on my account, what counts as being ruled out is contextual, in particular partly fixed by human considerations, including human interests and values. So for both epistemic and actual possibility there are, in different ways, facts about us that play a role in fixing the constraints. But for both, in large measure the constraints arise from objective considerations, that is, considerations that are entirely independent of us.

3.4   Extending Actual Possibility to What Is Presented with Other Means of Representation

My house is currently painted blue. I am considering having it repainted off-white but am unsure. The painter shows me a picture of my house painted off-white. This picture presents an actual possibility, of how my house would look in the future if painted off-white.

How do such actual possibilities, presented with pictures, compare with those given by statements? Actual possibilities presented with statements have success and failure conditions of truth and falsity. Other forms of representation will need similar success conditions to count as presenting actual possibilities. Pictures are not true or false.[22] Representational success for pictures is not truth, but bearing the similarity and resemblance characteristics, and sometimes other considerations, that are appropriately expected of the picture in the context.[23] So to say that a picture represents an actual but not yet actualized possibility is to say that realization of the contextually appropriate success conditions for the picture have not been ruled out.

This example shows us that what will count as not having been ruled out can vary significantly from the examples I discussed earlier. Perhaps the shade I would like is ruled out because the paint of that shade is not available – any chance of getting such paint is utterly discountable. Using some colors, for example, bright pink, will be utterly discountable when ruled out by city ordinance.

In the case of the picture of my house painted off-white, the indefiniteness involved with pictures is navigated in the ways described above for vagueness of statements. Given the picture of my house painted off-white, the attitudes, interests and values, and possibly other contextual considerations of the interested parties, determine the standards for settling the indefiniteness of the picture. With these standards in place, after completion of the painting, the appearance of the house will, or will not, count as realization of the actual possibility according to whether or not the contextual standards have been satisfied.

Models comprise an important class of non-linguistic representations. I have in mind a great many things that can count as models: physical models, the abstract model of the chemical formula for benzene, and the Lotka–Volterra model of predator-prey relations.[24]

Blueprints[25] provide a nice illustration. The lines of a blueprint mean nothing aside from the way a builder interprets them. There is always a range of options for interpretation – indeed this range is often narrowed by notations made on a blueprint during the construction process. When an intended target for an interpreted blueprint has not been ruled out, the blueprint-as-interpreted represents a real possibility. How might an option be ruled out? By the laws of physics – such a structure would collapse. As in the case of the color of my house, exclusion might be because of the unavailability of materials or because of city ordinance.

Let's consider abstract models that convey regularities or generalizations, usually conveyed with equations, where accompanying descriptions and illustrations function to specify how the equations are to be interpreted. Such models can be used as didactic illustrations, as when a model of the motion of a pendulum can be used to teach how pendula behave, without representing any past, present, or possible future pendulum. Perhaps such use of models can be thought of as presenting fictions. Abstract models can also be used to present generalizations. The Lotka-Volterra model gives a generalization about the pattern of change in the relative frequency of predators and prey. Such generalizations can then also function to present actual possibilities, as I will illustrate with the example of the Lotka–Volterra model.

Suppose that some initial distribution of predators and prey is actually possible. According to the equations the ensuing pattern is then likewise actually possible. The hedge, "according to the equations," covers two qualifications. First, the model, as such models very generally, is highly idealized, involving simplifying assumptions such as unlimited food supply for the prey. Insofar, what is described by the model will only be similar to actual possibilities, with the relevant kind of similarity set by the idealizations involved. Second, it has to be assumed that, within the relevant similarities, the equations correctly describe sequences of events. Insofar as the model is well confirmed, this is appropriately assumed.

All of the foregoing also applies to a sense in which such models can function to present actual possibilities of what could have happened: If an initial distribution was actually possible at some prior time, t, then the ensuing pattern was likewise then actually possibility, with both of the qualifications applying as in the case of the future.

3.5 Conclusion

Actual possibilities, and epistemic possibilities as well, importantly differ from other forms of possibility usually considered. In particular, metaphysical possibility is usually not constrained by specific circumstances special to the actual world. And likewise, for example, physical possibility if the physical possibility is understood with general laws as the only constraint. My analysis illuminates this difference. In ways absent for other kinds of alethic possibility, actual and epistemic possibility concern particular matters of fact peculiar to the actual world.

Notes

1 To be emphasized, it is only work on alethic possibilities that has most often proceeded in terms of whole possible worlds. Use of local has been the preference when working on deontic modalities and what one might call "objectual modalities," ones such as essences, powers and dispositions that concern alethic possibilities that arise from the identity and behavior of objects (See Vetter (2011) for summary). Work on epistemic possibilities varies between using local and global possibilities.

2 Where I have used "statement" some, such as Kmept (2014), work with propositions, a kind of content; or, as does Lewis (1986), with sentences, carriers of content. Sentences would need to be something like Quinian "eternal sentences," with unique contents and constant truth values. As long as one idealizes vagueness away, either will work, as will any other characterization of contents or of content carriers with unique contents and constant truth values. I will use "statement" as a cover term for readers' preferred option for carriers of content or for contents themselves. In section 3 I will deidealize vagueness and propositions will no longer do, so that treatment with contents would have to be more complex.

3 There has been very little written about actual possibly (under the name, "real possibility"), but there is a very extensive literature on abilities. The possibilities that arise when one is able to do something very much count as actual possibilities in my sense. While I believe that ability-possibilities will fit under my characterization, careful examination will have to wait.

4 See deRosset (2009) for some of the difficulties and further references.

5 When Lewis discusses ersatz possible worlds he works with sentences, so in my exposition of Lewis's analysis I will likewise use "sentence" rather than statement. Others, for example Kment (2014) develop ersatz possible worlds with propositions.

6 Kment (2017, section 3) expresses a similar approach to the relation among different varieties of possibility.

7 Often what is morally and legally permissible or required are characterized as modalities. These are modalities in quite a different sense. They characterize actions where the modalities under discussion here, the alethic modalities, characterize statements and by extension the events and states of affairs described by statements.

(See Kment, 2017, section 3.3) The alethic modalities are understood as having success conditions – normally truth, but see section 4 for a generalization.

8 Readers may be more familiar with the role of sets of possible worlds as, or as one way of representing, propositions. On my account it is particularly clear that there is no conflict as propositions are one way of characterizing local possibilities. See footnote 2 and further elaboration below.

9 Lewis also brings up problems of duplicate entities that I will not discuss as they have no bearing on the ideas I will present.

10 I would expect a similar reduction for epistemic possibility and perhaps other more specialized kinds such as physical, economical, etc. possibility.

11 Whether or not the problem of constitution will still require considerations of metaphysical possibility might depend on how one approaches that problem.

12 Nolan (2014) explains this issue and gives a general discussion of the entanglement of issues about modality and issues of mode of presentation in our representation of possibilities.

13 Given how I develop, below, the way the past and present constrain the future, the view need not presuppose indeterminism, but only indeterminism "in practice."

14 Additional relativity to an inertial frame can be built in by taking t to be the proper time of an inertial frame.

15 This kind of view will require truth value gaps for present statements about circumstances unsettled in the future. This issue is studied in the classic paper, Thomason. (1970) For a contemporary discussion, see Akama et al. (2008).

16 And even for mathematics some have questions.

17 Since standards can never be specified in complete detail, this understanding of constraint introduces an additional dimension of vagueness.

18 Actual and epistemic possibilities can be expressed using the subjective mood: Bill might, could buy bananas tomorrow; Mary might, could have bought bananas yesterday. But not logical, conceptual, metaphysical and other varieties of possibility. That there are talking donkeys is a logical, conceptual, metaphysical... possibility. But "there might, could be talking donkeys" I hear as straightforwardly false. Similarly for "The law of gravitation might, could be, or could have been, a $1/r^3$ law. I am tempted to speculate that this contrast is connected with the way actual and epistemic possibilties are tied to the actual world in ways in which other varieties of possibility are not.

19 A referee asks whether such newly discovered actual possibilities exist antecedently or are created when someone formulates the relevant statement. Since I take real possibilities to be either the statements that present them or the content thus presented, one's answer to this question will depend on one's answer to the same question applied to statements or contents.

20 An agent here can be an individual or a community. This and other details of epistemic possibility vary from author to author. What follows will apply to these varying analyses very broadly.

21 Of course, we determined many of the things that have occurred up to now, but these occurrences are then independent of us in the sense that we cannot go back and change what we did. Going forward, our actions are fixed and in this way independent of us.

22 Except in an older sense of "true" that means something similar to "faithful." To say that a picture is true to its subject is to say that it represents the subject faithfully, in contextually relevant respects.

23 And thus true or false in the older sense.

24 See Normann's "Prospective Modeling," this volume, for an especially interesting special case of models that present actual possibilities.

25 Blueprints have been replaced by computer-aided construction drawings, about which analogous comments apply.

References

Akama, Seiki; Nagata, Yasunori; and Yamada, Chikatoshi. 2008. "Three-Valued Temporal Logic Qt and Future Contingents." *Studia Logica*, 88(2): 215–31.

deRosset, Louis. 2009. "Possible Worlds: Modal Realism." *Philosophy Compass*, 4(6):998–1008.

Deutsch, Harry. 1990. "Real Possibility." *Nous*, 24(5): 751–55.

Gahér, František. 2003. "Logical, Scientific and Real Possibility." In *Possibility and Reality*, edited by Hans Rott and Horak Vitezslav, 169–86. Heusenstamm: Ontos Verlag.

Gibbs, Benjamin. 1970. "Real Possibility." *American Philosophical Quaterly*, 7(4): 340–348.

Katzav, Joel. 2014 "The Epistemology of Climate Models and Some of Its Implications for Climate Science and the Philosophy of Science." *Studies in History and Philosophy of Modern Physics*, 46(1): 228–38.

Kment, Boris 2014. *Modality and Explanatory Reasoning*. Oxford: Oxford University Press.

Kment, Boris. 2017 "Varieties of Modality." *Stanford Encyclopedia of Philosophy*: https://plato.stanford.edu/entries/modality-varieties/

Lewis, David. 1986. *On the Plurality of Worlds*. Oxford: Blackwell

Nolan, Daniel. 2014. "Hyperintensional Metaphysics." *Philosophical Studies* 171: 149–60.

Thomason, Richmond H. 1970. "Indeterminist Time and Truth-Value Gaps." *Theoria*, 86(3): 264–81.

Vetter, Barbara. 2011. "Recent Work: Modality without Possible Worlds." *Analysis*, 71(4): 742–54.

Williamson, Timothy. 2017. "Modality as a Subject for Science." *Res Philosophical* 94(3): 415–36.

Wirling, Ylwa Sjölin and Till Grüne-Yanoff. "Epistemic and Objective Possibility in Science." *British Journal for Philosophy of Science*. Forthcoming.

4 The Possibilities in Propensities
Emergence and Modality

Mauricio Suárez

4.1 Introduction to the Complex Nexus of Chance

A long tradition in the philosophy of probability is devoted to ontological issues. What is probability? What is the meaning of our statements involving probability? What are we referring to when we make the statement, for example, that "there is an even chance" that someone turns up, that "the probability that it rains tomorrow is 50%," or that the chance of a given radioactive isotope to decay in the next hour is "less than one in a million"? More specifically, and typically, is the referent of "probability" or "chance" in the mind or in the world?

Those theories of chance that provide a referent in the mind, such as *credences* or *degrees of belief*, have come to be known as "subjective" or "subjectivist." On these theories, all talk of probability is ultimately talk about states of mind, information, or knowledge of an agent or agents. Those that take probability to have a referent as *objective real chances* in the physical or material world are known as "objective" or "objectivist" theories. On such views, probability talk is ultimately about arrangements, setups, or sequences of outcomes in the real world. The debate about the subjective or objective nature of probability is of course very old, going back to the founding of the modern concept of probability in the 1660s (Hacking 1975).

Most philosophers nowadays are pluralists to some degree: they admit that probability talk is sometimes about credences, and sometimes about chances, and they consequently admit that there are at least two kinds of facts that make probability statements true. There are facts about credences—which make subjective probability statements true, but there are also facts about chances—and they make objective probability statements true. Hence, either there are, properly speaking, two different concepts under a name; or there is only one concept with a polar or dual aspect. Carnap (1945) famously embraced the first option, going as far as to disambiguate probability into two different notions, which he characteristically referred to as probability$_1$ and probability$_2$. Others were less sanguine. Thus, Ramsey (1926) admits that scientists legitimately use probabilities to study statistical populations, without regard for anyone's credences. Hacking (1975) adopts an ecumenical

DOI: 10.4324/9781003342816-6

approach preserving one integral concept of probability, but which he argues is essentially dual or two-faced. In other words, probability is simultaneously both subjective and objective. These are different forms of pluralism regarding probability—and they oppose any reductionist attempts to do away with one or another of the subjective and objective dimensions of probability.

While most philosophers of probability are pluralists regarding objective and subjective probability, monism is still a common understanding of each of these notions taken separately. That is, as regards probability$_1$ and probability$_2$, or subjective credence and objective chance, philosophers have continued by and large to hold reductionist views. Thus, subjective probability has been assimilated to the partial degrees of belief that are characteristic of subjective Bayesianism, although some defenders of epistemic probability continue to work in the logical interpretation tradition of Keynes (1929).[1] As regards objective probability, there are two schools, but they are mutually exclusive and presuppose objective probability, or chance, to be just one kind of stuff. The frequentist tradition takes this stuff to be frequencies, while the propensity tradition takes it to be propensities. Each takes objective probability to be just one thing.[2]

Nevertheless, this pluralism can be pushed further into the domain of objective probability. It is consistent to suppose that objective probability is not identifiable with either frequencies or propensities, but that in practical inquiry it is always related to both. That is, probability does little work on its own; but in the context of a model, and when allied with the cognate notions of propensity and frequency, it does a good deal of explanatory work. Thus, on the view that I shall be sketching, which I call the complex nexus (Suárez 2020), chance is really an intermingling of theoretical propensities, formal probabilities, and experimental frequency statistics within some articulate normative modelling practice. It is in the mangle of this modelling practice that chance is forged, and it is to this "mangle of practice" (Pickering 1995) that we must attend as philosophers if we are to comprehend it.

The relations between the three different notions are explanatory, with propensities grounding and providing sui generis explanations of single case probabilities, which in turn may subsume, and hence also explain, the observed frequencies. However, note that the explanations involved are rather thin, sui generis, and not necessarily causal, so there is no ensuing reductionism or inherent ontological hierarchy. In fortunate circumstances, where the underlying mechanisms are well known, propensities may indeed be employed to causally explain singular outcome events. However, this is rather rare in statistical modelling practice (it requires considerable background causal knowledge, which is rarely available). In this paper, I have little to say about the connection between propensities and frequencies anyway, and I shall focus rather on the relation between propensities and probabilities, which is where the main modal lessons in statistical modelling are to be found.

4.2 Probability and Propensity: Ontology or Modelling Methodology?

The ontological project to find the truth makers of objective probability gives rise, in the hands of propensity theorists, to the project of interpreting objective probability or chance as a propensity. At the heart of this project, lies a two-way identity thesis (Humphreys 2019, 186; Suárez 2013) regarding propensities and probabilities. The probability-to-propensity half of the identity thesis proclaims that all objective probabilities can be interpreted as propensities, while the propensity-to-probability half maintains that all propensities can be rendered as conditional probabilities. The problems with such a propensity *interpretation* of probability are well-known: It engenders what is known as "Humphreys' paradox." It is worth reviewing the paradox here, as well as why it compromises any attempt to provide a unified ontology for objective probability statements. As we shall see, the problems that emerge ultimately suggest an altogether different approach to the nature of chance in science.

The first probability-to-propensity half of the identity thesis had already been shown to be false by Wesley Salmon and others in the 1970s. It is, in fact, evident that not all objective probabilities—not even just objective conditional probabilities—can be interpreted as propensities. For any two types of correlated events A and B: Prob (A/B) > Prob (A/¬B). But this also means by inversion under Bayes' theorem that Prob (B/A) > Prob (B/¬ A). It is immediately obvious that many types of correlated events have only a propensity interpretation in one direction but not the other. In fact, it is most clear in precisely those cases where the propensity interpretation seems right in one direction that it cannot possibly be right in the inverse direction. For instance, take Salmon's example of the propensity of shooting to kill. In every context that one can imagine, the probability of death if shot is surely higher than otherwise: Prob (D/S) > Prob (D/¬S). This is because indeed shooting someone has a propensity to kill them. Yet, by inversion, the probability that someone may have been shot if they are dead is also higher than otherwise: Prob (S/D) > Prob (S/¬D). Yet, there is no conceivable propensity of death to retrospectively cause shooting. This just makes no sense: it is impossible to interpret that conditional probability, however objective, as a propensity. It is impossible precisely because the original non-inverted correlation has a clear propensity interpretation: It is reasonable to interpret the increased Prob (D/S) as the result of the causal propensity of shooting people to kill them. This is precisely the sort of objective causal propensity fact that underpins the laws controlling gun handling and possession that are in place in nearly every country in the world.

What is remarkable, and we owe it to Paul Humphreys, is the discovery that the other half of the identity thesis, the propensity-to-probability half, is also false. Up to that point, the project to universally interpret objective probability as propensity had something going for it. For if all propensities may be represented as conditional probabilities, and some objective conditional probabilities are interpretable as propensities, then the ontological project still holds some promise. It may be the case that propensity facts underlie every objective probability statement and provide the truthmakers for them. It may be that chance can be analysed away as propensity. It may be that—as Popper thought—the mere appropriate interpretation of some key conditional probabilities as propensities will solve the most intractable problems in, say quantum mechanics or evolutionary theory. It may, in other words, be the case that objective probability is ontologically nothing but relative propensity. Yet, none of this is to be, since, as Humphreys showed, the Kolmogorov probability calculus is just not the appropriate formal rendition for many, or perhaps most, propensities.

Humphreys proposed a thought experiment, a very common way to establish possibility claims in science. A brief sketch of the thought experiment is sufficient for our purposes.[3] Consider a source emitting one photon at a time t_1, reaching a half silver mirror at time t_2, and being transmitted at time t_3. It is then plausible to suppose that:

i Any photon that reaches the half-silver mirror has some finite (non-zero) propensity to be transmitted.
ii Any photon that is emitted has some propensity greater than zero but not one to reach the mirror.
iii Any photon that is emitted and fails to reach the mirror has propensity zero (i.e., it has no propensity) to be transmitted.

Humphreys (1985, 565ff.) rendered each of these claims in a conditional probability formulation, assuming that there is always a unique representation of propensity claims as statements of conditional probability. He then showed that these three formal conditions are inconsistent, modulo some caveats and assumptions, with the Kolmogorov axioms, including the fourth axiom for conditional probability (also known as the ratio analysis of conditional probability). The argument leads to the conclusion that, modulo those caveats and assumptions, there are conceivable propensities that cannot be represented as conditional (Kolmogorov) probabilities.

Humphreys' "paradox" is not so much a paradox as an argument to the effect that propensities and (Kolmogorov) probabilities are distinct. While there are some assumptions to Humphreys' proof of the contradiction that would ensue, they are rather innocuous if not trivial in this context. But if one is convinced by the pertinence of Humphreys' assumptions and by his reasoning—as I am—then it follows that either i) Kolmogorov is not the right formal calculus for probabilities, and the Kolmogorov axioms must be rejected as a definition of probability, or ii) propensities cannot be probabilities, and the identity thesis is false. Humphreys himself followed the first route out of the contradiction and suggested developing an alternative calculus such as Renyi's.[4] By contrast, I have for a decade now been urging that the best option is the outright rejection of the identity thesis *both ways*. It is of course possible that we both are (were) right. That is, it can certainly be the case that propensities have no probabilistic representation and, moreover, that the Kolmogorov axioms are not the appropriate formal calculus for probability (or, as many philosophers argue these days, the fourth axiom is at best an approximation to the definition of conditional probability).

If the identity thesis is false, and propensities are not probabilities, the question arises as to what the relation between them is. This question is no longer an ontological question about what probabilities are. It is rather a methodological question about how probabilities and propensities relate in practice. Such functional questions about concepts concern their workings in a cognitive practice, a disciplinary field, or some established working culture, rather than their definition or nature. Thus, questions regarding the functional relations between probabilities and propensities must be answered by reference to a practice, a methodology, or both. My suggestion is to look specifically at statistical modelling practice to figure out what the respective functions of probabilities and propensities are. I suggest specifically statistical modelling practice because it is here that probabilities play decisive cognitive functions in advancing knowledge, increasingly since the mid-nineteenth century.[5] We get a grip on all sorts of uncertainty through our statistical models, and probability is the key feature that earns those models their keep, endowing them with both explanatory and predictive value.

Note that this argument from practice overcomes purely theoretical objections to the explanatory power of dispositional notions, including what has come to be known as Molière's "dormitive virtue" objection. This is the thought that a dispositional explanation (e.g., of the dormitive effects of opium) trivially rephrases the explanandum in other terms.[6] On the contrary, statistical modelling practice only falls in line at this point with ordinary cognitive practice, where dispositional explanations are rather common. Thus, in everyday life we accept the fragility of a particular glass as part of the explanation of its breakage, the solubility of a particular sugar cube as part of the explanation of its dissolving in a hot drink, and so on. In most of these cases, under the usual interpretation, the explanans appeals to an abstract property, or set of abstract properties, that are responsible for such effects. While in many of these ordinary cases, such as fragility or solubility, there is the presumption that these abstract properties will reduce to some categorical basis in the physics or chemistry of the components of the given macroscopic objects, this is not always so. There are other everyday dispositions (generosity, kindness, etc.) that do not reduce in this way, and whatever dispositions there are in fundamental physics (Suárez 2007; 2015) certainly will not reduce, since there is no more fundamental categorical basis for them to do so.

4.3   Indexed Chance Functions and Their Outcome Spaces

To avoid Humphreys' paradox, the relation between propensities and the probabilities that they generate is best formally represented in terms of indexed chance functions.[7] Let us refer to a system's propensity to yield an outcome in a range {O}, with a given probability Prob (O) as Prop-to-O, or Prop_O for short. The identity thesis entices us to think of this propensity as a conditional probability over O: Prob (O/Prop_O). But, as we saw, this gives rise to "Humphreys trouble." My proposal is to resolve the tension not by changing the formal structure of the Prob function that determines the probabilities for the different values of O, but by abandoning the assumption that propensities are to go into the chance function that represents the probabilities they generate. Instead, I suggest thinking of propensities as literally giving rise to the chance functions that represent these probabilities. The propensities of a system are after all responsible for the single-case chances that it engenders.[8]

The move is in the spirit of the rejection of the identity thesis: since propensities are distinct from the probabilities they generate, they ought to be formally represented as generating those chance functions, but not as being subject to them. In other words, the obtaining or otherwise of the propensity cannot be an event in the sample or outcome space defined by the chance function. It must rather lie outside this sample space as the condition of possibility for the outcomes represented. This suggests writing the propensity of a system as an index that uniquely characterises the chance function as defined over the outcomes in the (sample) space: $\text{Prob}_{\text{Prop}}$ (O).

Thus, each propensity gives rise to one or more (depending on context) chance functions over a set of possible outcomes. Yet, critically, that chance function is not well defined over the propensity itself. In other words, the possession conditions for propensities and probabilities are distinct. The propensity enables or generates, and thus *grounds*, the probability that it gives rise to. But the probability does not symmetrically ground the propensity, since it does not enable or generate it. The characteristic asymmetry of propensities is thus most naturally represented formally as an "enabling," or "generating" index. Consider the simplest case of a coin toss on a table. If the coin is sufficiently thin, the landing area in the table sufficiently flat, and there are no air currents, or forces acting on either coin or table, or any other interfering factor, the only two outcomes within the range of significant possibility are "landing heads up" and "landing heads down." There is no third option: The coin cannot realistically fall on its edge. Moreover, if the coin is constructed to be fair (balanced, symmetric, not bent, etc.) and it is tossed with sufficient strength upwards, with a smooth distribution function over initial upwards and angular velocity over many tosses, then it is possible to prove that the two outcomes have equal probability. We can then define, for this fair coin, as tossed in the given context, free of any interfering factors, given the appropriate boundary conditions on smooth-behaving dynamics, a chance function over the two possible outcomes. The sample space is bivalent, merely the two outcomes, and the distribution function is flat, ascribing equal probability to each outcome:

$$Pr_{prop}(H) = Pr_{prop}(T) = \frac{1}{2}.^9$$

Propensities thus both *delimit and quantify the domain of the possible.* In other words, they determine what goes into the sample space and how probable it is. But how exactly do propensities achieve this, and what does it entail for the nature of the possibilities at issue? Philosophers have traditionally attempted to settle these questions, when they have addressed them at all, by means of studying the metaphysics of chance and causation (e.g., Mellor 2005; Mellor 2018; Stenwall, Persson, and Sahlin 2018). Yet, the questions are not strictly metaphysical, nor do they necessarily call for a study of the ontology of chance. They rather appear to be primarily questions about the functional relations operating in practice between two different concepts (propensity, possibility). In line with the methodological approach that was sketched in previous sections, these questions are to be answered here not by reference to ontology or metaphysics, but by reference to the practice of statistical model building.

The following two sections address each question in turn and show that they call for a functional and abstract characterization of propensities as precisely the sorts of properties that index objective chance functions. This, I argue, is the simplest neutral account of how propensities *ground* chances in general. Propensities are thus precisely *whatever properties of a system enable and generate, and thus ground, its objective chances.* This is a minimal functional characterization, yet it has significant implications for the second question regarding the nature of the possibilities in the outcome or sample space of the chance functions. Once the possible outcomes in a sample space are understood to be objectively grounded in propensities, at whatever level of description, the modal content of these possibilities can be seen to be highly constrained by the specific features of the system. We may refer to them as "material" possibilities because they are made possible not only logically, metaphysically, and nomologically, but also by the material conditions in which the relevant chance setup is placed. That is, the relevant modalities here are tailored to—and hence relative to—a very specific chance setup operating in highly locally constrained material conditions. This is therefore a kind of possibility more strictly and narrowly defined than mere logical, metaphysical, or even physical possibility.

4.4 The Emergence of Propensities

The minimal functional characterization of propensities renders them abstract placeholders for whatever properties of a system ground its objective chances.[10] Since systems exhibit chances at multiple levels of description, and since, on this view, there are no further ontological assumptions that under-pin propensities, it follows that there exist propensities at different levels of description, too. Not all propensities are "fundamental" properties. Some, if not most, are emergent. This is certainly true if it is the case that not all chances are fundamental, that is, if there are emergent chances.

Microphysics—the physics of the microscopic material world—is often taken to be the fundamental level of description to which other levels ought to ultimately reduce. Ontologically, at least, if all matter is ultimately made up of atoms, then atomic physics describes how all matter ultimately is. Reductionism has for some time now been a matter of contention both in philosophy and in science, where emergent complex properties are rife. Regarding the properties and laws of most systems in macroscopic physics; quantum, inorganic, and organic chemistry; the life sciences, from cellular and molecular biology all the way to organic evolutionary biology and ecology; never mind those in the cognitive sciences and social psychology: None are reducible to the properties and laws of physics.[11] There is thus no reason to expect the propensities of those systems at their different levels of description to be reducible to the propensities of micro-physical systems.[12] What is more, if we find that there exist objective chances in those fields that are not reducible to quantum or microphysical chances, then the minimal functional characterization entails that emergent propensities also obtain at those levels. It would follow that not all propensities are micro-physical, for not all objective chances pertain to microphysics.

Still, the statement remains conditional. Are there non-fundamental objective chances? Here is where ontology ceases to guide us, and I suggest that we take up guidance from the practice and methodology of statistical modelling instead. The practice seems to offer ample evidence that there are emergent chances in fields such as physical and 'chemical' chemistry, evolutionary biology and ecology, social psychology, and econometrics. Perhaps most remarkable is that there appear to be emergent macro-physical chances in classical mechanical systems too, despite the underlying deterministic dynamics that characterise Newtonian mechanics. In fact, many of our most established examples of games of chance arguably exhibit emergent mechanical chances. But how can this be, given the underlying deterministic dynamics that presumably govern many of these mechanical, chemical, biological, ecological, psychological, or econometric systems?[13]

Several authors over the last decade have made compelling arguments for emergent chance in a deterministic world. Some of them (Glynn 2010) require a prior commitment to a specific metaphysics of chance, such as Lewis' best system analysis. Others (Ismael 2016) more liberally argue for the emergence of chance from our open-ended experience as agents in a deterministic world. Here I shall focus on List and Pivato (2015; 2021), and List (2022), which are closer to my methodological stance since they argue that the modelling practice of complex systems alone warrants grounds for emergent chance—even if the underlying dynamics is deterministic.

The setting is complex systems analysis, which specifies *states* for the system as points in some state space X. The evolution of these states in *time* is given by a mapping onto a simple linear ordering relation T. A *history* of the system is then a path through state space, represented by a function $h{:}T \to X$, such that for each time t in T, h(t) is the system's state at time t. Different realms of possibility correspond on this picture with different constraints on the set of possible *histories* Ω. The largest set of histories is the set of all logically possible functions $h \in \mathcal{H}$: all mappings of any set of time points into any set of states. Thus, we prescribe logical possibility if we set $\Omega = \mathcal{H}$. List and Pivato (2021) regard this as too large a class for complex systems theory, and advocate instead the more constrained class of nomologically possible histories, that is, those histories possible relative to some laws.

Different laws of nature would impose different restrictions upon the possible evolutions of the system, thus generating proper subsets of *H*. We can assume that physical laws impose the weakest constraints on arbitrary systems, while other emergent laws impose stronger constraints. However, this assumption is entirely optional in a dynamical system, where the system can indeed be anything, described with any arbitrary degree of detail, and the laws that serve to constrain the space of histories can be operating autonomously at any level. Thus chemical, biological, psychological, and sociological possibilities need not correspond to monotonically increasingly stricter subsets of the set of logically possible histories. And, at any rate, as has already been suggested and will be emphasised in what follows, the relevant constraints may also reflect facts about the more local material conditions where chance setups operate.

A probability structure (a family of conditional probability distribution functions) can be defined upon the subsets E (also called "events") of the full set Ω of possible histories as follows. A probability function Pr_E is defined for each set E of histories $E = \{h_E\} \subseteq \mathcal{H}$, such that any other event E', or alternative set of histories, is ascribed the conditional probability $Pr_E(E') = \mathrm{Pr}(E'/E)$.[14] A temporally evolving system is then the pair consisting of the set Ω of possible histories, as defined under whatever constraints are relevant, together with the conditional probability structure $\{Pr_E\}_{E \subseteq \Omega}$. It is then possible to show that a fine-grained distribution function over the individual histories contained in E yields extreme (0 or 1) conditional probabilities only, while a coarse-grained distribution function over the entire set E yields non-extreme conditional probabilities. This can be intuitively interpreted as an objective chance emerging from underlying deterministic dynamics at the finer-grained level (List and Pivato 2015, 140ff). Thus, objective chances emerge at higher levels of description despite the ontological compositional fact that the coarser-grained events are just classes of the finer-grained events, and even though there is nothing indeterministic at the finer-grained level.[15]

Note that the probability distribution functions are formally indexed to the events E. However, the indexing here is not playing the role of a "generator" or "enabler" of the outcome space in the distribution function, unlike what, I argue, is the case for propensities. Hence, the set of histories does not ground the chance functions, and it is more convenient for our purposes to always leave it out of the subindex that contains the propensities while inserting the conditioned-upon event into the probability distribution function as follows: $Pr_{prop}(E'/E)$. Another important difference is that List and Pivato (2021) conceptualise histories as possible ways the entire world can be, in agreement with possible world semantics. This assumption in my view goes way beyond what any modelling practice provides warrant for. Instead, I am urging a deflationary reading of the set of histories as limited descriptions of the ways in which a particular system can evolve, under severe local material constraints, in response to the propensities that act upon or within it. No possible world semantics is required to understand how the chance setup operates within its environment, or the inferential work that it does for us in allowing us to write down the probability distribution functions that it generates or enables. This is true even if any possibility statements that we may deduce from such a framework may receive some sort of semantical interpretation—and possible worlds (of the *Ersatz* or non-*Ersatz* variety, see Lewis [1986] 2001), may always provide one. Yet, it can also be supposed that the truth conditions for those possibility statements refer also, or even exclusively, to the highly constrained material conditions present in the environments where the setups operate.[16]

At any rate, none of the lessons I would like to extract for propensities depends on what the appropriate semantical interpretation of those statements turns out to be. Under the abstract and functional characterization of propensities proposed, if there are emergent objective chances at higher levels of description then there are also propensities at those levels—even if there are no propensities at the lowest level to account for them (perhaps because the underlying dynamics at the lower level is deterministic). The reality of emergent objective chance requires us to postulate properties of the systems that are responsible for generating those chances. Whatever properties those systems have, at whatever level of description, they configure chance setups capable of generating those chances. Given our definitions, we know this to be true a priori, relative to the discovery of those emergent chances, and regardless of how much we know about the intricacies of the mechanisms that operate those chance setups.

4.5   The Modalities of Statistical Modeling

In the last section, I have argued that objective chances emerge at different levels of description. Even a deterministic physical universe would contain such chances at higher levels of chemical, biological, psychological, and sociological description. The universe may of course not be fully deterministic even at the physical level, in which case there are objective chances "all the way down." In the previous sections, I argued that the term 'propensity' is an abstract placeholder for the properties of a system that explain why it is a chance setup, that is, those properties of the system responsible for its generative or enabling objective chances. There is thus an intimate conceptual link between propensities and the outcome (sample) spaces in the chance or probability distribution functions that display those propensities. The propensities first generate the possibilities, then probabilistically quantify them. What does this tell us about the nature of the possibilities involved?

A debate rages nowadays over whether models describe objective or epistemic possibilities.[17] While not denying that some models in the cognitive sciences may merely represent agents' credences, I have assumed that there are objective chances—and that many scientific models in the statistical modelling sciences aim to capture some of these chances.[18] Now, it may be objected that in a deterministic physical world, any chance higher up the hierarchy is necessarily epistemic—it reflects exclusively our lack of knowledge regarding the initial conditions, the laws, or both. Yet, regardless of determinism, List and Pivato (2015) provide good arguments as to why the chances that result from coarse-graining at higher levels can be as objective as those further down the hierarchy. And a cursory inspection of statistical modelling techniques and activities across the sciences reveals that the presupposition of much of this work is to model objective chances.

A dynamical account of objective chance will at any rate approach the emergence of chance as a dynamical phenomenon—primarily the result of the operation of dynamical laws on particularly contrived systems. In this approach, the chance setup is not an entity or object standing aloof, with whatever properties it may possess at a given time. The chance setup is a complex and evolving dynamical process, where the relevant properties responsible for chanciness are the dynamical aspects of the evolution of the implicated components and their mutual interaction. The clearest example is provided by the methodology known as the Method of Arbitrary Functions (MAF), which has been defended by many of us in recent years. This is an approach to objective chance *in the macroscopic world* that, even within physics, dispels the worries about underlying determinism or, indeed, reduction.[19]

MAF shows that objective macroscopic chances may emerge as the result of a microscopically deterministic dynamical process. This takes a sufficiently smooth distribution over a range of relevant initial microscopic variables of the system to a probability distribution over some resulting macroscopic variables of the system at the end of the dynamical evolution. Its outstanding application is the lowly example of the toss of a fair coin, where the resulting flat distribution over the two possible outcomes has often been thought to be a trivial application of indifference. Keller (1986) demonstrates that, on the contrary, the flat distribution of a fair coin is rather the outcome of a very complex classical mechanical dynamical process where uncertainty is rife over the initial values of the upwards and angular velocity of the coin at the point of ejection. Effectively, this is a process that takes an initial distribution over microscopic initial conditions (the upwards (v) and angular (ω) velocities of the coin) and transforms it into a flat probability distribution over the only possible macroscopic outcomes at the resulting end (heads or tails, with equal probability: $\Pr(H) = \Pr(T) = 1/2$). But how does MAF carry out this "magic"? What justifies ascribing equal objective chances to each of the two outcomes of the coin toss? And where are these quantified objective possibilities coming from?

The key is the initial distribution over the microscopic variables v and ω. For the dynamical process to reliably generate the expected flat probability distribution over the expected outcome space, the initial distribution over the possible values of v and ω must be "smooth" (Strevens 2021, 14632ff.). The initial micro-probability distribution may just be coding in for our ignorance of the values of v and ω, in which case, the resulting macro-probability distribution over heads and tails is arguably epistemic as well. Or the initial micro-distribution may just collect statistics from repeated experiments, and the final macro-distribution exhibits the corresponding statistics for an actual or virtual ensemble of results. Neither of these options is very attractive since (i) both the subjective and the frequency interpretation of probability are plagued with problems, and (ii) they are not in the spirit of MAF as introduced by either Von Kries or Poincaré. But then, as already argued, there is no need to interpret this initial distribution for the connection with practice that is relevant here.[20]

The purpose of MAF is to rather display how a dynamical process by itself, however deterministic, can generate the outcome space and the objective chance distribution defined over it. There is no need for an interpretation of the initial micro-distribution (Rosenthal 2010), which can be taken to be the result of further dynamical processes that bring it about, as is the case, for instance, in perturbation theory (Strevens 2021). If the perturbations on the initial distribution of $\{v, \omega\}$ are also smooth, the dynamical process amplifies them, turning them into the objective chance distribution over the macro-variables. So, it is not the coin that possesses the propensities to land heads or tails, but the coin in its actual toss setup, given the constraints on its ejection, motion in the air, landing surface, etc. It is, in other words, the *entire* experimental set-up *as it evolves in time*, in accordance with the dynamical equations of motion, that generates the outcome space $O = \{H, T\}$, as well as the probability distribution $\Pr(H) = \Pr(T) = 1/2)$ defined over this space of possibilities.[21]

These unrealised material possibilities are thus also *real* in the sense of Deutsch (1990), *actual* in the sense of Teller (this volume), and *situated* in the sense of Ruyant (2021).[22] The options that appear in the outcome (sample) space are not merely logically and metaphysically possible—which of course they are—they are also made possible by the actual laws of nature as they apply in the specific circumstances in which the chance setups operate. With varying circumstances, or alternative setups, other logical and metaphysical possibilities become available, similarly consistent with the laws as they apply in very different material circumstances. Hence, these possibilities are relative also to singular facts pertaining to the operation of real existent chance setups, and the operations that they carry out. In analogy with Norton's (2021) material theory of induction, such possibilities are *material* since they are constrained by local matters of fact, and not merely by the logical principle of non-contradiction, metaphysical principles regarding what is conceivable or what follows from what is conceivable, or nomological statements regarding what is possible in accordance with natural laws. Material possibilities are additionally constrained by what the case is singularly in those precise locales where chance setups actually operate—they are hence relative to the very concrete material conditions in which propensities obtain and generate their range of quantified possibilities.

4.6   Conclusions

I have argued that propensities are abstract placeholders for the concrete properties of complex dynamical systems that generate objective chances, at whatever level of description. The Method of Arbitrary Functions shows that such dynamical processes can exist in fortunate circumstances regardless of underlying determinism. The objective possibilities that such dynamical processes bring about are local 'material' possibilities, which may or may not be ultimately realised, but which exist in virtue of very special dynamical processes acting in highly constrained environments. If the complex nexus of chance that is thus incorporated in the setup behaves smoothly and appropriately, both the material possibilities in the outcome space and the probabilities defined over the outcome space are genuinely objective.

Acknowledgements

Financial support is acknowledged from the Spanish Research Agency (AEI) projects PGC2018-099423-B-100 and PID2021-126416NB-I00. For comments I thank two referees, and audiences in Vienna (at the ERC "Modelling the Possible" seminar in May 2021), Madrid (at the "Modalities in Representation" workshop in May 2022), and Lisbon (at the Centre for Philosophy of Science, or CFCUL, seminar in September 2022). Special thanks to Tarja Knuuttila, Quentin Ruyant, and João Cordovil, respectively, for hosting me.

Much of my work on probability is indebted to Paul W. Humphreys (1950–2022), and I dedicate this chapter in his memory.

Notes

1 For subjective Bayesianism, see Ramsey (1926), Savage ([1954] 1972). De Finetti's (1972) radical subjectivism is perhaps most extreme in dispensing with any objective notion of probability altogether. In all these views the key to subjective probability is coherence, and the Ramsey-De Finetti theorem proves that partial degrees of belief must be probabilities if they are coherent. The logical interpretation, by contrast, embraces the notorious principle of indifference. Contemporary "objective Bayesians" are heirs to this tradition—see for instance Williamson (2010).

2 The frequency interpretation in its contemporary rendition was developed by Von Mises in his classic (1927) while having antecedents in Venn ([1876] 2006). The propensity interpretation originates in Peirce (1878), but it was Popper (1959) who made it famous. Later contributors include Gillies (2000), Mellor (2005), Humphreys (1985; 2004), Suárez (2013; 2014): In 2014, the four of us sat in a seminar room in London to discuss Paul's latest views—it is sad to think that half of us is now gone.

3 It is described in detail in many different places. See Suárez (2013; 2014) as well as, of course, Humphreys (1985, and 2004).

4 Renyi (1955). See Lyon (2014) for an exposition.

5 See Daston (1995), Gigerenzer et al. (1989), Hacking (1975).

6 In Suárez (2013), I took the bother to read carefully through Molière's delightful play *Le Malade Imaginaire* (Molière, 1673, see particularly the third interlude) and found the actual objection to be quite different. Molière's aim is rather to poke fun at pretentious pseudo-explanations in Latin by pompous medical practitioners: "Domandatur causam et rationem quare Opium facit dormire: A quoi respondeo, Quia est in eo Virtus dormitive." His is an objection that applies to any verbose attempt to substitute a pompous pseudo-explanation in Latin in place of any causal explanation in the vernacular, whether it be dispositional or categorical. There is nothing intrinsic about dispositional explanations that is objectionable; and Molière's dazzling diatribe is independent of whether dispositional properties are explanatorily, even causally, related to their manifestations.

7 See Suárez (2018) for the details of this approach and how it favourably compares to any alternative formulations.

8 See Gillies' (2000) excellent paper for the distinction between single-case and long-run propensities. Gillies' arguments for a long run and against single-case variants seem to me to rely on the identity thesis. By contrast, on my account propensities display single case chances.

9 The case of the coin has been studied intensely now, having been only properly described as recently as four decades ago. See the seminal contributions of Keller (1986), and Diaconis et al. (2007).

10 I am thus taking the grounding relation to provide the most generic form taken by all enablers or generators of an objective chance function without further specification. See Suárez (2018) for an extended discussion.

11 This is by now an old lesson, that there is no space to discuss here in full. Some of the landmarks include the key works in the Stanford school, led by Suppes (1978) or Dupré (1995); but in my view the most accomplished version of anti-reductionism is to be found in Mitchell (2003; 2012). These works show that ontological reductionism—the assumption that all material reality is ultimately composed of physical stuff—is compatible with the emergence of sui generis properties and laws at "higher" levels.

12 Such as the quantum propensities defended for different interpretations of quantum mechanics in (Suárez, 2007).

13 This is assuming that they are indeed deterministic systems, or that they can at any rate be described by means of deterministic models. At least in regards to evolutionary biology, these issues have been discussed extensively; see Mitchell (2003), Strevens (2016), Pence (2022). For similar claims regarding 'chemical chemistry' see Suárez and Sánchez-Gómez (2023).

14 I thank a referee for pointing out that these sets of histories may or not overlap.

15 See particularly List and Pivato (2015, 140ff.), and List (2022, section 6). An interesting response, challenging the coarse-graining procedure, is Kinney (2021). Even if List and Pivato's coarse-graining procedure fails, emergence can obtain, for a macro-system state need not be a mere coarse-graining of a micro-system's phase space state: it may live in an altogether different phase space.

16 There are different extant ways to provide a semantics in terms of local and contextual truth-conditions, including situations (Ruyant, 2021), real possibilities (Deutsch, 1990), or actual possibilities (Teller, this volume). My preference is to maintain a reference to "material possibilities," as the possibilities generated as the affordances locally instantiated by the operation of the chance setup within its working environment.

17 See Knuuttila (2021), Massimi (2019), Sjolin Wirling and Grüne-Yanoff (2021, forthcoming), Williamson (2016; 2018). Ylwa Sjölin and Till Grüne's chapter in this volume provides a nice summary.

18 There are, of course, ways to measure model uncertainty and to average over model ensembles to calculate overall uncertainty. The probabilities that appear in such procedures are prima facie epistemic, and there are interesting questions regarding the application of the multiplication rule to compute average uncertainty over objective chances in such cases. (I thank Tim Williamson for pointing out in conversation that the multiplication rule can fail; but this is a side-issue for the purposes of the present paper.)

19 The MAF originates in Von Kries and Poincaré. See Von Plato (1983) for a historical account. Reichenbach ([1916] 2008), Hopf (1932), Strevens (2013; 2021), Rosenthal (2010), Abrams (2012), Myrvold (2021), Suárez (2020) all endorse MAF to some degree, but on different grounds. For a dissenting view on the claim that MAF generates objective probabilities—and hence, on my view, objective possibilities too—see De Canson (2022).

20 See my defense of a minimalist approach to the interpretation of these probabilities—in line with Sober's (2010) no-theory theory—, for the initial distribution function in Suárez, 2020, 51ff.

21 The unrealized material possibilities are thus also actual possibilities, in the very concrete sense of Teller (this volume) or Stenwall et al. (2018). They are not merely permitted by the laws of physics but are actively brought about by the dynamical processes operating in the highly constrained conditions under which the experiment is run.

22 Deutsch (1990, 751) argues that judgements in medical practice regarding, for example, *real* possible cures for a given condition are not guided only by considerations of logical, metaphysical, or nomological possibility but also by whether such possibilities can or cannot realistically obtain in the given circumstances. Teller (this volume) generalizes to what he calls the *actual* possibilities studied by current science and argues cogently that no possible world semantics can provide an accurate semantics for possibilities so constrained. Ruyant (2021, chapter 4) further discusses the sense in which such possibilities are situated, that is, they are relative to local configurations of objects, and thus not merely to possible worlds (which are global descriptions of how the world can be), but to specific matters of fact obtaining in a given situation.

References

Abrams, Marshall. 2012. "Mechanistic Probability." *Synthese* 187(2): 343–75.

Carnap, Rudolf. 1945. "The Two Concepts of Probability: The Problem of Probability." *Philosophy and Phenomenological Research* 5(4): 513–32.

Daston, Lorraine. 1995. *Classical Probability in the Enlightenment.* Revised ed. Princeton, NJ: Princeton University Press.

De Canson, Chloé. 2022. "Objectivity and the Method of Arbitrary Functions." *British Journal for the Philosophy of Science* 73(3): 663–684.

Deutsch, Harry. 1990. "Real Possibility." *Nous* 24(5): 751–755.

Diaconis, Persi, Susan Holmes, and Richard Montgomery. 2007. "Dynamical Bias in the Coin Toss." *SIAM Review* 49(2): 211–35.

Dupré, John. 1995. *The Disorder of Things: Metaphysical Foundations of the Disunity of Science*. Revised ed. Cambridge, MA: Harvard University Press.

De Finetti, Bruno. 1972. *Probability, Induction and Statistics: The Art of Guessing*. New York: John Wiley and Sons.

Gigerenzer, Gerd, Zeno Swijtink, Theodore Porter, Lorraine Daston, John Beatty, and Lorenz Kruger. 1989. *The Empire of Chance: How Probability Changed Science and Everyday Life*. Ideas in Context. Cambridge: Cambridge University Press.

Gillies, Donald. 2000. *Philosophical Theories of Probability*. London; New York: Routledge.

Glynn, Luke. 2010. "Deterministic Chance." *British Journal for the Philosophy of Science* 61(1): 51–80.

Hacking, Ian. 1975. *The Emergence of Probability*. Cambridge: Cambridge University Press.

Hopf, Eberhard. 1932. "On Causality, Statistics and Probability." *Journal of Mathematics and Physics* 13: 51–102.

Humphreys, Paul. 1985. "Why Propensities Cannot Be Probabilities." *The Philosophical Review* 94(4): 557–70.

Humphreys, Paul. 2004. "Some Considerations on Conditional Chances." *British Journal for the Philosophy of Science* 55(4): 667–80.

Humphreys, Paul. 2019. *Philosophical Papers*. Oxford, New York: Oxford University Press.

Ismael, Jenann. 2016. *How Physics Makes Us Free*. Oxford, New York: Oxford University Press.

Keller, Joseph B. 1986. "The Probability of Heads." *The American Mathematical Monthly* 93(3): 191–97.

Keynes, John Maynard. 1929. *A Treatise on Probability*. London: Macmillan & Co.

Kinney, David. 2021. "Blocking an Argument for Emergent Chance." *Journal of Philosophical Logic* 50(5): 1057–77.

Knuuttila, Tarja. 2021. "Epistemic Artifacts and the Modal Dimension of Modeling." *European Journal for Philosophy of Science* 11(3): 65.

Lewis, David. 1986. *On the Plurality of Worlds*. 1st ed. Malden, MA: Wiley-Blackwell.

List, Christian. 2022. "The Naturalistic Case for Free Will." In *Levels of Reality in Science and Philosophy*, edited by Stavros Ioannidis, Vishne Gal, Hemmo Meir, and Shenker Orly. Cham: Springer.

List, Christian, and Marcus Pivato. 2015. "Emergent Chance." *The Philosophical Review* 124(1): 119–52.

List, Christian, and Marcus Pivato. 2021. "Dynamic and Stochastic Systems as a Framework for Metaphysics and the Philosophy of Science." *Synthese* 198(3): 2551–2612.

Lyon, Aidan. 2014. "From Kolmogorov to Popper, to Rényi: There's No Escaping Humphreys' Paradox (When Generalized)." In *Chance and Temporal Asymmetry*, edited by Alastair Wilson, 112–30. Oxford: Oxford University Press.

Massimi, Michela. 2019. "Two Kinds of Exploratory Models." *Philosophy of Science* 86(5): 869–81.

Mellor, Hugh. 2005. *Probability: A Philosophical Introduction*. London: Routledge.

Mellor, Hugh. 2018. "Propensities and Possibilities." *Metaphysica* 20(1): 1–3.

Mitchell, Sandra D. 2003. *Biological Complexity and Integrative Pluralism*. Cambridge Studies in Philosophy and Biology. Cambridge: Cambridge University Press.

Mitchell, Sandra D. 2012. "Emergence: Logical, Functional and Dynamical." *Synthese* 185(2): 171–86.

Molière (1673). *Le Malade Imaginaire: Commedie en Trois Acts*.

Myrvold, Wayne C. 2021. *Beyond Chance and Credence: A Theory of Hybrid Probabilities*. Oxford, New York: Oxford University Press.

Norton, John. 2021. *The Material Theory of Induction*. Calgary: Calgary University Press.

Peirce, Charles S. 1878. "Charles S. Peirce, Philosophical Writings | The Doctrine of Chances." 12 March 1878.

Pence, Charles H. 2022. *The Rise of Chance in Evolutionary Theory*. London: Academic Press.

Pickering, Andrew. 1995. *The Mangle of Practice: Time, Agency, and Science*. Chicago, IL: University of Chicago Press.

Popper, Karl R. 1959. "The Propensity Interpretation of Probability." *The British Journal for the Philosophy of Science* 10(37): 25–42.

Ramsey, Frank P. 1926. "Truth and Probability." In *Ramsey, The Foundations of Mathematics and Other Logical Essays*, edited by R.B. Braithwaite, Ch. VII, 156–98, London: Kegan, Paul, Trench, Trubner & Co.

Reichenbach, Hans. (1916) 2008. *The Concept of Probability in the Mathematical Representation of Reality*. Open Court.

Renyi, Alfred. 1955. "On a New Axiomatic Theory of Probability." In *Acta Mathematica Academiae Scientiarum Hungaricae* 6: 286–335.

Rosenthal, Jacob. 2010. "The Natural-Range Conception of Probability." In *Time, Chance, and Reduction: Philosophical Aspects of Statistical Mechanics*, edited by Andreas Hüttemann, and Ernst Gerhard, 71–91. Cambridge: Cambridge University Press.

Ruyant, Quentin. 2021. *Modal Empiricism*. Dordrecht: Synthese Library.

Savage, Leonard J. (1954) 1972. *The Foundations of Statistics*. 2nd ed. New York: Dover Publications.

Wirling, Sjölin, and Till Grüne-Yanoff. forthcoming. "Epistemic and Objective Possibility in Science." *The British Journal for the Philosophy of Science*.

Sjölin Wirling, Ylwa, and Till Grüne-Yanoff. 2021. "The Epistemology of Modal Modeling." *Philosophy Compass* 16(10), 1–11.

Sober, Elliott. 2010. "Evolutionary Theory and the Reality of Macro-Probabilities." In *The Place of Probability in Science*, edited by Ellery Eels and Fetzer James, 133–60. Dordrecht: Springer.

Stenwall, Robin, Johannes Persson, and Nils-Eric Sahlin. 2018. "A New Challenge for Objective Uncertainties and the Propensity Theorist." *Metaphysica* 19(2): 219–24.

Strevens, Michael. 2013. *Tychomancy: Inferring Probability from Causal Structure*. Cambridge, MA: Harvard University Press.

Strevens, Michael. 2016. "The Reference Class Problem in Evolutionary Biology: Distinguishing Selection from Drift." In *Chance in Evolution*, edited by Charles Pence, and Grant Ramsey. Chicago, IL: University of Chicago Press.

Strevens, Michael. 2021. "Dynamic Probability and the Problem of Initial Conditions." *Synthese* 199(5): 14617–39.

Suárez, Mauricio. 2007. "Quantum Propensities." *Studies in History and Philosophy of Science Part B: Studies in History and Philosophy of Modern Physics* 38(2): 418–38.

Suárez, Mauricio. 2013. "Propensities and Pragmatism." *The Journal of Philosophy* 110(2): 61–92.

Suárez, Mauricio. 2014. "A Critique of Empiricist Propensity Theories." *European Journal for Philosophy of Science* 4(2): 215–31.

Suárez, Mauricio. 2015. "Bohmian Dispositions." *Synthese* 192: 3203–28.

Suárez, Mauricio. 2018. "The Chances of Propensities." *British Journal for the Philosophy of Science* 69: 1155–1177.

Suárez, Mauricio. 2020. *Philosophy of Probability and Statistical Modelling*. Cambridge Elements in the Philosophy of Science. Cambridge: Cambridge University Press.

Suárez, Mauricio and Pedro Sánchez-Gómez. 2023. "Reactivity in Chemistry: The Propensity View", *Foundations of Chemistry*, 25: 369–380.

Suppes, Patrick. 1978. "The Plurality of Science." *PSA: Proceedings of the Biennial Meeting of the Philosophy of Science Association* 1978(2): 3–16.

Venn, John. (1876) 2006. *The Logic of Chance*. Mineola, NY: Dover Publications Inc.

Von Plato, Jan. 1983. "The Method of Arbitrary Functions." *The British Journal for the Philosophy of Science* 34(1): 37–47.

Williamson, Jon. 2010. *In Defence of Objective Bayesianism*. Illustrated ed. Oxford; New York: OUP Oxford.

Williamson, Timothy. 2016. "Modal Science." *Canadian Journal of Philosophy* 46(4–5): 453–92.

Williamson, Timothy. 2018. "Spaces of Possibility." *Royal Institute of Philosophy Supplements* 82(July): 189–204.

Part II
Possibility Spaces

5 Invariance, Modality, and Modeling

Andreas Hüttemann

5.1 Introduction

Scientific modelling can have many purposes. Generating modal knowledge and justifying modal claims are two such purposes. Such claims may concern, for instance, what is physically, biologically, and so on, conceivable, or they may concern possible explanations. In this chapter, I will focus on the relation between scientific modelling and *objective* or *de re* modal features of systems. More precisely, I will discuss how scientific modelling gives us knowledge about *possible* states of systems and about the ways in which the behaviour of target systems is *constrained*. I will argue that the concept of invariance is particularly helpful for exploring this relationship. After some stage setting (section 5.1), I will discuss the concept of invariance (of laws) and will argue that such invariances are typically empirically accessible (section 5.2). In section 5.3, I will argue that (some) modal features of the behaviour of systems can be understood in terms of invariance relations. Section 5.4 then concludes by discussing the connections between empirically accessible modal features of the behaviour of systems and some aspects of our modelling practices such as abstractions and idealizations.

5.2 Stage Setting: Models and Laws[1]

5.2.1 Models

In order to situate the following discussion, it will be useful to introduce some of the distinctions that have been discussed in recent literature on scientific modelling. Michael Weisberg stresses the flexibility of modelling by pointing to the variety of possible targets. Modelling can be used to study specific systems, for example, a specific fishery in the Adriatic Sea, clusters of targets, a generalized target (predator-prey-populations), or hypothetical targets such as species with more than two sexes. There is also targetless modelling when the features of models are studied without considering whether the model in question represents a target system (Weisberg 2013, chapters 5 and 7). In what follows, I will focus on models of generalized targets such

DOI: 10.4324/9781003342816-8

as harmonic oscillators, free-falling stones, or economies. (This chapter will thus be complementary to Michaela Massimi's paper (2019) on "Two kinds of exploratory models," which focuses on hypothetical models and targetless fictional models and their relation to modal knowledge.)

Another useful distinction has been introduced in a paper by Reutlinger, Hangleiter, and Hartmann (2018). Models are either embedded in empirically well-confirmed theories or they stand on their own and are thus autonomous. Quantum Mechanics and Newtonian mechanics are examples of theories within which models are embedded. Paradigm cases of embedded models are the harmonic oscillator, models of the hydrogen atom, and so on—the kind of models that Cartwright studied in *How the Laws of Physics Lie* (1983). These models typically serve the function of subsuming phenomena under abstract dynamic laws such as the Schrödinger equation. Autonomous models, by contrast, stand on their own. Well-known examples are Schelling's model of segregation and the Lotka-Volterra model. The focus of this chapter will be on embedded models.

5.2.2   Law Statements

With respect to laws of nature, their *content* (i.e., what the law statements state) can be distinguished from their *nomic status*. The latter will be discussed in section 5.3. Here I will focus on the content of law statements. Law statements (or theories) are not simply mathematical equations. Even if we know that in the equation $s = 1/2gt^2$ s stands for a path and t for a time, the equation cannot be taken to be Galileo's law of free fall. For instance, nobody takes Galileo's law to be disconfirmed by spheres rolling on a plane. Galileo's law is not simply a mathematical equation but the claim that the behaviour of a certain class of systems can be represented by the above equation:

Free-falling bodies behave according to the equation $s = \frac{1}{2} gt^2$.

It is essential for Galileo's law that it refers to free-falling bodies. They are the generalized target systems of the equation. For the purposes of this paper, I take the general form of a law statement to be the following:

(L) All physical systems of a certain kind K behave according to Σ

Σ is what I call the "law-predicate," which includes the mathematical equations that are meant to characterize the behaviour of the systems. (L) is ultimately of the time-honoured form "All Fs are Gs," but the latter hides all the interesting complexity in Σ that will be relevant for discussing the modal aspects of laws, models, and systems.

How are models and law statements related? Even though models have a plurality of possible functions (see Gelfert 2016 for an overview) within (L) models serve the function of representing a (generalized) target system. An example is the statement:

(H) All hydrogen atoms behave according to the Schrödinger equation with the Coulomb potential.

The hydrogen atoms are the generalized target systems. The Schrödinger equation with the Coulomb potential is meant to represent the behaviour of the system in question. The hydrogen atoms are modelled by the Coulomb potential. Modelling assumptions, for instance, idealizations and abstractions, are typically introduced when it comes to specifying the Hamilton operator. For instance, picking the Coulomb potential implies that certain variables are relevant while others are left out as irrelevant.

5.2.3 *Internal and External Generalizations*

Given the above characterization of law statements, we can draw a distinction between different kinds of generalizations that will play a role in what follows.

Law statements typically involve at least two different kinds of generalizations (see Scheibe 1991). A *system-external generalization* quantifies over systems. System-external generalizations *explicitly* occur in (L): "all systems of a certain kind K." *System-internal generalizations*, by contrast, are not explicitly mentioned in (L), but they are assumed. They quantify over the values of the variables that occur in the law equations. To illustrate this distinction: in Galileo's law (*Free-falling bodies* behave according to the equation $s = \frac{1}{2} g t^2$), we can distinguish (1) a generalization that quantifies over systems—the equation is said to pertain to all systems of a certain kind K, namely to free-falling bodies. (2) Generalizations that quantify over the values of the variables in the mathematical equation in Σ, in this case, s and t. The equation is meant to hold for all values of, say, the variable t (within a certain range). This is a system internal generalization.

It will turn out that many interesting aspects of the modality of laws are connected to internal generalizations.

Working with embedded models is an important aspect of scientific practice (at least in physics), and the question of how modelling practices in this context are related to modal features of the behaviour of the target systems is the issue I will address in the remainder of this chapter.

5.3 Invariance

5.3.1 *Different Kinds of Invariance*

In what follows, the notion of *invariance* will serve to link modelling assumptions such as idealizations and abstractions on the one hand and (objective) modal features of systems on the other. But what is an invariance? The basic idea is that something remains the same while certain changes or transformations do or could take place. Invariance is a modal notion, so in arguing that modal features of systems are to be understood in terms of invariances I am not engaged in the project of explaining modal notions away.

In the philosophical literature, invariance has been linked to notions such as truth and objectivity. The fact that in the special theory of relativity, the space-time-interval (as opposed to the spatial interval) is invariant under Lorentz transformations has been taken to show that this is an objective non-relative feature of the world (see, e.g., Nozick 2001). These implications of (some kinds of) invariance will not be the focus of this chapter. I will instead distinguish various kinds of invariances and explore their role in scientific practice.

Wigner, in a famous paper on invariance, observed:

> The world is very complicated and it is clearly impossible for the human mind to understand it completely. Man has therefore devised an artifice which permits the complicated nature of the world to be blamed on something which is called accidental and thus permits him to abstract a domain in which simple laws can be found. The complications are called initial conditions; the domain of regularities, laws of nature.
>
> (Wigner 1949, 521)

So, according to Wigner, the very notion of a law presupposes a distinction of something that changes on the one hand and the law(-equation) itself, which remains invariant, on the other.

Let us illustrate this notion of invariance with Newton's second law:

(N2): All bodies behave according to the equations $F = ma$.

Despite changes in initial conditions, that is, in the values of the force F, (N2) and thus in particular the law equation remains the same for all values of these initial conditions. The law and thus the internal generalization (as well as the external generalization) remain invariant.

Another example is Galileo's law:

Free-falling bodies behave according to the equation $s = \frac{1}{2}\, gt^2$.

Despite changes in initial conditions (e.g., the values of s), Galileo's law and thus in particular the law equation remain the same. In this case, however, the law equation is invariant only for small s compared to the diameter of the earth. Invariance, in this case, is domain-restricted, that is, not universal as in the case of Newton's second law. Furthermore, the domain restriction may depend on pragmatic considerations, for example, on the question with which precision we need to know s. The more precise we want to be, the more the domain of invariance will be restricted.

This provides us with a first kind of invariance:

Invariance *of the law equation* with respect to *the initial conditions*: The law equation holds independently of the values (within a certain domain) of the initial conditions.

(see Woodward 2018, section 3 for discussion)

What is characteristic of the first kind of invariance of laws is that the quantities or variables that change are characterized in terms of variables which are explicitly mentioned in the law predicate. There are, however, also changes with respect to quantities or features of reality that are not explicitly mentioned in the law equation. This leads to a second kind of invariance:

Invariance *of the law equation* with respect to *features of the target system that are not explicitly mentioned in the law equation*:

i same-level properties, for example, colour, shape, and mass in the case of free-falling bodies.
ii lower-level or constitutional properties, for example, the molecular structure of the gases in the case of the ideal gas law.

Closely related is a third kind of invariance:

Invariance *of the law equation* with respect to *changes* (not of the target system but) *in the behaviour of other systems*.

Newton's laws of motion and his law of gravity hold for the solar system whether or not other systems in the universe undergo changes. Another example is the ideal gas law. Its equation holds whether or not I am on time when I go to the dentist or whether or not a particular tree loses its leaves in New Mexico. Of course, the initial or boundary conditions of the system under consideration will change in the case of the solar system due to changes of other systems in the universe, but the law equations remain the same.

To sum up, the various types of invariance we have discussed so far (the list I presented is not meant to be exhaustive) differ with respect to the kind of changes that are envisaged.

The law equation of the particular system under investigation may remain invariant under changes with respect to:

- the initial conditions (first kind of invariance),
- features of the target system that are not explicitly mentioned in the law equation (second kind of invariance),
- the behaviour of systems elsewhere in the universe (third kind of invariance).

Let me briefly flag that there is one kind of invariance that, despite its prominence in the philosophy of physics literature, will not play an important role in what follows (for reasons to be indicated later).

Given the first kind of invariance, the law equation remains the same under changes of initial conditions. In general, however, the *solutions* of these law equations will change if the initial conditions change. Given Newton's second law, if a different force is applied, the law equation will remain the same; the solution of the equation will, however, in general, be different.

Changes in initial conditions that not only leave the law equations unchanged but also the solution of these equations lead to what might be called "transformational invariance":

Invariance not only of the law equation but also *of the solutions of the law equation* with respect to *certain changes of initial conditions* (which are called 'symmetry transformations' if they leave the solutions unchanged).

Simple examples are spatial transformations or velocity boosts in Newtonian mechanics. In this case, it is not only Newton's second law that remains the same but also the solutions of Newton's second law. (It does depend on the law equation(s) in question whether or not such symmetry transformations are allowed.) It is this kind of invariance that has been the focus of the philosophy of physics literature (Brading and Castellani 2003; Brading and Castellani 2007; Brading, Castellani, and Teh 2021; Wigner 1949).

5.3.2 *Invariance as an Empirically Accessible Relation*

Invariance relations are accessible by ordinary empirical methods. This will turn out to be relevant because it implies that other modal features of the behaviour of systems, which can be accounted for in terms of invariance relations, are empirically accessible too. Let me start with how we empirically investigate dependence claims. Suppose we are given claims like "The length of a metal rod depends on its temperature" or "The period of a simple pendulum depends on the length of the string L." Such dependence claims are (in principle) empirically accessible: we vary (in the first case) the temperature and figure out how the length of the rod changes as a result. Similarly, in the case of the simple pendulum's period and the length of its string.

Independence claims are empirically accessible in exactly the same way as dependence claims. Take, for example, the claim, "The period of a simple pendulum is independent of the mass of the pendulum." We now need to vary the mass of the pendulum and see how this affects the period. Provided other things have been kept equal, the period is independent of the mass if the period's values remain constant despite changes in the value of the mass. Of course, if we want to figure out dependence and independence claims for more complex systems in more complex settings, things will become more complicated. Background theories will play their part, and so on. But there is no reason to assume that independence claims are more difficult to ascertain than dependence claims or *vice versa*.

The essential point is that invariance claims are independence claims, and they are empirically accessible as other (in)dependence claims are. Whether or not Hooke's law or the Schrödinger equation continues to hold under changes of initial condition is something we can empirically check. In the case of Hooke's law (which states that the force F that pertains to a body attached to a spring extended by a distance x conforms to the equation $F = -kx$), it turns out that the law equation holds only for values of x that are small compared to the possible extension of the spring. Similarly, whether or not a certain law equation is invariant under changes of certain features of the target system is empirically accessible. In the case of Galilei's law, we can test whether or not colour, mass, and shape are irrelevant in free fall, and the same is true for changes in constitutive properties. Again, whether or not the behaviour of other systems leaves a law equation unchanged can be empirically accessed. The important point is that ascertaining whether or not law equations are invariant in any of the above-mentioned senses is not something that transcends our usual scientific methods.

5.3.3   *Realism about Invariance*

So far, we have talked about the invariance of laws or law equations. Within the frame of scientific realism (which I will assume here), we have good reasons to hold that the entities and properties quantified in well-corroborated laws or theories exist as mind-independent features of reality. As a consequence, invariances of laws—in general—translate into invariances of the behaviour of the target systems. Thus, for example, the fact that a law equation is invariant with respect to features of the target system that are not explicitly mentioned in the law equation can be translated into a claim about the target system: the behaviour of the target system (e.g., a falling body) is invariant with respect to features not explicitly mentioned in the relevant law equation (e.g., in colour of the body). This applies also to cases where the invariance relation pertains only to a restricted domain, as, for example, in the case of Hooke's law. Even if the boundaries of the domain are set by requirements on precision, which are ultimately due to pragmatic considerations, it is an objective and empirically determinable feature of the spring under consideration whether or not its behaviour (as described by Hooke's law) remains invariant within a certain domain for the elongation of the spring.

I hedged the claim that the translation from the invariance of laws to the invariance of the behaviour of systems holds by the "in general"-clause because there are specific reasons to be more cautious in the case of transformational invariance (the *solutions* of law equations are invariant with respect to certain kinds of symmetry transformations). There is an extended debate over whether or not this type of invariance is indeed indicative of invariances of the target system or whether it is due to mathematical surplus structures in the relevant theories. The question is whether, for instance, in the case of Newtonian mechanics the invariance of the solutions with respect to spatial transformation shows something about the behaviour of systems or whether it indicates that the mathematical structure of Newtonian mechanics allows one to distinguish situations that are in fact identical (see the essays in Brading and Castellani 2003 or Dewar 2019). So, according to this latter interpretation, symmetries like translational or rotational symmetries of space indicate that the theory has some mathematical surplus structure, which should not be interpreted realistically. In the case of Newton's Mechanics, absolute velocities or positions would have no physical meaning. I will not delve deeper into this debate. I have bracketed transformational invariance precisely because it is controversial as to whether it allows for a realist reading.

5.4 Nomological Modalities

Thus far, I have discussed the fact that laws and *a fortiori* systems that are characterized in terms of these laws may be invariant with respect to some changes. I have also argued that such invariance or independence relations are accessible by ordinary scientific methods. In this section, I argue that nomological necessity and nomological possibility—two modal features that are often associated with laws as well as the systems that are characterized in terms of these laws—can be understood in terms of invariance relations. This idea is not new; it has been put forward, among others, by Mitchell (2003, 140), Lange (2009) and Woodward (1992, 2018). The following section can be read as fleshing out Woodward's claim that "invariance-based accounts [of nomological necessity] provide a naturalistic, scientifically respectable, and non-mysterious treatment of what non-violability and physical necessity amount to" (Woodward 2018, 160).

To start, what qualifies as a modal feature of a system? A modal aspect of the behaviour of systems is an aspect that concerns not (only) the actual behaviour of systems but (also) possible behaviour or behaviour that takes place by some sort of necessity. To illustrate, laws tell us how systems might or would behave—provided certain conditions were to be met; that is, they tell us that this behaviour is possible. Laws furthermore tell us how the temporal evolution of systems is constrained, that is, that a certain necessary evolution is bound to happen. Nomologically modal features of systems are those modal features that are obtained simply by virtue of the fact that the behaviour of systems can be characterized in terms of statements of the form (L). Thus, laws attribute to the target systems, by virtue of the domain that the quantifier of the internal generalization is restricted to, a space of nomologically possible states. The law equation constrains, for example, how these states develop over time. Let us have a closer look at these modal claims.

5.4.1   *A Space of Possibilities*

By virtue of internal generalizations, laws attribute a space of possible states to systems. In the case of dynamical laws, it is assumed that the systems have a set of possible initial states. With respect to these states, we can distinguish two cases. Either the domain of quantification comprises all possible states (e.g., in the case of Newton's second law or the Schrödinger equation) or, as is the case of more specific laws, the domain of quantification comprises only a restricted range of states. Hooke's law, as we have seen above, holds only for a limited range of extensions of the spring.

The point to emphasize is that we are dealing with a modal claim because it is not only actual states or actual behaviour that the internal generalizations quantify. In fact, the internal generalization on its own does not even tell us which state of the system is the actual state.

The internal generalization's concern is possible behaviour only (whether actual or non-actual). Thus, the fact that internal generalizations come with a domain of values for variables requires the assumption that law statements attribute a space of possible (and mutually exclusive) states to systems. The fact that laws of nature attribute a space of possibilities to systems is essential for scientific practice (at least in physics) and the application of laws. It allows us to know what would happen if certain circumstances were to obtain (which we then might choose to bring about or to prevent to occur).

5.4.2   *Constraints*

Let me now turn to a second aspect of internal generalizations that is relevant for the examination of modal structure—the law equation. Law statements do not simply register the past, present, and future behaviour of systems; they describe how this behaviour is constrained. Internal generalizations put restrictions on the space of possible behaviour of systems by establishing relations between variables (i.e., law equations). These restrictions can either concern the synchronic co-possibility of values of variables that characterize the state of a system, as in the case of the ideal gas law, or the temporal evolution of the states of a system as in the case of the Schrödinger equation.

In the case of a synchronic law (law of co-existence), for example, the ideal gas law, the set of possible values for the variables p, V, and T is restricted to those that satisfy the equation $pV = v\,RT$. Thus, the possible states of the gas are constrained to a two-dimensional hypersurface of the three-dimensional space that is generated by the variables p, V, and T. The internal generalization does not only provide information about how the actual state of a system (if known) is constrained. In addition, it tells us how all possible states of the system are constrained, whether or not they are actual. That the systems are constrained means that those states not on the hypersurface are not accessible for the system. They are classified as states the system cannot possibly occupy, given the law equation, that is, as nomologically impossible states.

The fact that the gas satisfies the equation of the gas law allows a scientist or an engineer who can manipulate pressure and volume to *ensure* that the gas will have a certain temperature. Similarly, the engineer might want to prevent certain situations, such as preventing a gas from having a certain temperature. In such cases, she will rely on the fact that the law tells us that certain combinations of pressure, volume, and temperature will not occur; by setting pressure and volume appropriately, we can make sure that a certain temperature value will not be obtained.

The same holds for internal generalizations that describe the temporal evolution of a state of a system. Provided we prepare the system under consideration in a certain state and provided the equation in question is deterministic, we can ensure that at a later time, the system is in a certain state, and we can also prevent the system from being in certain other states. In the case of prevention, it is not only that given certain combinations of, say, pressure and volume, certain values for T simply do not occur. These values *cannot* occur. The use scientists and engineers make of internal generalizations in scientific practice is best understood by assuming that internal generalizations represent modal, that is, nomologically necessary relations.

5.4.3 *Nomological Modality and Invariance*

In this section, I will argue that modal notions such as nomological necessity or nomological possibility can be understood in terms of invariance relations. The point is not that modal notions can be understood in terms of a non-modal notion. Invariance, as introduced in section 5.2, characterizes something as remaining the same while certain changes or transformations do or could take place. It is defined with respect not only to actual but also to counterfactual changes. Invariance is thus clearly a modal notion. The project is not an attempt to eliminate or reduce modal notions but rather that of showing that nomological necessity and nomological possibility can be understood in terms of something that is empirically accessible.

How is nomological necessity related to invariance? Let us approach this issue by contrasting nomologically necessary laws with accidental generalizations. Take, for instance, Reichenbach's famous example of an accidental generalization: "All gold cubes are smaller than one cubic mile" and contrast it with Newton's second law. The former is a paradigm case of an accidental truth; that is, it is an accidental matter that it has not been rendered false. There is enough gold in the universe to build large gold cubes, it would have been costly but otherwise easy to do this. It simply never happens, neither in the past nor in the future, but it could have happened, for example, by actors intervening appropriately. By contrast, there are no interventions or other natural changes (more about natural changes in a minute) that could render Newton's second law false. It will hold, come what may. In other words, the difference between these examples for an accidental generalization and for a law is a difference that can be spelt out in terms of invariance: while Newton's law is invariant with respect to natural changes, an accidental generalization, such as Reichenbach's, is not.

Understanding nomological necessity in terms of invariance furthermore explains why we can rely on laws (e.g., when it comes to their application in technological contexts) while we cannot rely on accidental generalizations. We can rely on laws because they will not be rendered false, given a large range of circumstances. When we rely on Newtonian mechanics in building a bridge, we assume that the equations will continue to hold under a wide range of actual and possible conditions.

If we, as suggested, spell out the nomological necessity of laws (and *a fortiori* of the behaviour of the target systems) in terms of invariance, the fact that invariance relations may be domain-restricted translates to a domain restriction of nomological necessity.

Newton's second law or Schrödinger's equation may be invariant under any natural changes. Hooke's law, as we have seen, is not. Even Reichenbach's accidental generalization is invariant with respect to some changes. But that should not be considered to be a bug of the account. The traditional categories—accidental generalization on the one hand and law (nomologically necessary generalization) on the other—are the endpoints of a spectrum. This spectrum needs to be explored empirically. It may very well turn out that certain invariances (e.g., with respect to initial conditions) turn out to be more relevant for law status than others. But that is to some extent uncharted territory.

Let me briefly address a worry that has been raised (Psillos 2002, 185). I argued that Newton's law is invariant under *natural* changes. But it seems that we need to spell out natural changes as changes that are in accordance with laws of nature, that is, that are nomologically possible. The account thus appears to be circular. There are two things to be said about this. The account would indeed be circular if the project were one of reducing modal to non-modal facts. But that is not the project envisaged here. The point is to draw a difference between the gold-cube generalization being accidental and Newton's second law being nomologically necessary by relying on a class of natural changes that have been identified antecedently. (The same strategy is used in the causal interventionist literature *vis-à-vis* the objection that definitions of causes in terms of interventions are circular (see, e.g., Woodward 2003, 20–22.) Furthermore, it may very well be possible to characterize the class of what I have called here "natural changes" in terms that do not rely on the notion of law or nomological necessity (see Lange 2009, chapter 1, for such an attempt). However, for the purposes of this chapter, I do not need to commit myself to this position.

As we discussed above, laws ascribe a space of nomologically possible states to target systems by specifying a domain to which the quantifier of internal generalizations is restricted. In other words, nomologically possible states are those states of the target systems that are compatible with the law equation. Changing the system's state, if restricted to this domain of states, leaves the law equation invariant. Nomological possibility, when understood in terms of a space of possible states for a system, can thus be understood in terms of an invariance relation. The same holds if we use the term "nomological possibility" in a broader sense as referring to scenarios that are compatible with a whole set of laws. Again, compatibility with laws can here be understood as follows: switching from the actual world to such a scenario leaves the laws invariant. Thus, nomological possibility, both in a narrower as well as in a broader sense, can be accounted for in terms of invariance relations.

To sum up, what I have argued for is that our notions of nomological necessity and nomological possibility be explicated in terms of invariance or independence relations. While both the former as well as the latter are modal relations, the important point is that the latter are empirically accessible relations.

5.4.4 *Humeanism and Non-Humeanism*

Can we say something more about the nature of the invariance relations? One may, for instance, ask "What underpins these invariances?" (Bird 2007, 5). As is well-known, there are different accounts of nomological modalities in the literature on laws of nature. On the one hand, there are non-Humean accounts of laws which postulate via an inference to the best explanation that a *necessitation relation* obtains, which accounts for nomological modalities (e.g., Armstrong 1983). The idea is that this relation explains why a certain system is necessitated to behave as it actually does. Pieces of copper, for instance, cannot help but be excellent electric conductors because a second-order necessitation relation obtains between the property of being copper and that of being a good electric conductor. Others (dispositional essentialists, e.g., Bird 2005) postulate that it is of the essence of copper to be a good electrical conductor. Thus, again, a piece of copper in virtue of its essence cannot avoid being a good electrical conductor. Humeans, by contrast, typically don't believe in essences or necessitation relations. Ultimately, they think there is only the actual behaviour of systems and that there are the theories (maybe one *ideal* theory) about the actual behaviour that might serve as a basis for various counterfactual claims, which in turn underpin our modal talk (see, e.g., Lewis 1973).

The non-Humean is motivated by the idea that there needs to be something in nature, that is, *de re* necessities, to account for scientific practice. The Humean is motivated by the idea that we should refrain from postulating what is empirically inaccessible and thus rejects accounts such as Armstrong's or Bird's (see Earman and Roberts 2005). Both of these motivations seem plausible, and the invariance account allows you to take into account both: it argues that there are *de re* necessities which are this-worldly, immanent, and empirically accessible. That seems to be all we need.[2]

5.5 Modelling the Modal

We are now in the position to explain how (successful) scientific modelling leads to modal knowledge, where I take modal knowledge to be knowledge of (objective) modal features of the behaviour of target systems.

The construction of models has many purposes or functions (see, e.g., Gelfert 2019; Grüne-Yanoff 2013; Knuuttila 2021; Knuuttila and Loettgers 2017). Representing the behaviour of systems is (only) one of these—the one I have focused on here. When models are used for this purpose, the use of idealizations and abstractions may seem puzzling. I take idealizations and abstractions to be intentional misrepresentations of the properties or the behaviour of a target system, for example, by omission of properties of the systems or factors that contribute to its behaviour (abstraction) or by distorting the characterization of the behaviour or the properties of the system (idealization). The terms "idealization" and "abstraction" may refer both to the process or activity as well as to the product or result of a mis-representation. The fact that idealizations are *intentional* misrepresentations is important in order to distinguish idealizations from hypotheses that turn out to be false (Hüttemann 1997). An idealization has to be distinguished from the *hypothesis* that a certain body is indeed a point particle. Because idealizations are intentional misrepresentations (in contrast to the case of hypotheses that turn out to be false), the question for the rationale for ide-alizations arises.

Possible candidates for such a rationale can best be discussed in the con-text of an (apparent) paradox associated with idealizations. The paradox arises provided certain assumptions about the aim of science or at least the function of law statements such as (*L*) are made: if, for example, the aim of science is representation of the phenomena, the question arises of why scien-tists idealize. Similarly, if it is assumed that science aims at explanation and explanations invoke laws that need to be true, or if understanding is what is sought, and understanding presupposes veridicality (see discussion in Gelfert (2019) and Reutlinger et al. (2018)). If the aim of science presupposes that the descriptions of the behaviour of systems which are invoked to realize this aim are truthful, the question arises of why scientists idealize. What is the purpose or rationale of idealizations and abstractions?

Various approaches to deal with this paradox can be distinguished. First, there is the pragmatic approach. It sticks with the assumption that truth or representation is a necessary component of what science aims at and ac-counts for the use of idealizations as a compromise (see Strevens (2008, 298) for this characterization). We distort the characterization of target systems or leave out certain features in order to make the characterization math-ematically tractable or tractable by simple mathematics. Furthermore, it is assumed that we can always de-idealize and thus make our characterization of the target systems more truthful (see Knuuttila and Morgan (2019) and McMullin (1985) for a discussion of the prospects of de-idealization).

According to a second approach, truth is not a necessary condition for what science aims at (or for explanation in particular). One prominent example is Cartwright's simulacrum account of explanation, developed in her paper, "The truth does not explain much" (Cartwright 1983). Cartwright argues that explanatory power comes with unifying power rather than with truth only. As a consequence, laws (including the modelling assumptions that go into Σ) do not explain *in spite of* idealization but rather *in virtue of* idealizations. The tension is resolved by revising assumptions about what science aims at. A third approach distinguishes between the occurrent behaviour of systems on the one hand and an underlying more fundamental structure that gives rise to the occurrent behaviour on the other. While idealizations may misrepresent the occurrent behaviour of systems, they may nevertheless be instrumental in correctly identifying the underlying structure, for example, the dispositions that underly the overt behaviour of the systems. This approach has been discussed by Cartwright as well (Cartwright 1989; see also Hüttemann 1998; Hüttemann 2014). The tension between idealization and truth is resolved by revising what kind of truths science aims at (revising what laws/models are supposed to represent).

There is no reason to assume that the various kinds of idealizations and abstractions that have been distinguished and discussed in the literature (see, e.g., Hüttemann 1997, chapter 2; Weisberg 2007) can all be accounted for by the same approach. My aim in the remainder of this section is to argue for the claim that the use of some modelling practices indicates that scientists assume invariance claims about the behaviour of systems to hold. We employ idealizations and abstractions when we have reason to assume that certain invariance relations hold. This suggestion can be seen as falling under the third approach. Science is not (only) interested in the actual or occurrent behaviour of the target systems but rather in their modal profile: how the system might behave (provided the right conditions hold) and how its behaviour is constrained.

Let me illustrate this claim by discussing a simple example. Suppose the target of our modelling is the behaviour of a real physical pendulum. Let us furthermore suppose that we model the motion of the pendulum as a classical harmonic oscillator (Figure 5.1).

With F_{return} the return force, L the length of the pendulum, m its mass, T its period, and θ the angular displacement from the vertical equilibrium position

$$F_{\text{return}} = mg \sin\theta(t)$$

$$ma_{\text{return}} = mL\, d^2\theta(t)/dt^2$$

For small amplitudes $\sin\theta(t)$ can be replaced by θ, which yields

$$d^2\theta(t)/dt^2 + g/L\, \theta(t) = 0$$

and finally

$$T = 2\pi\sqrt{L/g}$$

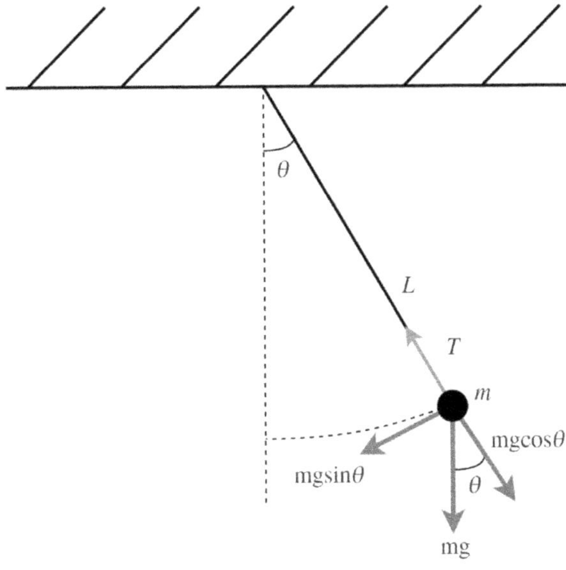

Figure 5.1 The harmonic oscillator.

If this model is empirically adequate (for small $\theta(t)$), it allows us to read off modal features of reality.

Why? First, the model is *abstract*. It leaves out features of the real pendulum and its environment. It does not mention the colour of the pendulum, the material it is made of, the exact form of the bob, the time of the day, the position of Jupiter, and so on. The fact that these things are not mentioned implies (for an empirically successful model) that the model is invariant with respect to these features. The equation with the force function provides an adequate account of the system, *whatever* the material, the colour, the time of the day, or the position of Jupiter. Abstractions in empirically successful models allow us to infer modal features of the target system.

Second, the model is idealized. It is assumed that the bob can effectively be treated as a point particle, even though it does have a finite extension. The fact that we can ascribe this feature to the bob, even though it does not have it, indicates that the extension of the bob is irrelevant to the behaviour of the pendulum (see Strevens 2008, Chapter 8). The law equation with the force function is—again—invariant with respect to this feature. Idealizations in empirically successful models allow us to infer modal features of the target system.

The following objection may be raised: models are empirically successful depending on pragmatic considerations, so that many models are approximate and simplifying yet very useful. But in those situations (which are ubiquitous in science), inferring modal features from the model invariances seems unjustified. This objection can be answered by pointing to the fact that even if one's measures of precision depend on pragmatic considerations and other purposes, it is an objective modal feature that the behaviour of the target system under consideration is or fails to be invariant relative to changes in a certain variable and provided the behaviour is characterized in this or that approximate way. It is, for instance, true that Galileo's law for free-falling bodies (and thus the fall of free-falling bodies) is invariant relative to changes in the height from which the body falls, given a certain accuracy of measurement.

Modelling practices assume that invariance relations hold. Conversely, if there is independent evidence (empirical or theoretical or both) for the obtaining of such invariance relations, the modelling practices in question may receive a *post facto* justification. Hüttemann, Kühn, and Terzidis (2015) have argued that the renormalization group approach to phase transitions provides a justification for abstractions that leave out the details of the systems' constitution.

5.6 Conclusion

Modelling the modal needs to be spelt out differently depending on the function of the models. In this chapter, I assumed that models sometimes have the function to represent target systems. Given this function, the role of idealizations and abstractions in modelling may seem puzzling. I argued that this puzzle can be resolved if what models and laws are meant to represent is not (only) the actual or occurrent behaviour but rather their modal profile. Idealizations and abstractions in empirically successful models allow us to infer modal features of the target systems—that their behaviour is invariant under certain natural changes.

Notes

1 This chapter uses and develops material from the first chapter of Hüttemann (2021).
2 In Hüttemann (2021, 74–78), I argue that while dispositions are an important ontological category to understand certain features of scientific practice, the modal nature of dispositions is not primitive but should rather be spelt out in terms of invariance relations.

References

Armstrong, David. M. 1983. *What Is a Law of Nature?* Cambridge: Cambridge University Press.
Bird, Alexander. 2005. "The Dispositionalist Conception of Laws." *Foundations of Science* 10: 353–70.
Bird, Alexander. 2007. *Nature's Metaphysics: Laws and Properties*. Oxford: Oxford University Press.
Brading, Katherine, and Elena Castellani eds. 2003. *Symmetries in Physics: Philosophical Reflections*. Cambridge: Cambridge University Press.

Brading, Katherine, and Elena Castellani. 2007. "Symmetries and Invariances in Classical Physics." In *Handbook of the Philosophy of Science. Philosophy of Physics*, edited by J. Butterfield, and J. Earman, 1331–67. Amsterdam, North Holland: Elsevier.

Brading, Katherine, Elena Castellani, and Nicholas Teh. 2021. "Symmetry and Symmetry Breaking." In *The Stanford Encyclopedia of Philosophy* (Fall 2021 Edition), edited by Edward N. Zalta (ed.). Stanford University. https://plato.stanford.edu/archives/fall2021/entries/symmetry-breaking/.

Cartwright, Nancy. 1983. *How the Laws of Physics Lie*. Oxford: Oxford University Press.

Cartwright, Nancy. 1989. *Nature's Capacities and Their Measurement*. Cambridge: Cambridge University Press.

Dewar, Neil. 2019. "Sophistication about Symmetries." *The British Journal for the Philosophy of Science* 70: 485–521.

Earman, John. 2002. "Laws, Symmetry, and Symmetry Breaking; Invariance, Conservation Principles, and Objectivity." (PSA 2002 Presidential Address)." *Philosophy of Science* 71(2004): 1227–41.

Earman, John, and John Roberts. 2005. "Contact with the Nomic: A Challenge for Deniers of Humean - Supervenience about Laws of Nature. Part I: Humean Supervenience." *Philosophy and Phenomenological Research* LXXI(1): 1–22.

Galilei, Galileo. 1954. Dialogues Concerning Two New Sciences. Transl. H. Crew and A. de Salvio. New York: Dover Publications.

Gelfert, Axel. 2016. *How to Do Things with Models*. Cham: Springer.

Gelfert, Axel. 2019. "Probing Possibilities: Toy Models, Minimal Models, and Exploratory Models." In *Model-Based Reasoning in Science and Technology*, edited by Matthieu Fontaine, Cristina Barés-Gómez, Francisco Salguero-Lamillar, Lorenzo Magnani, and Ángel Nepomuceno-Fernández, 3–19. Dordrecht: Springer.

Grüne-Yanoff, Till. 2013. "Appraising Models Nonrepresentationally." *Philosophy of Science* 80(5): 850–61.

Hüttemann, Andreas. 1997. *Idealisierungen Und Das Ziel Der Physik*. Berlin: de Gruyter.

Hüttemann, Andreas. 1998. "Laws and Dispositions." *Philosophy of Science* 65: 121–35.

Hüttemann, Andreas. 2014. "Ceteris Paribus Laws in Physics." *Erkenntnis* 79: 1715–28.

Hüttemann, Andreas. 2021. *Minimal Metaphysics for Scientific Practice*. Cambridge: Cambridge University Press.

Hüttemann, Andreas, Reimer Kühn, and Orestis Terzidis. 2015. "Stability, Emergence and Part-Whole-Reduction." In *Why More Is Different. Philosophical Issues in Condensed Matter Physics and Complex Systems*, edited by Brigitte Falkenburg, and Margret Morrison, 169–200. Springer: Dordrecht.

Knuuttila, Tarja. 2021. "Models, Fictions and Artifacts." In *Language and Scientific Research*, edited by Wenceslao J. Gonzalez, 199–22. Cham: Palgrave Macmillan.

Knuuttila, Tarja, and Andrea Loettgers. 2017. "Modelling as Indirect Representation? The Lotka–Volterra Model Revisited." *British Journal for the Philosophy of Science* 68(4): 1007–36.

Knuuttila, Tarja, and Mary S Morgan. 2019. "Deidealization: No Easy Reversals." *Philosophy of Science* 86(4): 641–61.

Lange, Marc. 2009. *Laws and Lawmakers: Science, Metaphysics, and the Laws of Nature*. Oxford: Oxford University Press.

Lewis, David. 1973. *Counterfactuals*. Oxford: Blackwell.

Massimi, Michaela. 2019. "Two Kinds of Exploratory Models." *Philosophy of Science* 86: 869–81.

McMullin, Ernan. 1985. "Galilean Idealization." *Studies in History and Philosophy of Science A* 16(3): 247–73.

Mitchell, Sandra. 2003. *Biological Complexity and Integrative Pluralism.* Cambridge: Cambridge University Press.

Nozick, Robert. 2001. *Invariances.* Boston: Belknap.

Psillos, Stathis. 2002. *Causation and Explanation.* Montreal: McGill-Queen's University Press.

Reutlinger, Alexander, Dominik Hangleiter, and Stephan Hartmann. 2018. "Understanding with (Toy) Models." *The British Journal for the Philosophy of Science* 69: 1069–99.

Scheibe, Erhard. 1991. "Predication and Physical Law." *Topoi* 10: 3–12.

Strevens, Michael. 2008. *Depth – An Account of Scientific Explanation.* Cambridge, MA: Harvard University Press.

Wigner, Eugene. 1949. "Invariance in Physical Theory." *Proceedings of the American Philosophical Society* 93(7): 521–6.

Weisberg, Michael. 2007. "Three Kinds of Idealization." *The Journal of Philosophy* 104: 639–59.

Weisberg, Michael. 2013. *Simulation and Similarity.* Oxford: Oxford University Press.

Woodward, James. 1992. "Realism about Laws." *Erkenntnis* 36: 181–218.

Woodward, James. 2003. *Making Things Happen.* Oxford: Oxford University Press.

Woodward, James. 2018. "Laws: An Invariance-Based Account." In *Laws of Nature,* edited by Walter Ott, and Lydia Patton, 158–80. Oxford: Oxford University Press.

6 Modeling the Biologically Possible

Evolvability as a Modal Concept

Marcel Weber

6.1 Introduction

An elephant with feathers doesn't seem to be equally impossible as a flying elephant. Laws of physics prohibit the latter but not the former. Yet an elephant with feathers would also be a strange creature because it is a mammal, not a bird or a dinosaur, and only the latter have evolved feathers. In fact, feathers are more recent in evolutionary history than the last common ancestor of mammals and birds. They are thought to be homologous to hair and scales. Mammals have evolved hair from the common primordial structures that gave rise to feathers in the dinosaur lineage leading up to contemporary avians; this is why the feathered elephant is biologically impossible. It would require both a reversal and a highly similar rerun of evolutionary turns that happened long ago, which is unlikely in the extreme (Beatty 1995, 2016). Hence our confidence that a feathered elephant is biologically impossible, at least relative to our actual evolutionary history.

While judgments of biological possibility are common in and outside of biology, the nature of these modalities has not been much studied by philosophers. A notable exception is Dennett (1995), who construed biological possibility as accessibility in a complete space of possible genomes that make up what he calls the "Library of Mendel." Max Hindermann (né Huber) has developed Dennett's idea into a full-blown modal logic (Huber 2017). Others have criticized the whole idea of trying to define a complete space of *all* biological possibilities (Maclaurin and Sterelny 2008), even if it is only relative to a given organism. However, this does not rule out possibility spaces that define a relevant set of possibilities given some explanatory goals, such as morphospaces in evolutionary biology. Morphospaces are representations of a range of possible forms that some biological structure, for example, a coiled snail shell, can take. The extent to which regions of such a space represent existing forms can give evolutionary biologists indications as to where to search for evolutionary mechanisms such as natural selection or developmental constraints, as shown by Huber (2017) and Maclaurin and Sterelny (2008).

DOI: 10.4324/9781003342816-9

While the modeling of modalities is increasingly being studied by philosophers of science (e.g., Ijäs and Koskinen 2021; Knuuttila 2021; Ladyman 1998; Sjölin Wirling and Grüne-Yanoff 2021a, 2021b), there have hardly been any attempts to clarify the modalities underlying biological concepts such as evolvability. The rest of this chapter will present such an attempt. In the following section, I will give some basics about evolvability and a biological field where it plays a central role, namely evolutionary developmental biology or "evo-devo." In section 6.3, I will give a more detailed account of a model system in which evolvability can be assessed. Section 6.4 will then present my account of evolvability as accessibility in genotype-phenotype map space. In section 6.5, I will analyze what kind of possibility might underlie biological models of evolvability. In the concluding section, I will draw out some general lessons from my analysis of this somewhat special case from evo-devo.

6.2 Evolvability Explanations

Evo-devo addresses questions such as: how can the same genetic mechanisms give rise to different organismal forms? How did specific developmental programs evolve? How do developmental processes affect evolutionary trajectories (Love 2020, 2024)? The concept of evolvability has been introduced to capture what is thought to be an intrinsic feature of a type of organism, namely its capacity to evolve (Alberch 1991; Brigandt 2015; Kirschner and Gerhart 1998; Love 2003; Villegas et al. 2023). This capacity is closely linked to the ability to produce heritable phenotypic variation such that it can sustain genetic modifications that remain stable over several generations (Kirschner and Gerhart 1998). Evolutionary forces such as natural selection or drift can only act on the gene variants that arise in a population, but many gene combinations may never occur because they don't allow the organism to develop to adulthood, that is, to the reproductive stage. Thus, evo-devo must attend to developmental mechanisms and seek to understand how genetic changes can modify these mechanisms such as to produce different viable phenotypes, which may then undergo natural selection or genetic drift. These mechanisms may show a bias for some forms or some regions of a morphospace, thus explaining certain evolutionary patterns non-adaptively (Brakefield 2006; Gould and Lewontin 1979). Of course, such biases can also combine with natural selection and/or drift to explain certain evolutionary patterns (Novick 2023). Such biases are also called "constraints" and come in different forms, including constraints on form and constraints on adaptation (Amundson 1994).

It is important to realize that there are not just one but several concepts of evolvability in biology. Love (2003) distinguishes between evolvability$_\mathrm{U}$ and evolvability$_\mathrm{R}$. The former, which is used mainly in quantitative evolutionary genetics, means the ability to respond to natural selection, which depends on heritability and additive genetic variance. It is considered to be a population property. The second term designates that which explains the differential evolutionary success of lineages, usually considered to be an intrinsic disposition of a type of organism. The literature contains many differing notions of evolvability (Hansen and Pélabon 2021), but they may be seen as all falling under one or the other side of Love's twofold distinction.

Philosophers attending to the concept of evolvability have invariably classified it as a *disposition*, analogous to fitness in the theory of natural selection (Brigandt et al. 2023). For example, Brown (2014) identifies evolvability with the probability $E = Pr_{x,b}(F_t)$ that a set of features F arise at future time t given the population x and its environment b at some starting point. Evolvability thus construed is a dispositional property, more precisely a *propensity* that results from the joint causal influences of factors internal to the organisms making up population x, given some environment b. The environment should be viewed as belonging to the manifestation conditions of the disposition, thus making evolvability an intrinsic property of a population (as most biologists insist).

A full appraisal of this account is beyond the scope of this chapter; my aim is rather to provide an alternative view. Nonetheless, I will point out some possible lacunae in the propensity view.

My approach in this chapter is to take seriously the idea that evolvability is essentially a *modal* concept that is used to make claims about what is evolutionarily possible. When we attend to the way in which the concept is used in biological practice, as we should (Brigandt 2015), we can see that it is often tied to a typical explanatory strategy, which I shall refer to as an *evolvability explanation*. This strategy takes as a starting point the assumption that evolutionary change, or at least some kinds of change, is *prima facie im*possible or has a low degree of biological possibility (if it comes in degrees). Then, an evolvability explanation postulates a mechanism that explains how a certain kind of change is possible after all. This explanatory structure is also known as "how-possibly explanation" (Dray 1957); however, diverging accounts of such explanations have been given (Grüne-Yanoff 2013; Reiner 1993; Verreault-Julien 2019). What matters here is that how-possibly explanations clearly involve modalities (Sjölin Wirling and Grüne-Yanoff 2021a).

Such a construal of evolvability explanations nicely fits the view that evolvability explanations provide a solution to an *evolvability problem*. This strategy is explained very clearly in Pavličev and G.P. Wagner (2012a). They begin by noting that organisms can only adapt to new environments if there are individuals who are "suited to survive under the new circumstances" (231). They then define "evolvability" as the "ability of a population to cope with the changing environment by adaptation." But adaptation by natural selection is only possible if random mutation delivers the "suited individuals." Now, random mutation can have "an incredible number of effects on the phenotype, and most of them will be deleterious under any circumstances, if not lethal" (Pavličev and Wagner 2012a). Thus, the following problems arise:

> How does such random genetic change produce the "right" kind of deviation often enough? How is change possible where multiple mutations are necessary but intermediate steps have no apparent advantage? How probable is adaptation if only some of the traits should be changed, without affecting those that are already in place?
> (Pavličev and G.P. Wagner 2012a, 232).

Thus, *prima facie*, it seems exceedingly unlikely that random mutation should be able to generate complex adaptations. This is a problem that Darwin had already been grappling with. The problem is that chances are there will likely be no "suitable individuals" alive, and if there are, they are not likely to have an advantage, so there is nothing that natural selection can do.

Now enter what I call evolvability explanations. Such explanations describe a mechanism or a set of mechanisms that can produce the necessary genetic variants for natural selection to act upon. Such mechanisms are frequently represented by using a so-called *genotype-phenotype* or *GP-map*, an important theoretical idea in evo-devo (Alberch 1991). This map specifies for a type of organism what phenotypes (usually from a range of forms that are chosen in view of a specific research problem) can be produced from what genotypes. The GP map is determined by developmental and physiological facts about the given species. More precisely, the map is a highly abstract summary of these facts. Pavličev and G.P. Wagner (2012a, 232) describe it as a "statistical summary" that abstracts away from the myriad of developmental and physiological processes that occur in a living organism. Framed in this way, the question becomes what kinds of GP-map structures give rise to high evolvability. Furthermore, there is the question of whether evolvability can itself evolve or evolve under selection, which I will leave out for now for the sake of simplicity, even though it is also a part of some evolvability explanations.

One proposed mechanism for high evolvability is that of a *modular* GP map, an idea originally due to Rupert Riedl (G.P. Wagner and Altenberg 1996). This means that the map is organized in such a way that pleiotropic effects mostly affect traits that form a complex or module with a distinct selected function (e.g., locomotion, visual perception, etc.) and fewer traits from a complex serving a different function. In other words, pleiotropy is not randomly distributed across all the traits of an organism.[1] This allows selection to act on the trait without affecting other traits and thus incurring fitness costs by pleiotropic effects. Modularity thus answers the third question raised by Pavličev and G.P. Wagner (2012a), to wit, how adaptation can change only some traits and leave those already in place intact.[2] Modularity removes a major theoretical obstacle to the possibility of adaptation and thus provides what I call an evolvability explanation, which is a solution to an evolvability problem. The extent to which it is responsible for adaptive change relative to other proposed mechanisms that can compensate for deleterious pleiotropic effects is subject to debate (Hansen 2003; Pavličev and G.P. Wagner 2012b).

Another classical idea is that of *robustness* (also called *canalization* in some older literature). When used in an ontic sense, robustness signifies a kind of invariance against perturbations. In biology, two important kinds of robustness are (1) invariance of the phenotype against environmental perturbations and (2) invariance of the phenotype against genetic mutation (A. Wagner 2012, 2013). I shall focus here on the second kind of robustness. It features centrally on an important kind of evolvability explanation. Biological systems at all levels of organizations are capable of genetic variation that does not manifest itself at the phenotypic level. There are several mechanisms behind this phenomenon. First, there are so-called "silent" point mutations in the coding region of genes. These do not alter the protein molecule encoded either because they are located in a non-coding region such as an intron (which is spliced out after transcription) or because they occur in the third position of a codon or base triplet, which, due to the redundancy of the genetic code, makes no difference with respect to the amino acid moiety encoded. Second, there are gene mutations that do change one or a few amino acids in the protein encoded, but without affecting the protein's shape, function, and stability. Third, some mutations are recessive, that is, they have no phenotypic effect when there is a second copy of the same gene present (in diploid organisms). Fourth, there are point mutations that neutralize each other when they occur in *trans* (i.e., in a distinct copy of the same gene present on a different chromosomal unit), a phenomenon known as "intra-allelic complementation." Fifth, there are molecules (protein or RNA) that are functionally redundant because there is another molecule that can take its function when their gene harbors a mutation affecting its function.

These five sources of robustness have been known for a long time, and they can increase evolvability by allowing genetic changes to accumulate without affecting the host. However, their effect on evolvability is a kind of side effect of the basic genetic mechanisms.

In addition, there are more specific molecular mechanisms that can increase robustness. A classic study in *Drosophila* genetics (Rutherford and Lindquist 1998) demonstrated the role of the heat-shock protein Hsp90 in increasing evolvability in fruit flies. Hsp90 is a so-called "molecular chaperone," which means that it helps proteins to fold into their most stable shape. A loss-of-function mutation in its gene leads to many misfolded or unfolded proteins (which tend to be sticky, hence the name "chaperone"—it prevents inappropriate protein-protein interactions). When the gene is mutated in fruit flies, laboratory populations show a significant increase in phenotypic variation, which is not due to genetic variation. Rather, it seems that functional Hsp90 masks a lot of genetic variation that is already present in the population by stabilizing mutant proteins. Thus, Hsp90 is part of a mechanism that directly increases robustness and hence evolvability. It allows mutant organisms to survive and even reproduce that harbor genetic mutations that may someday become valuable and hence selected for.

Robustness comes in different forms, and many mechanisms can increase robustness. In addition to the molecular mechanisms mentioned, there are also mechanisms at the level of gene regulatory circuit dynamics that are able to buffer mutations in non-coding regions (cis-regulatory elements) affecting transcription rates. Feedback circuits helping the developmental system to buffer perturbations abound in multicellular organisms (Siegal and Leu 2014). However, it seems that the principle is always the same: robustness solves the first and second evolvability problems according to Pavličev and G.P. Wagner, namely the problem of ensuring that useful mutations occur often enough in a population (and are also carried along into future generations even if they are potentially harmful or neutral), and that changes that require multiple mutational steps can occur.

I will discuss a specific robustness mechanism, namely so-called neutral networks, in the following section in more detail. For now, let us note that modularity and robustness are very general and abstract types of organizing principles that make adaptation (as well as evolutionary novelty) possible. They provide solutions to an evolvability problem and thus "possibilify" a process that seems *prima facie* impossible. I believe that this is a widespread type of explanatory reasoning also in biology[3] and that at least some explanations involving evolvability can also be understood in this manner.

I would also like to point out the abstract nature of such concepts as modularity and robustness. These concepts apply to a wide range of biological mechanisms (such as the Hsp90 mechanism and the neutral network mechanism to be discussed in the next section) and show what a series of quite distinct mechanisms have in common. In the present context, they make certain types of processes possible that seem impossible without them, given some theoretical assumptions. In the following section, I will take a closer look at an evolvability explanation of the robustness type.

6.3 Small RNAs as a Model of Evolvability

While in most cases vast regions of the GP map—even its dimensionality—remain unknown, there exist simple model systems where a GP map can be constructed. An example of such a system is small RNAs, that is, ribonucleic acids with a length of usually <100 base pairs. Not life forms of their own, small RNAs play various roles within all types of living cells. For example, some of them work as transfer RNAs in protein synthesis. They can form rather complex secondary structures due to Watson-Crick-type base pairing interactions within the same molecule. The secondary structure basically consists of a set of intramolecular base pairings (due to hydrogen bonds) and determines the RNA's three-dimensional shape as well as its

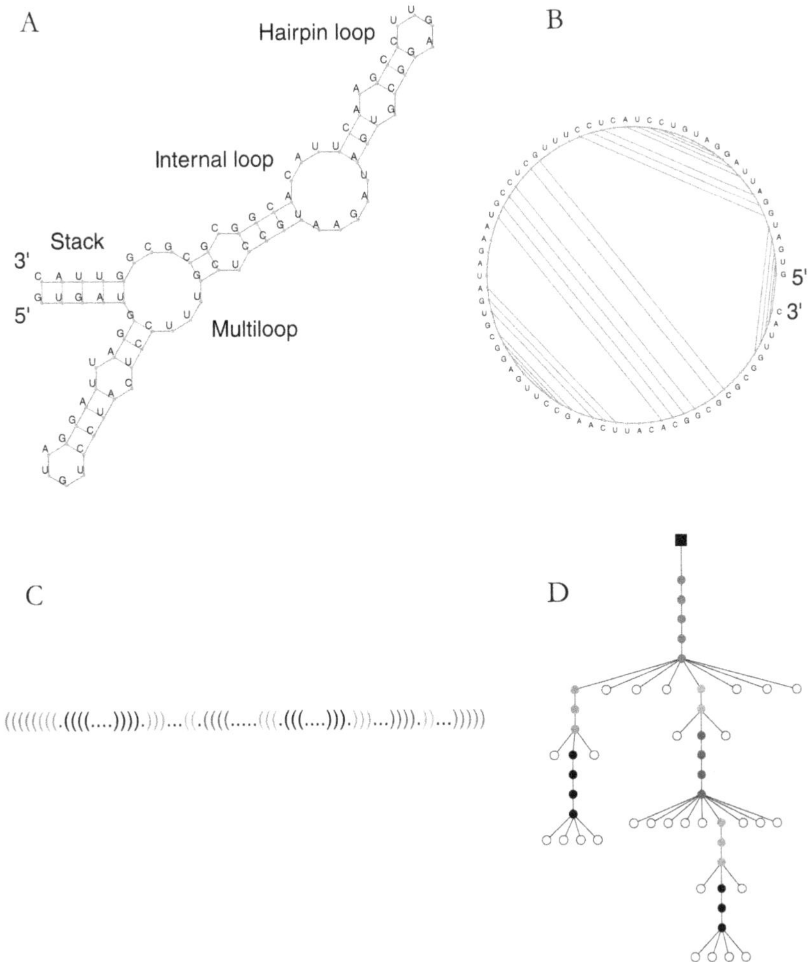

Figure 6.1 Secondary structure in small RNA molecules is defined by intramolecular base pairings that generate various loop-like structures. The molecule's three-dimensional shape is determined by the secondary structure. Many different sequences can have the same shape, and even slight variations in secondary structure do not necessarily alter the shape.

Source: Image reproduced with permission from Fontana (2002).

biological function. Typically, such RNAs form various hairpin- and loop-like structures (see Figure 6.1). In some cases, a huge number of different RNA sequences can give rise to the same three-dimensional shape, which may differ in free energy (i.e., thermodynamic stability).

Small RNAs are models for evolvability for the following reasons: (1) there is a *genotype-phenotype distinction*. The genotype is the nucleotide sequence of the RNA (or the DNA sequence of the gene that encodes it), while the phenotype is the secondary structure and the three-dimensional shape. The folding up of the molecules from a chain of ribonucleotides to a three-dimensional structure is analogous to the process of development. The relation of genotypes to phenotypes is many-one.[4] (2) RNAs are capable of evolving, that is, of being replicated and stable over generations with modifications. (3) There is a reliable mapping of genotypes into phenotypes, given suitable conditions (such as temperature and ionic strength). (4) This mapping is epistemically accessible for a large number of different sequences, thanks to powerful biocomputational tools, enabling researchers to predict secondary structures as well as three-dimensional shapes from pure RNA sequence information.

It is especially the fourth feature that distinguishes the small RNA model from whole organisms and that makes it suitable for studying evolvability. The epistemic accessibility of a vast number of genotype-phenotype relations allows for a kind of modal modeling, namely computing the GP map for a large number of possible sequences, most of which will never be realized. For illustration of the sheer magnitude, if one were to make just one single molecule of every RNA sequence with a length of 79 bp the aggregate of them would weigh more than the Earth (Manrubia et al. 2021)! Of course, only a small subset of this vast space of possibilities can be modeled. Fontana and Schuster (1998) built a computational model for studying so-called *neutral networks* in the space of possible genotypes. These are regions in genotype space that map into the same phenotype, as defined by a three-dimensional shape (see Figure 6.2).

The neutral network model provides a solution to the problem of how a molecule can evolve into a new shape (and potentially acquire a new function) by a process that requires several mutational steps. Throughout the process, that is, until the final mutational step, the molecule retains its old shape and therefore its old function (should it have one), thus providing an explanation for the evolvability of these molecules. Of course, the same result could be reached by following a different path in the sequence space, but paths that lie outside the neutral corridors might disrupt the molecule's function (and therefore stop the process short because the organism with disrupted small RNAs might not be viable). Thus, the small RNA model provides a simple and tractable model of evolvability. In the following section, I will examine the relationship between evolvability and biological possibility within the scope of this model.

6.4 Evolvability as Accessibility in Genotype-Phenotype Map Space

The neutral corridors in the GP-map space show which genotypes are accessible from any original genotype by a series of random genetic changes that are (1) not too unlikely to occur jointly in an actual population and (2) compatible with development and reproduction. Namely, those genotypes that

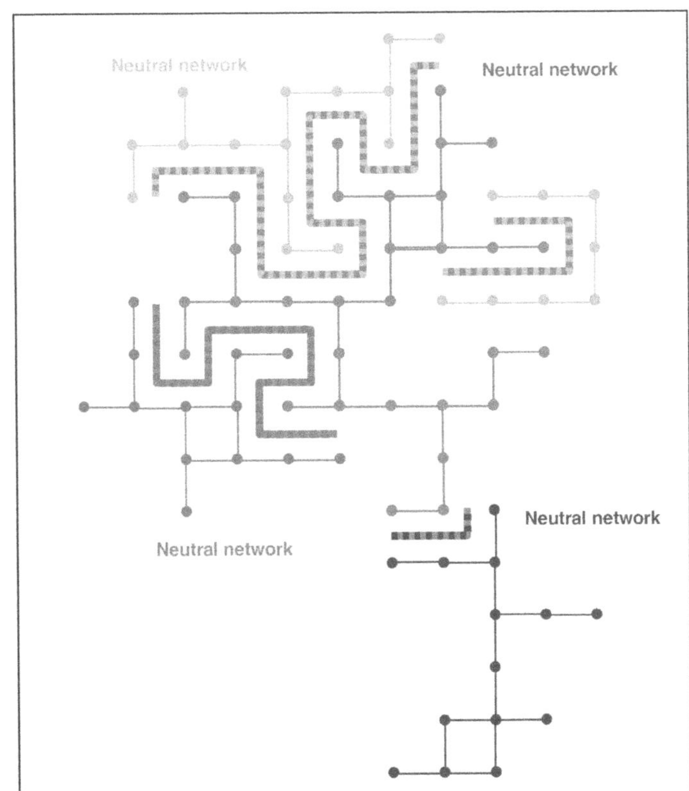

Figure 6.2 This map represents a simple genotype space for small RNAs. Each point in this space represents a possible RNA sequence. Neighboring points can be reached via a single one-nucleotide mutation. Each colored network in this space represents a series of changes that do not affect the molecule's shape and are therefore selectively neutral. Different colors represent different shapes, so a transition to a network shown in a different color here is not necessarily selectively neutral. This map shows that some transitions are likely because a random step out of a neutral network will take into a specific neighboring network (but not necessarily vice versa). Some changes are extremely unlikely because they would require several mutations to occur to transition from one network to another (see Figure 6.3).

Source: Image reproduced with permission from Fontana (2002).

are located in an adjacent region, with respect to the original genotype, of the GT map are accessible. I would like to suggest that this kind of accessibility can be used to define the relevant sense of possibility that underlies evolvability explanations in evo-devo. In order to see this, let us compare it to an account of accessibility that has been used to clarify the concept of biological possibility by way of a modal logic.

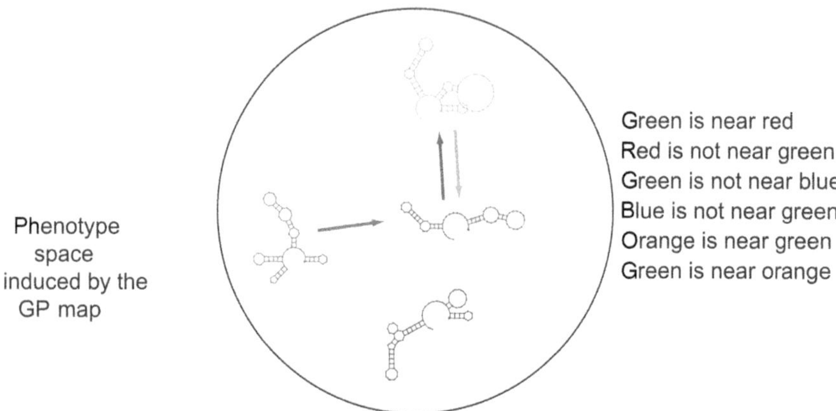

Figure 6.3 A GP map for the neutral network shown in Figure 6.2, showing the possible transitions between different phenotypes (i.e., different shapes of the RNA molecule).

Source: Image reproduced with permission from Fontana (2002).

Probably the most elaborate attempt of this kind so far has been provided by Max Hindermann (né Huber) in his PhD thesis (Huber 2017). Hindermann's modal logic refines and formalizes an idea presented by Dennett (1995). In order to conceptualize biological possibility, Dennett invented the "Library of Mendel" (LoM), a library that contains all the genomes that can be constructed from the four DNA bases A, T, C, and G (inspired by Jorge Luis Borges's "Library of Babel"). A "reader-constructor" maps genomes from the Library of Mendel to phenotypes. Biological possibility is then defined by Dennett in terms of an accessibility relation for genomes:

X is biologically possible if X is an instantiation of an accessible genome or a feature of its phenotypic products.

It is clear in Dennett's account that biological possibility is always relative to a given genome, *g*. A biological organism is possible at *g* to the extent in which it is the phenotypic product of a genome *g'* that is accessible from *g* (e.g., by a series of point mutations or sequence rearrangements). The more accessible *g'* is from *g*, the more possible its phenotypic product at *g*.

Dennett did not specify the relevant accessibility relation. This is where Huber's (2017) account comes in. He first reformulates the Library of Mendel as a relational structure (61):

The Library of Mendel is a relational structure $\langle \Sigma_M, R_M \rangle$ where the domain is the language of the Library of Mendel M and the binary relation is the accessibility relation R_M.

The language of the Library of Mendel consists of an alphabet containing the four nucleotide bases A, G, C, and T. Biological possibility is then defined in terms of satisfaction of the binary relation (61):

Some x is biologically possible at $g \in \Sigma_M$ if and only if there is some $g' \in \Sigma_M$ such that gR_Mg' and x is an instance of g' or a feature of the phenotypic products of g'

Finally, Huber provides an interpretation of the accessibility relation R_M:

For $g, g' \in \Sigma_M$, gR_Mg' if and only if there is a solution to a string editing problem with respect to g, g'.

A string editing problem is the problem of obtaining some string of symbols from another string by the least costly set of edit operations. For example, the string "AACTTC" can be obtained from the string "GGCTTC" by an edit operation that replaces all Gs in the string with As. The same sequence could also be obtained by first replacing all Cs with As, then changing back all As to Cs, and finally replacing all Gs with As. The latter edit operation would be more costly. For most cases, we can identify the number of edit steps needed with the cost; in other words, the cost of each step is identical. However, there might be cases where the cost varies with the kind of change introduced. For example, we could consider operations that cannot be brought about by an existing biological mechanism as being more costly. Alternatively, we could make the edit cost depend on the amount of metabolic energy needed. In any case, the solution to a string edit problem depends on the assumption of a cost function, and the space of biological possibility is going to be relative to such an assumption.

This formulation in terms of string editing allows Huber to use the string edit distance as a measure of possibility. Several such measures exist; for example, the *Levenshtein distance* is defined as the minimal number of operations needed to transform one string into another. As new genomes arise by mutation and recombination of existing genomes, the number of such changes needed to obtain a new genome from an existing one seems like a biologically relevant measure of accessibility.

Thus, in brief, according to Huber, the fewer mutational or recombinational steps are required to create a genome g' from an existing g, the more accessible and hence the more possible the latter is with respect to the former. This seems well in line with the intuition that, whatever biological possibility is, it must be relative to a given organism or lineage and it must come in degrees.

I think that Huber's account may capture some relevant sense of "biological possibility;" however, I will argue now that it is not the sense that underlies the concept of evolvability. There are several differences. First, the most obvious one is that Hindermann's accessibility relation—let's call it "H-accessibility" where "H" stands for Hindermann—refers only to nucleotide sequence, while

E-accessibility ("E" for "evolvability") refers to sequences plus phenotype. This already implies that different modal concepts must be in play here.[5] However, it should be noted that something like the string edit distance is also a component of E-accessibility. In the models at hand, biologists used not the Levenshtein distance but the *Hamming distance*, which is the number of individual positions in which two strings differ. In the small RNA model, they define a "neutral neighbor" as a sequence that gives rise to molecular shape α any sequence that gives rise to the same shape that can be produced by a single point mutation, that is, has a Hamming distance of 1. This neutral neighbor will itself have neutral neighbors, and so on, until we arrive at a sequence that gives rise to a different shape β. The distance between α and β in phenotype space can then be defined as the transition probability that a sequence that folds into β is produced from an α-folding sequence by a series of random point mutations, which is proportional to the number of genotypes that fold into shape α that are adjacent to genotypes that fold into β (Stadler et al. 2001, 258). Thus, distance in sequence space—here measured by the Hamming measure—is necessary but not sufficient because even two close locations in sequence space could be in non-adjacent regions of the GP map (see Figure 6.3 for an example).

A second difference is that the Levenshtein distance is necessarily *symmetrical* because any string edits that can be used to obtain a sequence G′ from sequence G can be carried out backward to produce sequence G from G′. By contrast, E-accessibility is not always symmetrical. In other words, there are evolutionary changes that are irreversible or that would take much more time than the corresponding change in the opposite direction (see again Figure 6.3 for an example). While this feature might seem counterintuitive, there are everyday examples that share this feature. Consider a map of Switzerland with Geneva sitting at the tip of the appendix in the extreme West of the country. The canton of Geneva is surrounded by France (it shares about 150 km of border with France and 4 km with the only neighboring Swiss canton of Vaud). Thus, when you step outside Geneva at a random location, you are very likely to find yourself in France. By contrast, when you step outside of France at a location randomly chosen on France's entire national border, you are, of course, not at all likely to be in Geneva. It's the same in the neutral networks. This kind of accessibility, too, is asymmetrical.[6]

Third, the Levenshtein distance can be used to define a *metric space* with the string edit distance as the relevant distance function. By contrast, E-accessibility is non-metric, which is already implied by the fact that it allows asymmetric relations. However, even though E-accessibility is not technically a distance measure, a binary relation of nearness can nonetheless be defined, for example, by considering some shape α and all the shapes that are accessible from it by random events above a certain likelihood threshold as a neighborhood (Fontana and Schuster 1998).

The choice of this threshold is not arbitrary but is informed by biological considerations, namely mutation rate, population size, or time frame (Stadler et al. 2001, 258). This has to do with the epistemic purposes of these models.

Their goal is to explain real evolutionary patterns. Therefore, they judge the possibility of certain changes under the actual conditions under which the evolutionary processes in question actually occur, which include the parameters just mentioned. Given enough time, a practically unlimited population size, and/or a sufficiently high mutation rate, any RNA shape could evolve from any other. But under real-world conditions, including limited time and population size, some transitions will be so unlikely as to have no evolutionary significance. It is these transitions that are judged to be inaccessible and hence impossible by biologists. Thus, judgments of possibility and impossibility depend on the explanatory purposes of these biological models.[7]

The last feature also points us to the limits of a purely probabilistic construal of evolvability such as Brown's (2014) propensity account. While probabilistic considerations do play a role in evolvability explanations, as we have seen, the propensity account does not fully account for biological practice because evolvability is often construed as a threshold property. Evo-devo researchers are interested in finding out "which evolutionary changes are possible or easy to achieve" (Hansen and Pélabon 2021). Probability considerations are often used alongside studying possibilities, so the two accounts are not mutually exclusive. But why would biologists talk about possibility if they were merely interested in how *likely* some changes are?

In the following section, I take a closer look at the relevant sense of modality.

6.5   What Kind of Modality?

In the previous sections, we have seen that evolvability and the underlying sense of possibility can be construed, at least in some cases, in terms of a kind of nearness in GP-map space. In this section, I will investigate further the nature of this space and compare the resulting sense of possibility with other conceptions of modality.

As we have seen, existing proposals concerning biological possibilities construe them as obeying a distance metric that defines a metric space or topology in the mathematical sense. This means that distances satisfy certain axioms, including symmetry and the triangle inequality: distance $(x, z) \leq$ distance (x, y) + distance (y, z). The Hamming and Levenshtein distances mentioned above define metric spaces; the Hamming distance if only point mutations are considered and the Levenshtein distance also for other mutational operations. No such metric distance is defined in the GP maps of our example. It contains only a much weaker ordering relation, namely to a so-called *pretopology* which is defined by a system of neighborhoods and allows asymmetries.[8]

The modalities generated by the RNA models are also different from the standard philosophical account of modality due to Lewis (1979). This account tries to define the notion of nearness of possible worlds by considering how similar such worlds are to the actual world (@). Lewis proposed to measure this similarity in terms of how big a violation of natural laws or "miracles" would be required to transform a world into actuality. If it takes a bigger miracle to turn world w into @ than world w^*, then w^* is more similar or closer to @ than w. The magnitude of miracles is to be assessed by using a system of weights and priorities according to which avoiding (1) "big, widespread, diverse violations of law" take priority over (2) the maximization of the "spatiotemporal region throughout which perfect match of particular facts prevails," which in turn has more weight than (3) avoiding "even small, localized, simple violations of law," while (4) securing "approximate similarity of particular fact, even in matters that concern us greatly" are of "little or no importance" (Lewis 1979, 472). These priorities are intended to create a metric ordering relation for judging the truth value of counterfactuals such as "had I not taken the milk off the stove, it would have boiled." This counterfactual is true because, in the most similar possible world where I do not take the milk off the stove, which differs from actuality in a matter of particular fact as in (4), the milk would have boiled. This is true because this world is more similar to actuality than a world where, all of a sudden while I was answering the door, the boiling temperature of milk jumped to 3000°C due to a localized violation of physical laws as in (3).

What is notable is that this account of possibility does not take into account whether there is an actual *process* or *mechanism* that could transform a world into actuality.[9] This is evident in Lewis's appeal to miracles. What biologists are more often interested in, by contrast, are such alternative scenarios for the realization of which there exist *naturally occurring biological processes* that *preserve the life of the organisms involved* (Weber 2017, forthcoming). In our RNA example, these two requirements are evident. The latter requirement is the reason why the biologists considered RNA genomes that fold into the same shape accessible because this means that the RNAs can preserve their function and thus sustain the life of their host organism. Furthermore, as we have seen, biologists considered evolutionarily relevant such genetic modifications that are sufficiently likely to occur by a small number of spontaneous mutations in a realistic population size, mutation rate, and time frame. Any other possible scenarios are not relevant for the epistemic purposes of this inquiry and can therefore be disregarded. This includes in particular neighboring possible worlds according to Lewis's criteria as well as the possibilities characterized by the Levenshtein distance. In the concluding section, I will draw some general lessons from the present analysis of the small RNA models for evolvability.

6.6   Conclusions

I suggest that several lessons can be drawn from this inquiry into how evolutionary biologists model the possible. First, I have argued that the concept of evolvability that is used in the type of evolutionary models considered here contains an underlying modality that is distinct from any other kind of modality that has previously been postulated by philosophers (but related to my notion of biological normality; see Weber 2017; forthcoming). Modality or biological possibility in this sense is a kind of accessibility in GP space, a relation which in some cases has unusual properties including asymmetry and the fact that it does, mathematically speaking, not form a topology but only a *pretopology*. This sets it apart from standard accounts of possibility, which posit metric modal spaces.

Second, the present analysis shows that judgments of possibility in biological models depend on the epistemic purposes of these models. Their purpose is to account for the phenomena, that is, evolutionary patterns such as directionality and punctuated equilibria. For this purpose, biologists working on the RNA model examined here needed to find out how to divide up the vast space of nearly endless variations of RNA sequence not only in regions that are close in terms of Hamming or Levenshtein distance—which are still vast—but which are also safe to travel for real evolving populations and can be traversed in a reasonable amount of time under the given constraints (mutation rate, population size). Thus, the explanandum phenomena and the evolutionary conditions of real populations determine what possibilities from the endless frontier of the Library of Mendel are *relevant*. Possibilities that are not relevant in this way do not enter into the picture and are therefore not subject to biological modeling. This includes also the modalities created (or picked out, if you prefer) by an accessibility metric such as David Lewis', which was constructed in order to account for the truth values of counterfactual conditionals, for example, in order to give a reductive account of causality (Lewis 1987). Philosophers of causality have largely abandoned the latter project (Woodward 2003). The extent to which the modal structures examined here might be able to underwrite justified counterfactual statements (not necessarily causal) about evolution[10] is an open question that is beyond the scope of this chapter.

Third, due to the special model system considered—small RNA molecules—it might be suggested that the biological possibilities examined here are not distinct from physical or "nomological" possibilities. After all, we are dealing with molecular species which behave in accordance with physical-chemical laws. I do not want to deny this, and I do not want to stay too attached to the terms "biological" and "physical;" however, it should be noted that the possibilities examined here are relative to taking a certain perspective, which is typical for biology. I am talking about a *functional* perspective. In the models examined here, this perspective manifests itself in the focus on shape-preserving transitions, which serves as a stand-in for function preservation. As I have argued, the biologically possible is constrained by the need to preserve the life of some organisms under the changes considered. This usually means that biologists will consider large regions of the modal space irrelevant because nonviable organisms will not have any influence on future generations. To use an extreme case as an illustration: if getting from one viable form to another would require evolution to transition through some lethal form or a form with meager chances of survival, then this is not an evolutionarily possible path and hence not a biological possibility, even though the resulting end form may itself be viable and the two forms may even be very close in genotype space. These constraints on possibility result from taking an evolutionary perspective and may not arise if one takes a purely physical or chemical perspective.

The general conclusions drawn here might also be applicable to more complex examples of evolvability from evo-devo. It must be admitted that the case of the small RNA models discussed here is quite special, not only because of the simplified nature and the availability of a complete GP map but also because neutral networks are not the only mechanism that makes for a highly evolvable system. As we have seen, neutral networks may be described as a realization of a broader phenomenon called robustness, which in this context means a kind of (phenotypic) invariance under perturbations. If a biological system is not able to buffer, to some extent, variations of its genome such as to allow for larger reconstructions it will never be able to give rise to novel forms (Pavličev and G.P. Wagner 2012a; Wagner 2013).[11] Thus, only a tiny bit of the vast space of possible genotypes can be realized by evolution. While our discussion in this chapter has focused on a particular type of model system, namely small RNA molecules, research using other computationally tractable systems such as protein folding, artificial life, or transcription factor binding has reached similar conclusions (Manrubia et al. 2021). There are thus reasons for thinking that the modal spaces for a broad range of biological systems resemble those that I have examined.

Notes

1 I owe this formulation to James DiFrisco.
2 See Herbert Simon's (1962) parable of the two watchmakers, one of whom is more efficient because he builds his watches in a modular fashion, allowing him to conserve modules already assembled. I am indebted to James DiFrisco for suggesting this analogy.
3 Here is an example from community ecology: we have *prima facie* reasons to think that only species with the same food and habitat requirements can survive in the same place because ecological models demonstrate that there must always be one species that outcompete all the others in a winner-takes-it-all type of competition. However, empirical evidence clearly shows that very similar species can and do peacefully co-exist in one and the same habitat. This paradox was resolved by postulating and verifying various coexistence mechanisms that prevent interspecific competition from running its course (Weber 1999). The case is similar to the classic how-possibly explanations in that we have some kind of a theorem according to which some phenomenon is impossible, and then mechanisms are suggested for how it can be possible after all.
4 In more complex cases, it is rather many-many, but here the genotype pretty much determines the phenotype, at least under normal conditions (temperature, ionic strength).
5 Thanks to Fabrice Correia for pointing this out to me.
6 The counterintuitive properties of the modal space may be attributable to its extremely high dimensionality. James DiFrisco also suggested to me an intuitive reason, namely that it is in general easier to break or obliterate a trait than to

build one. R.A. Fisher (1930) has illustrated this idea with the example of a microscope: the more adjustable knobs it has for focusing the image, the less likely it is that turning them randomly will produce a sharp image. He also proposed a geometrical argument involving fitness landscapes as to why mutations with small effects are more likely to be advantageous than mutations with large effects.

7 This does not imply that modal statements do not also depend on what is considered to be objectively the case, as Tarja Knuuttila has pointed out to me. There are possibilities that, while perhaps being objective, have no biological relevance (Weber, forthcoming).

8 A critical discussion of the various kinds of mathematical spaces used in evolutionary theorizing (sometimes metaphorically) can be found in Mitteroecker and Huttegger (2009).

9 Alistair Wilson pointed out to me that there are more recent possible world semantics that are attuned to actual physics, for example, Loewer (2012; 2020); however, I think my point that these accounts completely ignore biological processes stands.

10 The existence of such statements is controversial; see Beatty (1995).

11 James DiFrisco points out that robustness will also buffer against beneficial mutations, thus potentially preventing adaptive change. I am not aware of any general solutions to this problem. In the model discussed here, it is simply assumed that some possible leaps from one neutral corridor to another will be advantageous to the organism, although this is not represented in the model.

References

Alberch, Per. 1991. "From Genes to Phenotype: Dynamical Systems and Evolvability." *Genetica* 84(1): 5–11. https://doi.org/10.1007/BF00123979

Amundson, Ron. 1994. "Two Concepts of Constraint: Adaptationism and the Challenge from Developmental Biology." *Philosophy of Science* 61: 556–78.

Beatty, John. 1995. "The Evolutionary Contingency Thesis." In *Concepts, Theories, and Rationality in the Biological Sciences. The Second Pittsburgh-Konstanz Colloquium in the Philosophy of Science*, edited by Gereon Wolters, James G. Lennox, and Peter McLaughlin, 45–81. Konstanz/Pittsburgh: Universitätsverlag Konstanz/University of Pittsburgh Press.

Beatty, John. 2016. "What Are Narratives Good For?." *Studies in History and Philosophy of Science Part C: Studies in History and Philosophy of Biological and Biomedical Sciences* 58: 33–40. https://doi.org/10.1016/j.shpsc.2015.12.016

Brakefield, Paul M. 2006. "Evo-Devo and Constraints on Selection." *Trends in Ecology & Evolution* 21(7): 362–8. https://doi.org/10.1016/j.tree.2006.05.001

Brigandt, Ingo. 2015. "From Developmental Constraint to Evolvability: How Concepts Figure in Explanation and Disciplinary Identity." In *Conceptual Change in Biology: Scientific and Philosophical Perspectives on Evolution and Development*, edited by Alan C. Love, 305–25. Dordrecht: Springer Netherlands. https://doi.org/10.1007/978-94-017-9412-1_14

Brigandt, Ingo, Cristina Villegas, Alan C Love, and Laura Nuño de la Rosa. 2023. "Evolvability as a Disposition: Philosophical Distinctions, Scientific Implications." In *Evolvability: A Unifying Concept in Evolutionary Biology?* edited by Thomas F. Hansen, David Houle, Mihaela Pavličev, and Christophe Pélabon, 55–72. Cambridge, MA: MIT Press.

Brown, Rachael L. 2014. "What Evolvability Really Is." *The British Journal for the Philosophy of Science* 65(3): 549–72. https://www.jstor.org/stable/26398395. Accessed 17 January 2023

Dennett, Daniel C. 1995. *Darwin's Dangerous Idea: Evolution and the Meaning of Life*. New York: Simon and Schuster.

Dray, William. 1957. *Laws and Explanation in History*. London: Oxford University Press.

Fisher, Ronald A. 1930. *The Genetical Theory of Natural Selection*. Oxford: Clarendon.

Fontana, Walter, and Peter Schuster. 1998. "Shaping Space: The Possible and the Attainable in RNA Genotype-Phenotype Mapping." *Journal of Theoretical Biology* 194(4): 491–515. https://doi.org/10.1006/jtbi.1998.0771

Fontana, Walter. 2002. "Modelling 'Evo-devo' With RNA." *BioEssays: News and Reviews in Molecular, Cellular and Developmental Biology* 24(12). https://doi.org/10.1002/bies.10190

Gould, Stephen J, and Richard C Lewontin. 1979. "The Spandrels of San Marco and the Panglossian Paradigm: A Critique of the Adaptionist Programme." *Proceedings of the Royal Society of London* B 205: 581–98.

Grüne-Yanoff, Till. 2013. "Appraising Models Nonrepresentationally." *Philosophy of Science* 80(5): 850–61. https://doi.org/10.1086/673893

Hansen, Thomas F. 2003. "Is Modularity Necessary for Evolvability? Remarks on the Relationship between Pleiotropy and Evolvability." *Bio Systems* 69(2–3): 83–94. https://doi.org/10.1016/s0303-2647(02)00132-6

Hansen, Thomas F, and Christophe Pélabon. 2021. "Evolvability: A Quantitative-Genetics Perspective." *Annual Review of Ecology, Evolution, and Systematics* 52(1): 153–75. https://doi.org/10.1146/annurev-ecolsys-011121-021241

Huber, Max. 2017. *"Biological Modalities."* PhD Thesis, University of Geneva. https://archive-ouverte.unige.ch/unige:93135

Ijäs, Tero, and Rami Koskinen. 2021. "Exploring Biological Possibility Through Synthetic Biology." *European Journal for Philosophy of Science* 11(2): 1–17. https://doi.org/10.1007/s13194-021-00364-7

Kirschner, Marc, and John Gerhart. 1998. "Evolvability." *Proceedings of the National Academy of Sciences of the United States of America* 95(15): 8420–7. https://doi.org/10.1073/pnas.95.15.8420

Knuuttila, Tarja. 2021. "Epistemic Artifacts and the Modal Dimension of Modeling." *European Journal for Philosophy of Science* 11(3): 1–18. https://doi.org/10.1007/s13194-021-00374-5

Ladyman, James. 1998. "What Is Structural Realism?" *Studies In History and Philosophy of Science* 29: 409–24.

Lewis, David. 1979. "Counterfactual Dependence and Time's Arrow." *Noûs* 13(4): 455–76. https://doi.org/10.2307/2215339

Lewis, David. 1987. "Causation." *Philosophical Papers II*: 159–213. https://doi.org/10.1093/0195036468.003.0006

Loewer, Barry. 2012. "Two Accounts of Laws and Time." *Philosophical Studies* 160(1): 115–37. https://doi.org/10.1007/s11098-012-9911-x

Loewer, Barry. 2020. "The Package Deal Account of Laws and Properties." *Synthese* 199(1–2): 1065–89. https://doi.org/10.1007/s11229-020-02765-2

Love, Alan C. 2003. "Evolvability, Dispositions, and Intrinsicality." *Philosophy of Science* 70: 1015–27.

Love, Alan C. 2020. "Developmental Biology." In *The Stanford Encyclopedia of Philosophy* (Spring 2020), edited by Edward N. Zalta. Stanford, CA: Metaphysics Research Lab, Stanford University. https://plato.stanford.edu/archives/spr2020/entries/biology-developmental/. Accessed 27 July 2020.

Love, Alan C. 2024. *Evolution and Development: Conceptual Issues*. Cambridge: Cambridge University Press.

Maclaurin, James, and Kim Sterelny. 2008. *What Is Biodiversity?* Chicago: University of Chicago Press.

Manrubia, Susanna, José A Cuesta, Jacobo Aguirre, Sebastian E Ahnert, Lee Altenberg, and Alejandro V Cano, et al. 2021. "From Genotypes to Organisms: State-of-the-Art and Perspectives of a Cornerstone in Evolutionary Dynamics." *Physics of Life Reviews* 38: 55–106. https://doi.org/10.1016/j.plrev.2021.03.004

Mitteroecker, Philipp, and Simon M Huttegger. 2009. "The Concept of Morphospaces in Evolutionary and Developmental Biology: Mathematics and Metaphors." *Biological Theory* 4(1): 54–67. https://doi.org/10.1162/biot.2009.4.1.54

Novick, Rose. 2023. *Structure and Function*. Cambridge: Cambridge University Press. https://doi.org/10.1017/9781009028745

Pavlicev, Mihaela, and Günter P Wagner. 2012a. "Coming to Grips with Evolvability." *Evolution: Education and Outreach* 5(2): 231–44. https://doi.org/10.1007/s12052-012-0430-1

Pavlicev, Mihaela, and Günter P Wagner. 2012b. "A Model of Developmental Evolution: Selection, Pleiotropy and Compensation." *Trends in Ecology & Evolution* 27(6): 316–22. https://doi.org/10.1016/j.tree.2012.01.016

Reiner, Richard. 1993. "Necessary Conditions and Explaining How-Possibly." *The Philosophical Quarterly* 43(170): 58–69. https://doi.org/10.2307/2219941

Rutherford, Suzanne L, and Susan Lindquist. 1998. "Hsp90 as a Capacitor for Morphological Evolution." *Nature* 396(6709): 336–42. https://doi.org/10.1038/24550

Siegal, Marc L, and Jun-Yi Leu. 2014. "On the Nature and Evolutionary Impact of Phenotypic Robustness Mechanisms." *Annual Review of Ecology, Evolution, and Systematics* 45: 496–517. https://doi.org/10.1146/annurev-ecolsys-120213-091705

Simon, Herbert. 1962. "The Architecture of Complexity." *Proceedings of the American Philosophical Society* 10(6): 467–82.

Sjölin Wirling, Ylwa, and Till Grüne-Yanoff. 2021a. "The Epistemology of Modal Modeling." *Philosophy Compass* 16(10): e12775. https://doi.org/10.1111/phc3.12775

Sjölin Wirling, Ylwa, and Till Grüne-Yanoff. 2021b. "Epistemic and Objective Possibility in Science." *The British Journal for the Philosophy of Science*. https://doi.org/10.1086/716925

Stadler, Bärbel M. R, Peter F Stadler, Günter P Wagner, and Walter Fontana. 2001. "The Topology of the Possible: Formal Spaces Underlying Patterns of Evolutionary Change." *Journal of Theoretical Biology* 213(2): 241–74. https://doi.org/10.1006/jtbi.2001.2423

Verreault-Julien, Philippe. 2019. "How Could Models Possibly Provide How-Possibly Explanations?" *Studies in History and Philosophy of Science Part A* 73: 22–33. https://doi.org/10.1016/j.shpsa.2018.06.008

Villegas, Cristina, Alan C Love, Laura Nuño de la Rosa, Ingo Brigandt, and Günter P Wagner. 2023. "Conceptual Roles of Evolvability Across Evolutionary Biology: Between Diversity and Unification." In *Evolvability: A Unifying Concept in Evolutionary Biology*, edited by Thomas F. Hansen, David Houle, Mihaela Pavlicev, and Christophe Pélabon, 35–54. Cambridge, MA: MIT Press.

Wagner, Andreas. 2012. "The Role of Robustness in Phenotypic Adaptation and Innovation." *Proceedings of the Royal Society B: Biological Sciences* 27(1732): 1249–58. https://doi.org/10.1098/rspb.2011.2293

Wagner, Andreas. 2013. *Robustness and Evolvability in Living Systems*. Princeton, NJ: Princeton University Press.

Wagner, Günter P, and Lee Altenberg. 1996. "Perspective: Complex Adaptations and the Evolution of Evolvability." *Evolution* 50(3): 967–76. https://doi.org/10.2307/2410639

Weber, Marcel. 1999. "The Aim and Structure of Ecological Theory." *Philosophy of Science* 66: 71–93.

Weber, Marcel. 2017. "Which Kind of Causal Specificity Matters Biologically?" *Philosophy of Science* 84(3): 574–85. https://doi.org/10.1086/692148

Weber, Marcel. forthcoming. "Causal Selection Versus Causal Parity in Biology: Relevant Counterfactuals and Biologically Normal Interventions." In *Philosophical Perspectives on Causal Reasoning in Biology*, edited by Brian J. Hanley, and C. Kenneth Waters. Minneapolis: University of Minnesota Press.

Woodward, James. 2003. *Making Things Happen: A Theory of Causal Explanation.* Oxford: Oxford University Press.

7 The Combinatorial Possibilities of Synthetic Biology

Tarja Knuuttila and Andrea Loettgers

7.1 Introduction

Synthetic biology aspires toward the biologically possible by reimagining and building novel biological parts, devices, and organisms. It has heavily relied on the idea that it is possible to (re)engineer biology for useful purposes by recombining biological components. Indeed, one of the best-known projects of synthetic biology is the goal of standardizing biological parts, achieving so-called BioBricks that could be combined in a Lego-like fashion into new biological designs with useful novel functions.[1] Alongside standardization, the goal of (re)wiring has been formative for synthetic biology. The guiding idea is to build biological circuits from well-characterized biological parts based on general design principles, describing, for instance, different control mechanisms. Such general design principles are often transferred from other scientific domains, especially from engineering.

The publication of the first two synthetic genetic circuits in *Nature* (2000) is often regarded as the beginning of synthetic biology proper. These synthetic genetic circuits were constructed from biological components that were wired according to a select circuit design, borrowed from engineering, to perform a particular function, and then implanted into living cells. One was a synthetic toggle switch (Cameron, Bashor, and Collins 2014), and the other, the repressilator, was modeled on the ring oscillator, a familiar component in electronics (Elowitz and Leibler 2000). The repressilator implements a negative feedback loop between transcriptional repressors and can create oscillations in protein levels. It was later shown that the addition of a positive feedback loop to the negative feedback loop would increase the robustness and tunability of oscillations (Stricker et al. 2008). Results such as these led to the expectation that synthetic biologists could rewire and reprogram organisms for useful ends such as, for example, producing novel therapeutics, materials, and biofuels.

Although much of synthetic biology still relies on tedious trial-and-error experimentation in a laboratory, the possibility of constructing novel genetic circuits and implementing them within living cells showed that biology might be rationally engineered, even allowing for the introduction of "foreign"

DOI: 10.4324/9781003342816-10

design principles from engineering into biology. Synthetic biologists might be able to engineer leaner and more effective designs than the structures resulting from natural evolution. This aim of synthetic biology—to extend beyond natural evolution toward possible biology—has its clearest expression precisely in the engineering-oriented program of rationally designing and assembling novel, non-natural biological systems and parts.

How should we analyze the modal character of synthetic biology, which extends beyond natural evolution? Drawing on the combinatorial nature of synthetic biology's engineering agenda, we use insights from David Armstrong's combinatorial theory of possibility to investigate the development of synthetic genetic circuits. Armstrong's theory approaches possibilities in terms of the combination of actual elements of the world that appears well-suited to the construction of synthetic genetic circuits. We argue that apart from the principle of combinatorialism, two other features of Armstrong's theory are crucial for synthetic biology: structural universals and the requirement of their instantiation. The search for general design principles, which we conceptualize as structural universals, is one of the primary goals of understanding biological organization. It is one thing to be able to mathematically model a particular possible system and quite another to actually succeed in building a functional biological construct in the laboratory and implement it in a living biological organism. Synthetic biology studies whether various design principles, often transferred from other disciplines and modeled in systems biology, can in fact be realized in existing biological systems.

We proceed as follows: we begin by revisiting how synthetic biologists have conceived of potential biology in terms of a space of biological possibilities where novel synthetic systems lie elsewhere than on the path of evolutionary explorations (section 7.2). In section 7.3, we will look at Daniel Dennett and Andreas Wagner's use of the idea of all possible DNA sequences, also called the Library of Mendel, to conceptualize the space of biological possibilities. We argue that although combinatorial in nature, the idea of such a vast library does not capture the particular challenge of constructing synthetic genetic circuits. Instead, we apply Armstrong's combinatorial theory of possibility (section 7.4). In section 7.5, we study structural universals, their exploration, and instantiation in systems and synthetic biology. Section 7.6 concludes.

7.2   Extending the Space of the (Biologically) Possible through Synthetic Biology

In a programmatic article published in *Nature*, two leading synthetic biologists, Michael Elowitz and Wendell Lim, discuss the importance of the "expansion of biology from a discipline that focuses on natural organisms to one that includes potential organisms" (Elowitz and Lim 2010, 899). They urge biologists to follow the examples of physics and chemistry which study "the physical and chemical principles that govern what can or cannot be,

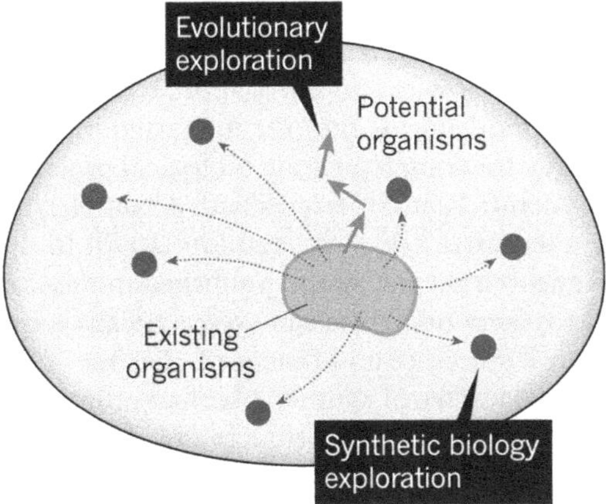

Figure 7.1 The expansion of biology from natural organisms to potential organisms. *Source*: Elowitz and Lim (2010, 890).

in natural and artificial systems" (Elowitz and Lim 2010). This analogy highlights the notion that more general principles might also govern biology and that the utilization of these principles could provide access to unexplored areas of the space of biological possibilities—consequently extending biological research beyond naturally evolved organisms and their evolutionary paths. Figure 7.1, taken from Elowitz and Lim's article, encapsulates the idea of expanding biology through synthetic biology. The image shows how synthetic biology could take biology into directions different than those of natural evolution and, more extensively, explore the space of biological possibilities. While evolutionary exploration follows a gradual path, synthetic biology explorations offer shortcuts to other areas within the space of possibilities not reachable by evolution (indicated by dashed lines).[2]

How should this space of naturally evolved and potential organisms be understood? Elowitz and Lim do not explicate their conception of the space of biological possibilities, nor what they mean by evolutionary exploration. They focus on how synthetic biology explorations into possible biology might also be beneficial for understanding actual biology. Among the many benefits are the insights the lab-constructed synthetic genetic circuits provide on natural circuit-design principles that would be hard to come by using the conventional method of perturbing natural systems. Synthetic biology approaches are also helpful for developmental biology in finding out, for example, what kinds of circuits would be sufficient to program the development of organisms (Elowitz and Lim 2010, 890). Moreover, synthetic biology supplies a complementary approach for the study of metabolic networks, where instead of studying metabolic pathways in distinct organisms, it can combine enzymes from many species to explore what kinds of metabolic networks are possible (Elowitz and Lim 2010).

As the examples mentioned above show, synthetic biologists are not aiming to construct networks that would necessarily be comparable to actual, naturally evolved networks. Elowitz and Lim's approach to building synthetic circuits is more akin to modeling as they are interested in more "basic science" questions. They argue for complementing biological practice, attuned toward the details of idiosyncratic, actual systems with a "simpler, more modular and more predictable alternative" (Elowitz and Lim 2010) In the same vein, Lim and colleagues argue in a review article published in the same year, that "by building minimal toy networks, one can systematically explore the relationship between network structure and function" (Bashor et al. 2010).

Given the modal character of synthetic biology, the oft-repeated criticism that organisms are not engineered artifacts (e.g., Nicholson 2019) seems, at least partially, beside the point. The implementation of design principles from engineering can also be viewed as an investigation into the extent to which biological systems are adaptable enough to allow for organizational structures that are simpler than those of natural systems, and not outcomes of evolution. Following Herbert Simon, synthetic biology could be considered as part of the sciences of the artificial, where artifacts are designed according to what "ought to be, that is, in order to obtain goals and to function" (Simon 2019, 5). Consequently, the parts and systems designed and constructed in synthetic biology could be taken as artifacts, and their engineering relies on the existence of well understood and probed design principles.

7.3   The Library of Mendel

As we saw in the previous section, Elowitz and Lim envision synthetic biology explorations through the notion of a possibility space that contains all biological possibilities. Referencing Borges' "Library of Babel," Daniel Dennett and Andreas Wagner have articulated a space of the biologically possible as a "Library of Mendel" that is basically a combinatorial space consisting of descriptions of genomes, whose standard codes in known living systems consist of only four characters (A, C, G, and T). Dennett defines biological possibility in the following way: "x is biologically possible if and only if x is an instantiation of an accessible genome or a feature of its phenotypic products" (Dennett 1995, 118).[3] Similarly, Wagner has invoked the metaphor of the library in depicting the spaces of evolutionary explorations (e.g., Wagner 2014). He considers "genotype libraries," such as the libraries of protein genotypes, regulatory genotypes, and metabolic genotypes, a "giant realm of possibility" (Wagner 2019, 40). Evolution explores such libraries through evolving populations.

One problem of casting biological possibility in terms of such libraries is their hyper-astronomical size. Due to the combinatorial explosion caused by the possible combinations of, for example, amino acids forming proteins, their number is far bigger than the estimated number of hydrogen atoms in the universe. The problem with such gigantic numbers is not just their incomprehensible enormity, but also the time available in the real world.

The time required for the evolutionary process to visit all possible states would be longer than the universe's existence. This raises the critical question of how life could have evolved and enabled such a wide range of living forms on Earth.[4] Apart from the vastness of the library, Dennett points out how the biological possibilities in such a library are in fact disappearing in it: the overwhelming majority of the "text" in such a library is gibberish, containing only "a vanishingly thin thread" of biologically meaningful "useful" text that could lead to viable organisms.

While libraries like this, as such, capture but an aspect of biological possibility, they address a key component of synthetic biologists' understanding of it. What matters is the combinatorial composition of elements within such libraries and how such a combinatorial approach ties to the concept of rational design: creating new biological functions through assembling standardized biological parts. However, due to the aim of their standardization, BioBricks are not simply strings of DNA—as the Library of Mendel would have it—but already artificial constructs. Shetty, Endy, and Knight have spelled this out clearly:

> We define a biological part to be a natural nucleic acid sequence that encodes a definable biological function, and a standard biological part to be a biological part that has been refined in order to conform to one or more defined technical standards.
>
> (Shetty, Endy, and Knight 2008)

Since the introduction of the BrioBricks standard in 2003, other standards for synthetic biological parts have been proposed.

In addition to the goal of standardization, synthetic biology faces the problem of assembly, that is, how biological parts can be wired together into genetic circuits able to produce desired behaviors and products. The notion of general design principles, applied to biology and engineered systems alike, has been the guiding heuristic of the study of synthetic genetic circuits. Indeed, the basic science-oriented field of synthetic biology that experiments on different circuit designs investigates what kinds of design principles can be implemented within biological systems, thus also studying the limits of the engineering principles in biology (Knuuttila and Loettgers 2013; 2014; 2021). While systems biology has studied different kinds of design principles abstractly with modeling, within synthetic biology, such abstract designs have provided blueprints for constructing functional biological circuits. For synthetic biology, the actual construction of such circuits and their integration into living cells is crucial. Within a cell, the circuit is exposed to interactions and constraints of its cellular environment.

In what follows, we study these aspects of synthetic biology through David Armstrong's combinatorial theory of possibility. It offers analytic resources to capture not just the combinatoriality of the synthetic biology program, but also the importance it places on the assembly of functional circuits as well as their biological implementation. The questions of assembly and implementation can be cast in Armstrongian terms as those concerning structural universals and their instantiation.

7.4   Armstrong's Combinatorial Theory of Possibility

David Armstrong's combinatorial theory of possibility (Armstrong 1986; 1989) is based on combinations of states of affairs that consist of a particular having a property or a relation, between particulars (Armstrong 1997, 1).[5] These properties, like redness or the charge of a particle, and the distinct relations between such particulars, are universals that particulars instantiate. Starting from this combinatorialist premise, Armstrong's account is naturalist in that it seeks to reduce the possible to the actual. He has characterized his naturalism as that of "actual-world chauvinism" (Armstrong 1986, 56) relating "the very idea of possibility to the idea of combinations—all the combinations which respect a simple form—of given, actual elements" (Armstrong 1986, 575). In other words, recombination is the key to modality, as possibilities arise from the reassembly of the existing parts of the world. Such combinations also cover the notions of contraction and expansion, which are important when it comes to evolutionary processes and synthetic biology.

The actual world of a combinatorialist consists of (1) individuals (particulars) that can be characterized by (2) a set of properties and relations, and (3) the distribution of these properties and relations, which exhaustively specifies which individuals have what properties and in what kind of relationships they stand with one another (Kim 1986).[6] Once individuals and their attributes (properties and relations) are established, all possibilities can be generated from them through combination and recombination.

While Armstrong perceives the world through this realistic naturalism, in terms of actual individuals, properties, and relations forming a single spatiotemporal system, he cautions against viewing it as "a tinker-toy construction from three different parts" (Armstrong 1986, 577). Instead, these "elements" should be regarded as abstractions from states of affairs; only by "selective attention may [they] be considered apart from the states of affairs in which they figure" (Armstrong 1986, 57). According to Armstrong's metaphysics, the world is an immense collection of states of affairs that consist of particulars and universals. The simplest existing thing will be a simple state of affairs that consists of a simple particular instantiating a simple property. Each universal, such as redness, squareness, or roundness, is "one that runs through many" being identical in different particulars. Armstrong denies the independent existence of universals, such as Platonic forms. Universals are nothing without particulars; likewise, particulars are nothing without universals.

How, then, are individuals and universals related? Armstrong illustrates his solution to the problem by considering how the size of a thing stands in relation to its shape: "Size and shape are inseparable in particulars, yet they are not related. At the same time, they are distinguishable, and particular size and shape vary independently" (Armstrong 1978, 10). This quote expresses Armstrong's immanent realism, according to which, for each universal U, at least N particulars exist such that they are U. In addition to this "Principle of Instantiation," there is the "Rejection of Bare Particulars," which states that for each particular, x, there exists at least one universal, U, such that x is U (Mumford 2007, 29).

A particular can instantiate several universals, but the question is whether Armstrong adheres to a fixed-base ontology of simple individuals that do not have any proper parts. For instance, Kim (1986) thinks that Armstrong would need to commit to such a base. Armstrong allows, however, that it is a contingent matter: whether it is structures all the way down and whether any given individuals, properties, and relations are indefinitely complex (Kim 1986, 586). For our purposes, it is essential to note that the combinatorial Armstrongian world is structured, most likely in highly intricate ways—and that Armstrong thinks that natural science and empirical investigation should have the last word on what kind of real universals there are.

In addition to monadic universals, there are higher-order universals, such as relations between universals. Armstrong distinguishes between conjunctive and structural properties. For instance, F & G would be a conjunctive property, where F and G are themselves properties. The Library of Mendel, with its collection of all conceivable genomes (and their phenotypic outcomes), appears to be an example of a space composed of such conjunctive properties. Structural properties are more complex, involving relations in which any number of universals might be in various types of relationships with one another. Like monadic universals, structural universals are properties that different individuals can instantiate.

One problem in Armstrong's account that has generated much philosophical discussion is that there are no "alien" universals (Mumford 2007; Sider 2005)—that is, there are no universals that do not occur in the natural world. While Armstrong allows for expanding the world by more individuals, the same does not apply to universals.[7] The prohibition of "alien universals" follows from Armstrong's actualism: some actual individuals need to instantiate the building blocks of worlds. Irrespective of the philosophical problems concerning "alien universals," it seems interesting to ask whether the program of constructing synthetic genetic circuits would, for example, involve the investigation of alien universals not yet instantiated by any natural entity or system. It does not seem, however, that the practice of synthetic biology is committed to alien universals: it transfers designs from engineering and physics to biology, considering them more general design principles on the one hand, and seeks to develop such designs toward a more "biology-inspired" direction on the other (Goel, McAdams, and Stone 2013).

Scientific examples are needed to get a more concrete idea of what structural universals are. The molecule methane provides a stock example of structural universals within the philosophical discussion. Methane comprises four hydrogen and one carbon atom, forming covalent bonds between them.

Fisher explains methane's status as a structural universal as follows: "If methane is instantiated, the molecule that instantiates it has five spatiotemporal parts. These parts must instantiate certain universals: four of the five parts instantiate hydrogen, and the remaining part instantiates carbon" (Fisher 2018, 2). The compositional structure of structural universals has remained a subject of philosophical discussion. For instance, Lewis has argued against structural universals, among other things because methane has only one universal called "hydrogen" and not four (Lewis 1986). The consequent philosophical discussion has offered various theories that attempt to explain the compositional nature of structural universals (see Fisher 2018). Regardless of how this problem may be solved, the notion of structural universals is crucial for understanding the synthetic biology agenda and the assumption of general design principles in biology.

Along with the notion of structural universals, Armstrong's emphasis on instantiation appears critical for understanding synthetic biology. As already discussed, Armstrong considers properties such as roundness as universal because different individuals can instantiate them. One universal can run through many individuals. The property of being round cannot exist by itself, but only in combination with an individual. States of affairs for Armstrong are particular bearings, such as a plate being round. The world, according to Armstrong, consists of nothing else than such states of affairs. They are composite objects, consisting of a particular (this plate) and a universal (roundness). The states of affairs can be causally linked by laws that are instantiated in particular causal transactions by virtue of universals that are nomically connected (Mumford 2007, 53). There will also be different orders of state of affairs. The higher-order states of affairs constrain the states of affairs below them. Since the states of affairs are spatiotemporal, they will have a spatiotemporal relationship with the other states of affairs. Armstrong assumes, moreover, that all states of affairs are causally linked to each other, drawn together in a causal net of chains and resulting in branches.[8]

While systems biology and synthetic biology both address the structural universals, the latter is also especially engaged in investigating their instantiation. Through engineering synthetic systems, synthetic biology studies whether some structural universals abstractly conceived by mathematical models can be implemented in—or instantiated by—biological organisms and their parts. The science Armstrong has in mind is physics, where it is conceivable, at least to some degree, that laws provide the relations between the universals. In contrast to physics, there are few laws in biology. What could be considered structural universals in biology? We suggest that general design principles in biology could qualify as Armstrongian structural universals, comparable to physical laws.

7.5   The Study of Structural Universals in Biology

The search for, and exploration of, organizational or general design principles is not specific to synthetic biology, but concerns biology and biochemistry more generally (Montévil and Mossio 2015; Moreno and Mossio 2015;

Mossio 2024). Biochemist Michael Savageau defines a design principle as "a rule that characterizes some biological feature exhibited by a class of systems such that discovery of the rule allows one not only to understand known instances but also to predict new instances within the class" (Savageau 2001, 142). The conceptualization of biological systems in terms of general design principles and attempts to formulate and search for such organizational principles has a long lineage in the biological sciences, especially in the areas leaning toward systems theory, physics, and engineering sciences (Bertalanffy 1969; Hartwell et al. 1999; Wiener 1961).

Since Jacob and Monod's (1961) operon model of gene regulation, biologists have entertained the idea that genetic control might be governed by negative feedback. Various feedback or feedforward loops commonly employed in engineering have been considered so generic that they apply to biology as well. Going a step further, such basic design principles could enable the development of increasingly complex organizational structures. This combinatorial vision is constitutive of synthetic biology and its program of exploring biological possibilities and producing novel functional biological parts and systems.

In terms of Armstrong's combinatorial theory of possibility, biological functions[9] could be conceptualized as (second-order) properties instantiated by circuits that consist of individuals and their properties arranged in specific ways. For example, transcription factors (individuals) arranged in a feedback loop give rise to the structural universal of homeostasis. Unlike the methane example, the structural-functional relationships in the biological feedback systems are not static; rather, the dynamics between network elements are required for the instantiation of biological functions.

Below, we will take a look at two different research strategies to study biological networks that are based on a combinatorial approach. The first aims to identify recurring structures in biological organisms using genomic data, whereas the second, the synthetic approach, seeks to engineer novel biological constructs from genes and proteins. The synthetic strategy differs from model-based approaches, which impose a presumed organizational structure based on theoretical considerations that are heavily influenced by concepts and model templates transferred from physics, chemistry, engineering, and complexity theory.[10] Focusing on material instantiations of dynamic structures and their products brings the synthetic biology approach closer to exploring objective possibilities, at the cost of being directly exposed to biology's complexity.

7.5.1 *Identifying Design Principles from Databases*

The work of the systems biologist, Uri Alon, on network motifs is one of the first attempts to identify network structures from biological data. According to Alon "[t]he structure of biological circuits—the precise way that their components are wired together—provides them with special dynamical features" (Alon 2006, 2). Network motifs are, for Alon, the basic connectivity structures within biological circuits. In scanning through gene regulatory data, Alon and his group discovered connectivity structures that occur far

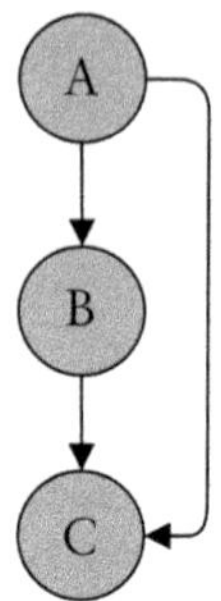

Figure 7.2 Example of a network motif in the form of a feedforward network.

more frequently than would be expected if they occurred by chance (Milo et al. 2002). Alon's strategy relates to Elowitz and Lim's suggestion of using recurrent pathways in biological systems to investigate the design principles of biological organization. Instead of researching them directly in biological systems, Alon relied on data. Alon and his colleagues created a computer algorithm that searched for recurrent patterns in the gene transcription data of *E. coli* bacteria.

Figure 7.2 depicts one of the recurring patterns that Alon and his coworkers discovered throughout their investigation. This particular network motif is a feedforward loop, which is defined as "pairs of source and target nodes that have two or more internally disjoint connecting paths" (Berka 2012, 75). In a biological environment, such loops can, for example, speed up the production of proteins. Figure 7.2 depicts a network with three spatiotemporal parts, with A and B representing transcription factors that bind to receptor C. A feedforward network is a structure in which A directly controls C or through an intermediate transcription factor B.

A network motif such as the feedforward loop is notable for its abstract character. It can depict one of the underlying motifs of any kind of more complex network, making it a structural universal that could be instantiated by different kinds of material entities. Alon, too, sees it this way. Network motifs are recurrent, statistically significant subgraphs or patterns in a larger graph. Networks of all kinds—natural, engineered, and social—studied in different scientific disciplines can be represented as graphs composed of smaller subgraphs.

A precursor for the idea of network motifs can be found from the work of Holland and Leinhardt (1974). While they studied social networks, introducing methods to identify different types of subgraphs and testing whether the frequency of those subgraphs is higher than expected in random networks, Alon generalized their approach to cover gene regulation. Alon (2007) sought also to offer a biological justification for the particular network motifs he and his coworkers identified from the data. He attributed some specific biological functions to different motifs, claiming more generally that "[e]volution seems to have converged on the same motifs in different systems and different organisms, suggesting that they are selected for again and again, on the basis of their biological functions" (Alon 2006, 253). Furthermore, he referred to the ability of each network motif to "carry out defined information-processing functions" (Alon 2006), the notion of information-processing itself obviously extending also to other kinds of systems than biological ones and pointing toward the generality of the notion of a network motif.

Consequently, Alon appears to see the network motif as a (structural) universal that runs across various types of networks. As such, a network motif lacks biological or other specificity in the form of actual components and processes, conveying instead a generic relationship between a structure and a function that can be realized in materially distinct systems. Alon's approach amounts to reverse engineering, that is, proceeding from the biology-specific data on genes and transcription factors to general structures. The dynamics of such structures can be mathematically modeled, thus providing blueprints for forward engineering approaches, for example, the construction of synthetic networks that explore the biological realizability of such structural universals.

7.5.2 The Synthetic Biology Approach for Exploring Biological Possibilities

Whereas Alon sought to identify actual patterns from data that are, in principle, applicable to various kinds of possible systems, synthetic biology allows for probing the realizability of different kinds of design principles in biological systems.[11] As synthetic biologists Mukherji and van Oudenaarden have put it: "An important aim of synthetic biology is to uncover the design principles of natural biological systems through the rational design of gene and protein circuits" (Mukherji and van Oudenaarden 2009, 895). The pioneering work that Elowitz and his coworkers did on the repressilator, the other one of the two first synthetic genetic circuits, provides a good example of such a rational design approach.

The repressilator (Elowitz and Leibler 2000) is a network of three genes organized in a ring, inhibiting/repressing each other's expression. It is a molecular analog of the ring oscillator, which is an electronic circuit made up of an odd number of NOT gates arranged in a ring with an output that oscillates between two voltage levels. Elowitz and Leibler sought to see if the repressilator exhibited similar dynamics. In the repressilator, each of the gene's transcription factors (proteins) binds to the transcription site of its nearby gene, inhibiting its transcription. This arrangement can exhibit oscillations in protein production.

The success of the repressilator in producing oscillations has broader biological implications, in addition to being a significant feat of bioengineering. Biological control is assumed to occur through various types of oscillations and many biological phenomena are rhythmic or cyclical in nature, such as the circadian clock, body temperature, and metabolic processes. In contrast to the mathematical model, which served as a template for building the repressilator, this synthetic model has the same materiality as naturally evolved biological systems. As a result, it is subject to the same constraints as biological systems and their components, even though the repressilator was built from genes that are not found in such a combination in any known natural system. The molecular components of the repressilator were chosen from various biological mechanisms to optimize the model's properties of interest, such as the strength of the oscillations at the protein level, and to make the genetic circuit as independent as possible from the rest of the cell.

Being constructed from biological components and integrated into the *E. coli* bacterium, the repressilator system was obviously biological in contrast to models produced in other media, such as the original mathematical model. This was crucial for its modal significance. Although the biochemical interactions in the cell are largely unknown, this embedment, as Waters (2012) has pointed out, "avoids having to understand the details of the complexity, not by assuming that complexity is irrelevant but by incorporating the complexity in the models." The repressilator showed that it was possible to implement a simple oscillatory system from engineering within a biological system. This raised high hopes concerning the engineering of biology for useful ends.

As for the actual construction process of the repressilator, the first step was to select a circuit design and create a mathematical model. Synthetic biologists commonly refer to such a mathematical model as the "blueprint"; it is needed to study the properties of the synthetic system, informing its construction. In the following stage, the synthetic repressilator was built in the lab and integrated into a live bacterial cell that continued to divide into a colony of cells, all of them hosting the repressilator. The researchers were able to transfer the lab-created repressilator plasmid into *E. coli* bacteria using the bacteria's capacity to absorb extra-chromosomal DNA from their surroundings. Furthermore, to make the oscillations visible, green fluorescent protein (GFP) was fused to one of the three genes that served as a "reporter." Through fluorescence microscopy, the researchers could then observe the gene's protein level oscillations.

Positively surprised, the scientists found that the repressilator indeed produced oscillations, though they were noisy (contrary to what the underlying mathematical model predicted). Several questions surfaced. First, could noise be conducive to biological systems? Second, are regular oscillations even possible in a stochastic environment like a cell? Third, how do the stochastic fluctuations observed in the repressilator connect to other forms of noise that occur independently of those stochastic fluctuations? The Elowiz lab addressed these problems in subsequent research by both mathematical modeling and building further synthetic constructs. In addition to the synthetic model, an electronic version of the repressilator was constructed (Buldú et al. 2007). The comparison between the electronic and synthetic versions of the repressilator is interesting in view of universals being instantiated, or multiply realized (cf. Koskinen 2019), by different material systems. The researchers reasoned that an electronic network provides a good analog of the repressilator for the study of robust oscillations, since, as the researchers put it, "this system is subject to electronic noise and time delays associated with its operation [...] since its parameters depend on the actual values of capacitances and resistors [...]" (Mason et al. 2004, 709). The researchers found that in the electronic repressilator, robust oscillations were possible in the presence of noise. Clearly, the material instantiation of the network made a difference, even though the underlying network design was basically the same.

The previous two decades have proven synthetic biology more difficult than anticipated (Endy 2005; Knuuttila and Loettgers 2013). The biochemical properties of biological parts and simple synthetic structures, as well as their interactions with the rest of the cell, are often not well understood and make it challenging to apply design ideas borrowed from engineering (Knuuttila and Loettgers 2014). Despite the difficulties, synthetic biologists still pursue a combinatorial engineering approach to biology. It is important to keep in mind, though, that the investigation of network architectures transferred from engineering to biology does not imply that synthetic biologists regard them as universals shared by both engineering and biology. Evolutionary design principles are unlikely to be very similar to design principles of engineering, yet biology may be adaptable enough to allow for an engineering-inspired construction approach. Moreover, the development of synthetic systems does in fact test the assumed flexibility of biological organisms and consequent possibilities of engineering them (Knuuttila and Loettgers 2013).

7.6 Conclusion

In this chapter, we have investigated the modal character of synthetic biology by considering it through Armstrong's combinatorial theory of possibility. Synthetic biology's combinatorial nature is evident from its overarching engineering approach. In comparison, its aim of exploring general design principles within living biological organisms has gained less attention. We have suggested that the general design principles studied by synthetic biology could be considered structural universals. For Armstrong, structural universals as such are mere abstractions in need of instantiation by actual particulars. Accordingly, while general design principles can be studied abstractly by modeling, synthetic biology studies whether such principles and their consequent functions can in fact be realized by biological organisms. We have argued that synthetic biology does not mistake biology for engineering but rather uses engineered circuits to probe naturally evolved circuits and biological possibilities more generally. As Michael Elowitz has put it: "Natural circuits can inspire more effective synthetic designs, while synthetic circuits provoke unique questions about the natural circuits with which they interact" (Michael Elowitz).[12]

To be sure, the combinatorial theory of possibility has, as a philosophical theory, several problems that we have largely side-stepped, and things become even more complicated when applying this theory to scientific research (and not just to some examples of distinct molecules like methane). One caveat of applying Armstrong's theory to actual scientific research is its grounding in logical analysis. Consequently, interpreting actual scientific research through the concepts of the combinatorial theory of possibility may strike problematic. We are at a loss, for instance, for providing a scientific interpretation of the idea that any and all recombinations of simple individuals, properties, and relations are possible (e.g., Armstrong 1986). Clearly, the philosophical attempt at accommodating our modal language does not necessarily provide the soundest scientific intuitions. We still believe, however forced our attempt to highlight the modal features of synthetic biology in Armstrong's terms may be, that Armstrong would have supported such an endeavor. Apart from being a naturalist, he was an empiricist, who contended that "except for the primitive verities of ordinary experience, it is natural science that gives us whatever detailed knowledge we have of the world" (Armstrong 1997, 5).

Acknowledgment

This project received funding from the European Research Council (ERC) under the European Union's Horizon 2020 research and innovation program (grant agreement no. 818772). We wish to thank Marcel Weber and Rami Koskinen for their comments on this work.

Notes

1 BioBricks conform to the restriction-enzyme assembly standard that cuts DNA at specific sites, catalogued in *The Registry of Standard Biological Parts*. The Registry was founded in 2003 at the Massachusetts Institute of Technology.
2 The space depicted in this sketch comes closest to the morphological space.
3 It is noteworthy that Dennett allows for alternative genetic systems as well, which would lead to other spaces of possibilities. See Koskinen (2019) for a study of the modal character of synthetically created alternative genetic alphabets.
4 Wagner has a two-fold answer to the problem of big numbers. First, the evolutionary explorations happen at the level of populations, whose members differ from each other, yet share the same phenotype. The libraries are structured as enormously complex genotype networks that are related to different phenotypes. Populations explore the genotype networks of these libraries by moving through genotype networks in small steps, for example, by way of mutations. Second, the neighborhoods in these libraries are diverse. As the networks are so large, populations drift within the genotype space widely, getting into different neighborhoods. For a discussion, see Chapter 3 in Wagner (2014).
5 A particular having a property, or a relation between two particulars are first-order states of affairs. In second-order states of affair, first-order states of affairs have properties or stand in relation, or first-order properties and relations have properties or stand in relation (e.g., *redness being a color*), and similarly for higher orders (Meinertsen 2023). See also below.
6 See also (Knuuttila and Loettgers 2022).
7 Later on, in his *World of States of Affairs* (1997), Armstrong attempted to accommodate possibilities involving alien universals (for discussion, see Schneider 2001).

8 In *World of States of Affairs* (1997), Armstrong embraced a new view of instantiation: the idea that particulars and universals are partially identical. The new view is not consistent with his earlier contingentist ideas of combinatorial freedom. As Mumford puts it: "[Armstrong] no longer has any ground, therefore, for invoking the possibility that a universal have a further, additional instantiation to the one it has, nor that a particular could instantiate one further universal over and above the ones that it does" (Mumford 2007, 193). As a result, Armstrong appears to have left behind his earlier theory of combinatorial *possibility* entering instead into the realm of necessity.

9 We treat functions here simply as activities that biological entities do by themselves. See (Wouters 2003) for different notions of biological function. We thank Marcel Weber for pointing out Wouters' paper.

10 For model templates, see Knuuttila and Loettgers (2023).

11 See Ijäs and Koskinen (2021), who argue that synthetic biology can be used to explore biological possibilities through two mutually complementary strategies: the *design* of new-to-nature functional systems and the material *redesign* of extant functions.

12 https://www.elowitz.caltech.edu/research.

References

Alon, Uri. 2006. *An Introduction to Systems Biology: Design Principles of Biological Circuits*. 1st ed. Boca Raton, FL: Taylor & Francis Inc.

Alon, Uri. 2007. "Network Motifs: Theory and Experimental Approaches." *Nature Reviews Genetics* 8(6): 450–61. https://doi.org/10.1038/nrg2102.

Armstrong, D. M. 1978. *A Theory of Universals: Volume 2: Universals and Scientific Realism*. Cambridge University Press.

Armstrong, D. M. 1986. "The Nature of Possibility." *Canadian Journal of Philosophy* 16(4): 575–94.

Armstrong, D. M. 1989. *A Combinatorial Theory of Possibility*. Cambridge University Press.

Armstrong, D. M. 1997. *A World of States of Affairs*. Cambridge Studies in Philosophie. Cambridge University Press.

Bashor, Caleb J., Andrew A. Horwitz, Sergio G. Peisajovich, and Wendell A. Lim. 2010. "Rewiring Cells: Synthetic Biology as a Tool to Interrogate the Organizational Principles of Living Systems." *Annual Review of Biophysics* 39(1): 515–37. https://doi.org/10.1146/annurev.biophys.050708.133652.

Berka, Tobias. 2012. "The Generalized Feed-Forward Loop Motif: Definition, Detection and Statistical Significance." *Procedia Computer Science*, Proceedings of the 3rd International Conference on Computational Systems-Biology and Bioinformatics 11(January): 75–87. https://doi.org/10.1016/j.procs.2012.09.009.

Bertalanffy, Ludwig von. 1969. *General System Theory: Foundations, Development, Applications*. New York: G. Braziller. https://doi.org/10.1109/TSMC.1974.4309376.

Buldú, Javier M., Jordi García-Ojalvo, Alexandre Wagemakers, and Miguel A. F. Sanjuán. 2007. "Electronic Design of Synthetic Genetic Networks." *International Journal of Bifurcation and Chaos* 17(10): 3507–11. https://doi.org/10.1142/S0218127407019275.

Cameron, D. Ewen, Caleb J. Bashor, and James J. Collins. 2014. "A Brief History of Synthetic Biology." *Nature Reviews Microbiology* 12(5): 381–90.

Dennett, Daniel C. 1995. *Darwin's Dangerous Idea: Evolution and the Meanings of Life*. New York: Simon & Schuster.

Elowitz, Michael, and Stanislas Leibler. 2000. "A Synthetic Oscillatory Network of Transcriptional Regulators." *Nature* 403(6767): 335–38. https://doi.org/10.1038/35002125.

Elowitz, Michael, and Wendell A. Lim. 2010. "Build Life to Understand It." *Nature* 468(7326): 889–90. https://doi.org/10.1038/468889a.

Endy, Drew. 2005. "Foundations for Engineering Biology." *Nature* 438(7067): 449–53. https://doi.org/10.1038/nature04342.

Fisher, A. R. J. 2018. "Structural Universals." *Philosophy Compass* 13(10): e12518. https://doi.org/10.1111/phc3.12518.

Goel, Ashok K., Daniel A. McAdams, and Robert B. Stone. 2013. *Biologically Inspired Design: Computational Methods and Tools.* Springer Science & Business Media.

Hartwell, Leland H., John J. Hopfield, Stanislas Leibler, and Andrew W. Murray. 1999. "From Molecular to Modular Cell Biology." *Nature* 402(6761): C47–52. https://doi.org/10.1038/35011540.

Holland, Paul W, and Samuel Leinhardt. 1974. "The Statistical Analysis of Local Structure in Social Networks." Working Paper. Working Paper Series. National Bureau of Economic Research. https://doi.org/10.3386/w0044

Ijäs, Tero, and Rami Koskinen. 2021. "Exploring Biological Possibility through Synthetic Biology." *European Journal for Philosophy of Science* 11: 39.

Jacob, François, and Jacques Monod. 1961. "Genetic Regulatory Mechanisms in the Synthesis of Proteins." *Journal of Molecular Biology* 3(3): 318–56. https://doi.org/10.1016/S0022-2836(61)80072-7.

Kim, Jaegwon. 1986. "Possible Worlds and Armstrong's Combinatorialism." *Canadian Journal of Philosophy* 16(4): 595–612.

Knuuttila, Tarja, and Andrea Loettgers. 2013. "Basic Science through Engineering? Synthetic Modeling and the Idea of Biology-Inspired Engineering." *Studies in History and Philosophy of Science Part C: Studies in History and Philosophy of Biological and Biomedical Sciences,* Philosophical Perspectives on Synthetic Biology 44(2): 158–69. https://doi.org/10.1016/j.shpsc.2013.03.011.

Knuuttila, Tarja, and Andrea Loettgers. 2014. "Varieties of Noise: Analogical Reasoning in Synthetic Biology." *Studies in History and Philosophy of Science Part A* 48(December): 76–88. https://doi.org/10.1016/j.shpsa.2014.05.006.

Knuuttila, Tarja, and Andrea Loettgers. 2021. "Biological Control Variously Materialized: Modeling, Experimentation and Exploration in Multiple Media." *Perspectives on Science* 29(4): 468–92.

Knuuttila, Tarja, and Andrea Loettgers. 2022. "(Un)Easily Possible Synthetic Biology." *Philosophy of Science* 89(5): 908–17. https://doi.org/10.1017/psa.2022.60.

Knuuttila, Tarja, and Andrea Loettgers. 2023. "Model Templates: Transdisciplinary Application and Entanglement." *Synthese* 201(6): 200. https://doi.org/10.1007/s11229-023-04178-3.

Koskinen, Rami. 2019. "Multiple Realizability and Biological Modality." *Philosophy of Science* 86(5): 1123–33. https://doi.org/10.1086/705478.

Lewis, David K. 1986. *On the Plurality of Worlds.* Oxford; New York: Blackwell Publishers.

Mason, Jonathan, Paul S. Linsay, J. J. Collins, and Leon Glass. 2004. "Evolving Complex Dynamics in Electronic Models of Genetic Networks." *Chaos: An Interdisciplinary Journal of Nonlinear Science* 14(3): 707–15. https://doi.org/10.1063/1.1786683.

Meinertsen, Bo R. 2023. "Immanent Realism and States of Affairs." In *The Routledge Handbook of Properties,* edited by A. R. J. Fisher and Anna-Sofia Maurin. London: Routledge.

Milo, R., S. Shen-Orr, S. Itzkovitz, N. Kashtan, D. Chklovskii, and U. Alon. 2002. "Network Motifs: Simple Building Blocks of Complex Networks." *Science* 298(5594): 824–27. https://doi.org/10.1126/science.298.5594.824.

Montévil, Maël, and Matteo Mossio. 2015. "Biological Organisation as Closure of Constraints." *Journal of Theoretical Biology* 372(May): 179–91. https://doi.org/10.1016/j.jtbi.2015.02.029.

Moreno, Alvaro, and Matteo Mossio. 2015. *Biological Autonomy*. Vol. 12. In *History, Philosophy and Theory of the Life Sciences*, edited by Philippe Huneman, Thomas A. C. Reydon, and Charles T. Wolfe. Dordrecht: Springer Netherlands. https://doi.org/10.1007/978-94-017-9837-2.

Mossio, Matteo, ed. 2024. *Organization in Biology*. Cham: Springer Nature. https://doi.org/10.1007/978-3-031-38968-9.

Mukherji, Shankar, and Alexander van Oudenaarden. 2009. "Synthetic Biology: Understanding Biological Design from Synthetic Circuits." *Nature Reviews Genetics* 10(12): 859–71. https://doi.org/10.1038/nrg2697.

Mumford, Stephen. 2007. *David Armstrong*. Stocksfield: Routledge.

Nicholson, Daniel J. 2019. "Is the Cell Really a Machine?" *Journal of Theoretical Biology* 477: 108–26. https://doi.org/10.1016/j.jtbi.2019.06.002.

Savageau, Michael A. 2001. "Design Principles for Elementary Gene Circuits: Elements, Methods, and Examples." *Chaos: An Interdisciplinary Journal of Nonlinear Science* 11(1): 142–59. https://doi.org/10.1063/1.1349892.

Schneider, Susan. 2001. "Alien Individuals, Alien Universals, and Armstrong's Combinatorial Theory of Possibility." *The Southern Journal of Philosophy* 39(4): 575–93.

Shetty, Reshma P., Drew Endy, and Thomas F. Knight. 2008. "Engineering BioBrick Vectors from BioBrick Parts." *Journal of Biological Engineering* 2(1): 5. https://doi.org/10.1186/1754-1611-2-5.

Sider, Theodore. 2005. "Another Look at Armstrong's Combinatorialism." *Nous* 39(4): 679–95. https://doi.org/10.1111/j.0029-4624.2005.00544.x.

Simon, Herbert A. 2019. *The Sciences of the Artificial*. Cambridge, MA: MIT Press.

Stricker, Jesse, Scott Cookson, Matthew R. Bennett, William H. Mather, Lev S. Tsimring, and Jeff Hasty. 2008. "A Fast, Robust and Tunable Synthetic Gene Oscillator." *Nature* 456(7221): 516–19. https://doi.org/10.1038/nature07389.

Wagner, Andreas. 2014. *Arrival of the Fittest: Solving Evolution's Greatest Puzzle*. New York: Penguin.

Wagner, Andreas. 2019. *Life Finds a Way: What Evolution Teaches Us About Creativity*. London: Hachette UK.

Waters, Kenneth C. 2012. "Experimental Modeling as a Form of Theoretical Modeling." Paper presented at the Philosophy of Science Association 23rd Meeting, San Diego.

Wiener, Norbert. 1961. *Cybernetics: Or Control and Communication in the Animal and the Machine*. 2nd ed. Cambridge, MA: MIT Press Ltd.

Wouters, Arno G. 2003. "Four Notions of Biological Function." *Studies in History and Philosophy of Science Part C: Studies in History and Philosophy of Biological and Biomedical Sciences* 34(4): 633–68. https://doi.org/10.1016/j.shpsc.2003.09.006.

8 Beyond Networks

Explaining Dynamics in the Natural and Social Sciences

James DiFrisco, Johannes Jaeger*, and Andrea Loettgers**

8.1 Introduction

Over the past few decades, the study of networks has spread like wildfire across the natural and social sciences. Network theory originated in formal graph theory and, for quite some time, remained confined to this specific domain of mathematics. At around the same time, network-based models were being employed in condensed-matter physics. A classic example is the Ising model of ferromagnetism, which consists of binary ("up" or "down") magnetic moments arranged on a regular spatial lattice that interact with their immediate neighbors and undergo phase transitions to an aligned global configuration at some critical temperature via a random assortment of up or down spins (Ising 1925).

By the end of the 20th century, with large amounts of data becoming available across various scientific disciplines, network models entered other areas of investigation. Small-world and scale-free networks became common model templates that were easily transferred between disciplines (Barabási and Albert 1999; Watts and Strogatz 1998; reviewed in Barabási and Oltvai 2004; Caldarelli 2007; Newman 2003; Watts 1999, among others). Such models are now regularly used in the study of metabolism, gene regulation, computer networks, power grids, networks of actors and mathematicians, the spread of disease, and the structure of language and communication. Structural analysis has been particularly useful for explaining the robustness of networks against perturbation and relatedly the percolation of perturbations through a network.

Network models, by their very structure, appear to exhibit a surprising degree of generality, often in fields that famously lack it. Therein lies both their strength and their weakness: these models operate at a high level of abstraction, which makes them independent of the (often very complex and unknown) mechanistic details of the systems they are supposed to represent. The Ising model, for example, can be used as a general model of phase transitions in networks precisely because of its high level of abstraction and lack

* All authors contributed equally to this chapter.

DOI: 10.4324/9781003342816-11

of context-specific detail. The downside is that structural models of networks only represent certain generic dispositional properties of systems (such as robustness) and ignore the causal factors that generate those phenomena as well as their dynamical, time-dependent realizations.

This imposes under-appreciated limitations on the applicability and scope of structural network models, as well as on their explanatory and predictive power. In particular, such models should not be used to explain phenomena that depend on system dynamics, broadly, or those phenomena for which the particular underlying causal mechanisms make a difference. In addition, even some generic structural properties, such as robustness, depend more on dynamics than is usually acknowledged. We will use network percolation in different contexts as one prominent example that illustrates this point. In summary, insights gained from structural network models should always be taken with a grain of salt. Yet, this basic fact eludes not only lay people but also many experts in the field, which leaves them susceptible to what Whitehead (1925) called the "fallacy of misplaced concreteness," or the inappropriate reification of abstract notions.

Often, static structural models of networks are only the first step toward dynamical models of a complex system. And yet, even if the analysis of a network is extended to its dynamic aspects, network-based approaches have important limitations. In particular, the analysis of dynamical systems is historically centered on linearization around steady states or asymptotic equilibria, ignoring relevant nonlinear *transients*, or system behaviors that differ from its long-term, asymptotic, or equilibrium behavior. Furthermore, the structure and boundary conditions of a system are usually assumed to remain constant over time. When a time-variable structure *is* considered, its dynamics are usually prescribed by the modeler, from the outside. And yet, many time-invariant dynamical systems change their structure endogenously. We have called the challenge of representing such time-dependent dynamic properties of systems *the problem of diachronicity* (DiFrisco and Jaeger 2019). The central issue here is the fundamental tension between simplicity and tractability and the complex nature of the systems we are studying.

Each time network models and structural analysis get applied to a new field, a transfer of model templates occurs between the original and the new context of application. A model template is an "abstract conceptual idea, which aims to capture the intertwinement of a mathematical structure and associated computational tools with theoretical concepts that, taken together, depict a general mechanism that is potentially applicable to any subject or field displaying particular patterns of interactions" (Knuuttila and Loettgers 2014, 295). The generic ontology of model templates—in the case of the Ising model, consisting of magnetic moments that are binary variables located on a regular lattice with coupling restricted to immediate neighbors, together with concepts taken from statistical physics—suggests the kinds of systems to which a template can be applied.

Taking this into account, network models are very successful model templates *not* because they provide accurate representations of the target systems to which they are applied, but because of their generic ontology and associated conceptual framework that can be easily used for modeling a wide range of different systems (Knuuttila and Loettgers 2023). Analogously, we use linear analysis to study dynamical systems, not because we believe that complex systems are predominantly at steady state, but because this provides us with mathematical tools to analyze the system in a rigorous manner. When the wider context of complexity is forgotten, however, we run the risk of blinding ourselves to the idealizations of our models and the generic nature of the explanations they provide. We observe that this problem is widespread in network science and other system-theoretic approaches across scientific disciplines.

Both structural network models and dynamical systems can be thought of as defining and generating abstract *spaces of possibilities*, parameterized by the modeled system's features, that is, the specific formalisms, variables, and parameters that those models are built upon. In this way, model templates constrain and shape the space of possibilities of the models that are based on them. In general, the space of possibilities[1] of static network models is a very low-dimensional space compared to the space of possibilities in a dynamical model targeting the same system. Static network models include formal variables and parameters like connectivity, degree distributions, or the number of nodes, and tend to exclude spatiotemporal features other than the inter-node relationships that define the topology of the network. By contrast, dynamical models include temporal features (parametrized by rates of change, time scales, and relaxation times) and may also include spatial features. This grants them larger possibility spaces (called "configuration spaces" in dynamical systems theory), higher resolution on possible system states, and thus more determinate information about actual (and possible) occurrences in space and time.

In what follows, we will briefly outline the history of the Ising model as a case study, with a specific focus on the various network models it inspired. Then we will analyze the processes of model template transfer that led to the application of these related network models across disciplines, from cognitive neuroscience to cell and developmental biology, ecology, sociology, to economics. We will focus on the particular mismatch between abstract and idealized network-level models and the dynamic causal mechanisms they represent. In each case, the space of possible states of a static network model cannot capture the dynamic aspects of the modeled system because there is no simple or straightforward relationship between the structure (topology) and the dynamical behavior of a complex network. Each example will be illustrated using examples from the relevant field of study.

8.2 The Origin of Network Models in Condensed-Matter Physics

One of the most successful network models in physics is the Ising model, developed during the 1920s by Ernst Ising. In the original version of the model

(Ising 1925), magnetic moments of atoms were modeled as binary variables, taking values of +1 ("up") or −1 ("down"), which are arranged in a linear string. Magnetic moments only interact with their immediate neighbors in the string. Above a critical temperature T_c, all magnetic moments point in random up or down directions due to thermal fluctuations. At temperatures below T_c, the interactions between magnetic moments start to overcome thermal motion, and the magnetic moments begin to align. Thus, at the critical temperature, the system undergoes a phase transition and becomes ferromagnetic. In the original one-dimensional Ising model, such a phase transition could not be observed. However, in 1942, Lars Onsager showed that phase transitions occur in the two-dimensional case. Following this discovery, the properties and behavior of the model were explored systematically, by changing magnetic moments and interactions between them (Sherrington and Kirkpatrick 1975). In this context, novel possibilities of performing simulations, enabled by technological progress in computer technology, played a key role.

8.3   Neural and Cognitive Networks

Physicists like John Hopfield recognized the possibility of applying Ising-like models to neural networks in the late 1970s and early 1980s. While constructing a model of associative memory, (Hopfield 1982) made use of a spin-glass model, from which he transferred conceptual ideas, structure, ontology, and mathematical methodology to neuroscience. The most important conceptual idea introduced this way to neuroscience is *cooperativity*, describing a particular quality of interaction between the components of a system that gives rise to specific kinds of observable behavior. In the case of the Ising model, this is illustrated by the phase transition from random to globally aligned spins. The interaction is modeled abstractly, removed from the specific mechanisms that cause phase transitions in magnetic systems.

Spin-glass models have their origin in the Ising model but differ from it in that they replace strict next-neighbor with infinite long-range interactions and feature coupling between the magnetic moments that decrease with distance. Even though the Hopfield model neglects many of the details of actual neural systems, it nonetheless has been a great success. This is largely due to the tractability of the model and the fact that the model allows for learning and the retrieval of patterns. Since then, recurrent Hopfield neural networks (Hopfield 1984) have come to form the very foundation of the *connectionist* approach to cognitive neuroscience as well as deep learning in artificial intelligence research. The conceptual strength of such connectionist models lies in the simple representation of network structure, in particular, the abstract representation of neural interactions: positive for excitation, negative for inhibition (and zero for no interaction).

Connectionist network models are particularly suited to modeling and implementing learning processes. In this context, the emphasis lies entirely on network structure: the transformation of input to output depends on the topology of the neural network that mediates it. Learning is implemented by algorithms that alter this wiring and the strength of interactions until an optimal structure is found. In this view, the difference between cognitive processes is explained by differences in the structure of the underlying neural networks.

The main issue here (as in many other examples) is that the map between network structure and system dynamics is degenerate. Many different structures can realize the same process, and many different processes can occur on the same structure depending on the broader context. In other words, network structure alone grossly underdetermines the behavior of the system, and is therefore inadequate as an explanation for specific cognitive processes, since it does not even address the dynamics of the system. Building on the problem of diachronicity, we call this the *problem of correspondence* between network structure and network function (DiFrisco and Jaeger 2019). It implies the basic failure of static structural models to capture essential causal factors that are required to generate the behavior of a neural or cognitive system. These factors are intrinsically dynamic and often exert their influence only transiently. The possibility (configuration) space of a static network model simply does not contain them, and in this respect fails to connect to its target phenomena.

8.4 Regulatory Networks in Cell and Developmental Biology

Recurrent Hopfield neural networks have also found their way into the study of regulatory networks in cell and developmental biology (Mjolsness, Sharp, and Reinitz 1991). Due to their ability to "learn" to reproduce patterns, they are ideally suited as network models for reverse engineering. In this approach, the structure of a gene regulatory network is inferred by fitting the model to data that represents the spatiotemporal dynamics of gene expression by the system (see, e.g., Jaeger and Crombach 2012). During this optimization process, the structure of the network is adjusted in a way that is analogous to machine learning and the connectionist approach to cognitive neuroscience, until the model output resembles the observed expression dynamics as closely as possible. What results from this procedure is a network model that "has learned to" reproduce the observed expression data. This approach not only yields a specific network structure, but the dynamical model of the regulatory network allows us to connect specific interactions and/or subnetworks to specific expression features in the data (e.g., Jaeger et al. 2004; Verd et al. 2018).

This explicitly dynamic approach can be contrasted to more general approaches that focus on network structure only. Famous examples are the work of Barabási and colleagues on global network analysis (Barabási and Oltvai 2004) or that of Uri Alon and colleagues on local network motifs (Alon 2007). Also worth mentioning are the gene regulatory networks claimed to implement a genetic program by Eric Davidson and colleagues (e.g., Davidson et al. 2002; Levine and Davidson 2005). Other examples abound. The general idea is always the same: by pinning down the structure of a network, we supposedly are able to infer some of the underlying system's generic properties. For instance, the widespread occurrence of power laws in the degree distributions of metabolic and gene regulatory networks suggests that these networks are robust to mutational and environmental perturbations unless the perturbations affect one of the rare network hubs—exceptionally highly connected network nodes from the fat tail of the power-law distribution. Another example is provided by the claim that particular small network motifs are "enriched" (and unexpectedly frequent) in regulatory systems since they are tied to specific functions, such as filtering or amplifying noisy input, or providing feedback control.

The fundamental limitation of all these approaches, again, lies in the simple fact that structure does not determine function (the problem of correspondence). The dynamics of a process not only depend on the interaction of network components but also on other factors, for example, diffusion or boundary conditions such as the shape of the (often growing or changing) spatial domain on which they occur. Moreover, the dynamics of local network motifs depend on the context of the larger network they are embedded in. The demarcation of subnetworks always depends on the specific interactions that are selected as relevant out of a wider system of interactional complexity (Wimsatt 1972). Another issue is the chosen degree of abstraction. The model templates that structural network models of metabolism and gene regulation are based on are so widely applicable exactly because they abstract not only from molecular mechanistic details but also from dynamics in general.

Such abstraction can lead to various problems, especially when mechanistic claims are made on the basis of evidence from a purely structural model. One particularly salient example is the claim that power-law degree distributions indicate a specific mechanism of network growth—the preferential attachment mechanism first proposed by Barabási and Albert (1999). As it turns out, power-law signatures are generic, occurring in a wide class of complex systems that exhibit simple path- or history-dependence in their dynamics (Corominas-Murtra, Hanel, and Thurner 2015; Perkins et al. 2014). At the same time, power-law distributions are extremely hard to detect unambiguously from data (Broido and Clauset 2019; Clauset, Shalizi, and Newman 2009). Therefore, based on finding such a signature, virtually no mechanistic inferences can be drawn with any certainty about the processes underlying the network or how they came to be.

But even uncontroversially successful claims to robustness must be scrutinized. Stuart Kauffman has famously claimed that living systems must exist in a particular dynamical regime that he called *"the edge of chaos"* (Kauffman 1993). This is not about the mathematical theory of deterministic chaos, which reveals that the behavior of many complex systems is highly sensitive to their particular initial conditions, but rather indicates a state of self-organized criticality, in which the dynamics of biological regulatory processes reside close to critical thresholds (such as those observed in the Ising model) that separate highly ordered and disordered dynamical regimes of the system. As a key characteristic, systems at the edge of chaos exhibit "islands of order" in "a sea of disorder."

Importantly, structural or environmental perturbations to any system that resides in such a stable island regime usually remain local due to the dynamic self-maintaining organization of the system. Contrast this with notions of percolation through network structure, where the propagation of perturbations is modeled exclusively based on structural properties, such as local modularity in network connections or global small-world-ness of a network (the property that every component of the network can be reached from any other in a small number of steps). Evidently, these two notions of error propagation differ fundamentally from each other, since the latter does not take into account any dynamical properties of the system, such as varying time scales (delays) or feedback regulation, that may be able to buffer fluctuations. Expressed in more technical terms: network robustness, as used in structural analysis, is not at all equivalent to the kind of *structural stability* conveying robust behavior that can only be identified through proper dynamical analysis.

Connectionist network models based on recurrent Hopfield neural networks (that are fit to quantitative expression data and validated against empirical evidence, as described above) allow us to bridge the gap between abstract studies of network structure and dynamics in simulated ensembles of network models, and the empirical investigation of gene regulation at the molecular and genetic level. And yet, they still depend on the idealization of regulatory network structure, which is treated as static over the time course of a biological regulatory process. In reality, however, such structures can be highly transient, depending on a diffuse assembly of stochastic molecular processes that are often actively involved in determining the behavior of the system. This limits the applicability of network models in these areas, with network edges becoming ill-defined, "fuzzy" and transient under closer scrutiny.

8.5 Ecological Networks and Ecosystems

Network concepts have long been used in ecology, but the arrival of mathematical network theory has led to a deluge of network modeling tools in contemporary ecology (see, e.g., Borrett, Moody, and Edelmann 2014). Network models are used to capture qualitative patterns of connectivity and interaction among individuals and populations. A paradigmatic example is the trophic food web. This describes what eats what, or the general topology that channels the flow of primary production and detritus from autotrophs[2] to various levels of heterotrophs. The formal tools of network theory can be used to predict generic properties for different network topologies. For example, it is widely appreciated that *modularly* structured networks tend to be more stable to random perturbations than non-modular networks because perturbations within a module are less likely to propagate to other modules (cf. Simon 1962). Similarly, the disruption of a hub, or highly connected node such as a "keystone species," is more likely to percolate to many other nodes than the disruption of a peripheral, or less connected, node (see also section 8.4).

As we have already seen, static network models like trophic webs generally abstract from time and flatten a dynamic set of real-world interactions into a static structure of qualitative connections. It is therefore natural to suppose that networks may be a more useful modeling tool to the extent that the system being modeled is *time-invariant*—that is, the interactions targeted by the model do not depend in a significant way on timing, rates, relaxation times, temporal fluctuation, and so on. Real-world trophic exchanges (networked food chains) would initially seem to be a good candidate for time-invariant representation in an ecological context. However, the connectivity and even directionality of trophic networks can change as organisms undergo development. The marine and freshwater ecosystems that cover much of the earth are dominated by "life history omnivores"—species that feed from different trophic levels at different life stages. Many fish, for example, feed on phytoplankton after hatching and then on other fish as adults—including the very piscivorous fish that ate them while young (Montoya, Pimm, and Solé 2006). When developmental timing is asynchronous, there will be multiple trophic networks at the same time, potentially with different stability properties, and disentangling them will require reincorporating time into the model.

Related to the feature of time-invariance, network models designed to capture system-level stability or robustness generally picture node variables as binary or "on/off" switches—like a gene knockout, electrical power station failure, or species extinction. Network models are often enriched with concepts from percolation theory (e.g., Callaway et al. 2000), which investigates generic network behaviors under removal or switch-like perturbations of nodes. As with time-invariance, this suggests that network models may be most applicable to systems whose components exhibit discrete "on/off" behavior. In ecology, an example would be the extinction or removal of a population from an ecosystem.

Even though extinction is a discrete event, the processes leading to it and deriving from it are inherently time-dependent. Therefore, the ecological role of a population can be difficult to capture as a discrete "on/off" phenomenon. Unlike the near-instantaneous percolation of failures in electrical networks, the effect of extinction of one population on another population takes time to manifest. In the phenomenon of "extinction debt" (Tilman et al. 1994), for example, habitat destruction has a delayed effect on species diversity—and this occurs even in linear systems (Hastings 2004, 42). The situation is more dramatic in a nonlinear context. Population biologists have long appreciated how nonlinear coupling between populations means that small-graded differences in population size, interaction rates, and generation times can have disproportionately large effects on ecosystem dynamics.

In order to be able to capture phenomena like extinction debt, models must take into account that real-world interactions between populations are dynamic processes that take place on multiple time scales. Their stability is not an instant all-or-nothing affair pertaining to the persistent identity of an abstract network, as described in network models of robustness, but is a composite of different stability properties of different components on varying time scales. This is fundamental for understanding ecological dynamics. As Alan Hastings (2010, 3472) writes:

> Various internal timescales in ecosystems arise from the inherently diverse endogenous timescales over which different biological processes operate and the vastly different timescales of interacting species [...] The dynamics of ecological systems can only be understood by combining exogeneous and the endogenous dynamics on appropriate timescales.

Interestingly, time scale separation is not only a passive consequence of combining sub-systems with different endogenous time scales (e.g., different generation times of interacting species) but can also be a "driving force" behind their observed dynamics (Hastings 2010; cf. DiFrisco 2017). Dynamics on multiple time scales are expected to give rise to transients, which, as explained above, are behaviors of a system that are distinct from its long-term, asymptotic, or equilibrium behavior (Hastings 2004). Coupled predator-prey systems, for instance, exhibit transients, and epidemics and outbreaks *are* transients. Transients are only definable and detectable on the basis of considerations of time scale, and so they cannot be captured by network models at all. Moreover, they are also refractory to modeling strategies of linearization around steady states or asymptotic equilibria. We will revisit this important issue below.

Other aspects of time-dependence in ecological systems could be adduced here that can only be captured through dynamical models. For example, stochasticity and stochastic forcing pertain to the spatiotemporal fluctuation length in a system (stochasticity is distinct from *probabilistic* percolation, which is still discrete), and oscillation, synchronization, and phase-coupling are time-dependent. From this brief overview, it should be clear that network models face rather severe limitations for understanding ecological dynamics and ecological stability.

8.6 Entanglement of Model Templates from Physics and Economics

In this section, we will close the circle by coming back to model templates and the application of the Ising model in economics. This example shows how two models after having been developed independently from each other become entangled as templates. Searching for new research fields in which to apply their methodology, physicists started to apply variants of the Ising model to problems in economics, sociology, and neuroscience around the 1990s (see Sornette 2014). In fact, the application of spin-glass models, such as that of Sherrington-Kirkpatrick, became quite common outside of physics. One specific example is the mutual entanglement of the Ising model and the Schelling model. Schelling's model is commonly regarded as providing a tool for exploring the social phenomenon of segregation, for example, racial segregation in mixed urban neighborhoods (Schelling 1971; 1978). Both models capture generic structural properties of networks, which in the understanding of many scientists, grant them interdisciplinary applicability. This is thought to be the case even though they have been constructed for modeling vastly different and seemingly unconnected problems—ferromagnetism and racial segregation in urban neighborhoods. Together, they nicely illustrate the evolution and mutual interactions of model templates in a complex historical and sociological scientific context (Knuuttila and Loettgers 2016).

The situation of the Ising and Schelling models differs from the straightforward one-way historical transfer and application of spin-glass models to cognitive science and biology described above. It is a case of convergent, rather than divergent model evolution. In fact, physicists remained unaware of Schelling's model for almost three decades. When they finally noticed it, they immediately recognized the structural similarity of the two models. Qualitatively equivalent cooperative interactions between micro-components drive vastly different macro-phenomena in each model: segregation in the case of Schelling, and ferromagnetic phase transitions in the case of Ising. Once this conceptual equivalence was established, the history of the two model templates became entangled through shared computational methods, abstract concepts, and tools for analysis (cf. Knuuttila and Loettgers 2023; Schelling 2006).

The following example illustrates this point. In both models, binary variables are arranged on a two-dimensional lattice, representing spins or individuals as its vertices. In the case of Ising, each spin has its own temperature T. In the original Schelling model, the temperature parameter T in the Ising model becomes the tolerance T of each individual toward other individuals and has been treated as constant. This again corresponds to the special case when $T = 0$ in the Ising model. Using computer simulations, physicists investigated Ising systems that show dynamic phase transitions due to increasing temperature. By transferring these methods to the social sciences, physicists aimed to explore potential threshold effects in increasing or decreasing the tolerance of individuals in the context of segregation. Depending on the orientation/social identity of the neighboring spins/individuals, variables (vertices) in the network will flip to align or, in the case of the individuals of Schelling, their tolerance toward neighbors will increase or decrease.

In addition to this shared behavior, there is a general tendency to lose sensitivity over time: decreasing magnetization in the case of Ising, or attenuating tolerance in the case of Schelling, which can lead to spontaneous segregation (the formation of ghettos), a phenomenon which is sensitive to the dynamic rate of forgetting. From this behavior of the model, one can draw the inference that "life-long reinforcing of tolerance is needed to avoid ghetto formation" (Stauffer 2008). This example highlights that dynamical properties of the system (rates of forgetting) are crucial for the explanation of the observed phenomenon. Unfortunately, evidence for such dynamic mechanisms is even harder to come by in the social sciences than in biology, a situation which would be important to remedy, in our view.

This example shows that including a dynamic perspective qualitatively extends the range of possible model behaviors and, at the same time, in an interesting way entangles the two models. The starting point for this entanglement is the abstract structural similarity between the two model templates, which becomes extended by including the dynamical element, related computational and mathematical methods, and concepts affiliated with the Ising model, to the context of the Schelling model (e.g., Knuuttila and Loettgers 2023).

8.7   Possibilities and Limitations in Static and Dynamic Network Models

The examples of static and dynamic network models we have introduced in the previous sections illustrate a number of interesting characteristics of the practice of modeling. First, it should be evident that formalized possibility spaces do not exist prior to their corresponding models. Instead, abstract spaces of possibilities are *generated* in the process of identifying a natural system with its relevant features and then encoding these features into a formalized model. And the generation of a possibility space can also proceed the other way around when scientists apply a model template to a new phenomenon. The dimensions of the resulting possibility space are determined not only by some features of the natural system represented in the model but also crucially by the properties and practical aspects of the model templates involved. The latter include not only particular mathematical formalisms (such as the Hopfield recurrent neural networks discussed above), but also analysis tools (like the study of degree distributions in static networks, or linear analysis at steady states for dynamical models), plus the conceptual frameworks required for embedding the model in its specific context and interpreting it in an appropriate manner (Knuuttila and Loettgers 2023). An example of the latter is the conceptual analogies employed in examining the consequences of the Ising vs. the Schelling model.

In this view, possibility spaces generated by the application of a model are central for assessing a model's capability to address successfully some aspects of its target phenomena. Unfortunately, this boundedness of the possibility space to the methods and aims of the model in question tends to be ignored by practitioners, especially if the model leads to some empirical insight, as is the case for the study of the robustness of networks in the instance of random perturbation. These models can point toward a number of illustrative cases where they *do* have some explanatory power, especially compared to the situation of having *no* model. It is easy to forget to ask whether they are missing unconsidered cases of robustness that may be equally or more important but that are harder to access in terms of the empirical data and the modeling methodologies required. This question can however be evaluated by considering the constraints on the possibility spaces generated by a model. Are they well-motivated by the scientific problem at hand? Or are they imposed, in a historically contingent manner, by the template transfers underlying the application of a model in an unfamiliar context? How can the use of the model template be justified in this particular application?

Let us illustrate such an analysis with two examples. One from the domain of static network models, the other one revealing some limitations of dynamical systems modeling in turn. As we have pointed out, there are often time-variant aspects of a system that are crucial for explaining its behavior. In the context of static network models, we have seen that robustness to perturbation can originate from structural properties—specific degree distributions that prevent perturbations from percolating, but also other features like the presence of redundant network paths or feedback regulation. In contrast, robustness through structural stability, that is, the behavior of a dynamic system governed by attractors that are able to buffer and course-correct following perturbations, only occurs in models that include time-dependent variables.

The possibility spaces of static models simply preclude structural stability as a possible mechanism for robustness. Therefore, such models run the danger of misinterpreting the phenomenon of robustness due to their limited range of possible mechanisms. For example, modelers may infer that a system known to be robust must have a modular network architecture when its robustness is instead based on structural stability or compensatory negative feedback. As we have mentioned, it is risky to draw any mechanistic conclusions from highly abstract structural properties of networks, since there is no simple or straightforward connection between such properties and the observable behavior of the system.

Expanding the space of possibilities by including dynamics is a first step in the right direction to remedy such overextension of limited models. However, it is important to note that current dynamical models suffer from their own set of limitations. The first, already mentioned above, is that our methods for constructing and analyzing dynamical systems often require the imposition of assumptions guided by mathematical tractability rather than biological realism. For example, when modeling nonlinear processes acting on multiple time scales (i.e., most biological processes), one can deploy idealizations that allow treating certain processes as instantaneous or constant due to separation of time scales. According to the *rapid equilibrium assumption*, a faster process reaches stable equilibrium effectively instantaneously relative to a slower process, such that the state variables describing the fast process are no longer dependent on time. Similarly, the *quasi-steady state assumption* holds that a process instantaneously reaches the steady state it would reach if the other variables were constant (Ingalls 2013).

Both approaches can improve the analytic tractability of a dynamical model by providing a "reduced model" in which differential equations are replaced by algebraic ones and state variables by algebraic expressions. In effect, these approaches idealize away some of the time dependence of natural biological processes. Rapid equilibrium and steady-state assumptions are essential to our foundational biochemical models of gene regulation, such as Michaelis-Menten kinetics (Gunawardena 2014; Michaelis and Menten 1913) and binding cooperativity (Hill 1910). Although there are some methods for estimating the error such approaches introduce (Fenichel 1979; Segel and Slemrod 1989), they often leave the modeler in a state of uncertainty as to their biological accuracy.

Stability analysis of dynamical systems is also based on steady-state assumptions. Steady states can be evaluated as stable or unstable based on linear approximations of the equations describing the system's behavior near the steady states. The idea that a system converges to a steady-state profile is a fairly safe assumption for systems like closed chemical reaction networks, but not for the open systems more often of interest for development and evolution. When studying biological or social systems, we must carefully examine, case by case, whether the fact that regulatory processes often remain not only far from thermodynamic equilibrium, but also far from any kind of metastable steady state has an important effect on our analysis and conclusions.

Another serious limitation of dynamical models is the following: even though they allow the state of the system to transition over time, they rarely involve temporal (especially state-dependent) changes in parameter values or in the very rules encoded in the model (e.g., the functional form of the equations that constitute a dynamical systems model). Yet, such change is ultimately unavoidable in biological systems. We have touched upon this in the section about ecological models, where trophic exchanges may depend on the developmental stage of an individual, or on the maturity of an ecological succession. Similarly, the rules governing developmental or cellular regulatory processes only remain constant over limited time windows, and phenomena such as Hebbian learning can lead to a substantial rewiring of neural network processes over extended periods of time. When our dynamical models incorporate such longer-term structural changes, the latter often are imposed as boundary conditions on the model from outside. What we need for the study of many biological processes, however, are models where such change can happen based on model-intrinsic factors. At present, this important aspect of living systems is simply outside the possibility spaces of most of our models, whether explicitly dynamical or not.

8.8 Concluding Remarks

The success and spread of network models in the landscape of contemporary science can be attributed (a) to their capacity to provide model templates that enable transdisciplinary application and (b) the computational and mathematical tractability of the models. The given structure of model templates shapes and constrains, in a historically contingent way, the space of possibilities captured by the models we use in a wide range of scientific contexts.

There is an obvious and strong cognitive appeal to the idea that phenomena as diverse as ferromagnetism, metabolism, ecological stability, social segregation, and many more, may share structural principles that can be captured by the same model template. In this chapter, we have examined the modeling capacities of network formalisms in the face of real-world complex dynamical phenomena. A wide variety of examples converge on the conclusion that network models show significant limitations in their capacity to describe, explain, and predict phenomena of interest in the study of complex systems. When network models fall short in this way, it is generally due to the problems of diachronicity—that is, the time-dependence of system behaviors and properties—and correspondence between network structure and dynamics (as elaborated in detail in DiFrisco and Jaeger 2019). With network models, the possibility spaces of states of the real-world system are of much higher dimension and complexity than the corresponding possibility space of the model, because the behavior and structure of natural and social systems are intrinsically time-dependent, while the models are not. To phrase this central conclusion more positively, in identifying the sources of the breakdown of network models, we can begin to reach a better understanding of their proper domain of applicability. This is important for guiding researchers to use network models for situations that match their strengths, rather than perpetuating the illusion of understanding through an uncritical process of interdisciplinary template transfer.

Acknowledgments

This chapter has been supported by funding from the John Templeton Foundation project "Pushing the Boundaries" (grant ID: 62581) as well as the European Research Council (ERC) under the European Union's Horizon 2020 research and innovation program (grant agreement no. 818772).

J.D. gratefully acknowledges support from the Francis Crick Institute, which receives its core funding from Cancer Research UK (CC2240), the UK Medical Research Council (CC2240), and the Wellcome Trust (CC2240).

Notes

1 In this chapter, the phrase "space of possibilities" is used to refer to this notion in a general sense. The phrase "possibility space" is used when pertaining to a model.
2 An autotroph is an organism that can produce its own food using light, water, carbon dioxide, or other chemicals. Heterotrophs cannot produce their own food, occupying higher levels in a food chain.

References

Alon, Uri. 2007. *An Introduction to Systems Biology: Design Principles of Biological Circuits*. Boca Raton, FL: Chapman & Hall/CRC Press.

Barabási, Albert-László, and Réka Albert. 1999. "Emergence of Scaling in Random Networks." *Science* 286(5439): 509–12. https://doi.org/10.1126/science.286.5439.509.

Barabási, Albert-László, and Zoltán N. Oltvai. 2004. "Network Biology: Understanding the Cell's Functional Organization." *Nature Reviews Genetics* 5(2): 101–13. https://doi.org/10.1038/nrg1272.

Borrett, Stuart R., James Moody, and Achim Edelmann. 2014. "The Rise of Network Ecology: Maps of the Topic Diversity and Scientific Collaboration." *Ecological Modelling*, Systems Ecology: A Network Perspective and Retrospective 293(December): 111–27. https://doi.org/10.1016/j.ecolmodel.2014.02.019.

Broido, Anna D., and Aaron Clauset. 2019. "Scale-Free Networks Are Rare." *Nature Communications* 10(1): 1–10. https://doi.org/10.1038/s41467-019-08746-5.

Caldarelli, Guido. 2007. *Scale-Free Networks: Complex Webs in Nature and Technology*. Oxford: Oxford University Press.

Callaway, Duncan S., M. E. J. Newman, Steven H. Strogatz, and Duncan J. Watts. 2000. "Network Robustness and Fragility: Percolation on Random Graphs." *Physical Review Letters* 85(25): 5468–71. https://doi.org/10.1103/PhysRevLett.85.5468.

Clauset, Aaron, Cosma Rohilla Shalizi, and M. E. J. Newman. 2009. "Power-Law Distributions in Empirical Data." *SIAM Review* 51(4): 661–703. https://doi.org/10.1137/070710111.

Corominas-Murtra, Bernat, Rudolf Hanel, and Stefan Thurner. 2015. "Understanding Scaling Through History-Dependent Processes with Collapsing Sample Space." *Proceedings of the National Academy of Sciences* 112(17): 5348–53. https://doi.org/10.1073/pnas.1420946112.

Davidson, Eric H., Jonathan P. Rast, Paola Oliveri, Andrew Ransick, Cristina Calestani, Chiou Hwa Yuh, Takuya Minokawa, et al. 2002. "A Genomic Regulatory Network for Development." *Science* 295: 1669–53. https://doi.org/10.1126/science.1069883.

DiFrisco, James. 2017. "Time Scales and Levels of Organization." *Erkenntnis* 82(4): 795–818. https://doi.org/10.1007/s10670-016-9844-4.

DiFrisco, James, and Johannes Jaeger. 2019. "Beyond Networks: Mechanism and Process in Evo-Devo." *Biology & Philosophy* 34(6): 54. https://doi.org/10.1007/s10539-019-9716-9.

Fenichel, Neil. 1979. "Geometric Singular Perturbation Theory for Ordinary Differential Equations." *Journal of Differential Equations* 31(1): 53–98. https://doi.org/10.1016/0022-0396(79)90152-9.

Gunawardena, Jeremy. 2014. "Time-Scale Separation – Michaelis and Menten's Old Idea, Still Bearing Fruit." *FEBS Journal* 281(2): 473–88. https://doi.org/10.1111/febs.12532.

Hastings, Alan. 2004. "Transients: The Key to Long-Term Ecological Understanding?" *Trends in Ecology & Evolution* 19(1): 39–45. https://doi.org/10.1016/j.tree.2003.09.007.

Hastings, Alan. 2010. "Timescales, Dynamics, and Ecological Understanding." *Ecology* 91(12): 3471–80. https://doi.org/10.1890/10-0776.1.

Hill, A. V. 1910. "A New Mathematical Treatment of Changes of Ionic Concentration in Muscle and Nerve under the Action of Electric Currents, with a Theory as to Their Mode of Excitation." *The Journal of Physiology* 40(3): 190–224. https://doi.org/10.1113/jphysiol.1910.sp001366.

Hopfield, J. J. 1982. "Neural Networks and Physical Systems with Emergent Collective Computational Abilities." *Proceedings of the National Academy of Sciences* 79(8): 2554–58. https://doi.org/10.1073/pnas.79.8.2554.

Hopfield, J. J. 1984. "Neurons with Graded Response Have Collective Computational Properties Like Those of Two-State Neurons." *Proceedings of the National Academy of Sciences* 81(10): 3088–92. https://doi.org/10.1073/pnas.81.10.3088.

Ingalls, Brain. 2013. *Mathematical Modeling in Systems Biology: An Introduction.* Cambridge, Massachusetts: The MIT Press.

Ising, Ernst. 1925. "Beitrag zur Theorie des Ferromagnetismus." *Zeitschrift für Physik* 31(1): 253–58. https://doi.org/10.1007/BF02980577.

Jaeger, Johannes, and Anton Crombach. 2012. "Life's Attractors." In *Advances in Experimental Medicine and Biology*, edited by Orkun S. Soyer, 93–119. New York, NY: Springer.

Jaeger, Johannes, Svetlana Surkova, Maxim Blagov, Hilde Janssens, David Kosman, Konstantin N. Kozlov, Manu, et al. 2004. "Dynamic Control of Positional Information in the Early Drosophila Embryo." *Nature* 430 (6997): 368–71. https://doi.org/10.1038/nature02678.

Kauffman, Stuart A. 1993. *The Origins of Order: Self-Organization and Selection in Evolution.* Oxford: Oxford University Press.

Knuuttila, Tarja, and Andrea Loettgers. 2014. "Magnets, Spins, and Neurons: The Dissemination of Model Templates Across Disciplines." *Monist* 97(3): 280–300. https://doi.org/10.5840/monist201497319.

Knuuttila, Tarja, and Andrea Loettgers. 2016. "Model Templates Within and between Disciplines: From Magnets to Gases – and Socio-Economic Systems." *European Journal for Philosophy of Science* 6(3): 377–400. https://doi.org/10.1007/s13194-016-0145-1.

Knuuttila, Tarja, and Andrea Loettgers. 2023. "Model Templates: Transdisciplinary Application and Entanglement." *Synthese* 201(6): 200. https://doi.org/10.1007/s11229-023-04178-3.

Levine, Michael, and Eric H. Davidson. 2005. "Gene Regulatory Networks for Development." *Proceedings of the National Academy of Sciences* 102(14): 4936–42. https://doi.org/10.1073/pnas.0408031102.

Michaelis, L., and Maud L. Menten. 1913. "Die Kinetik Der Invertinwirkung." *Biochemistry* 49(2): 333–69.

Mjolsness, Eric, David H. Sharp, and John Reinitz. 1991. "A Connectionist Model of Development." *Journal of Theoretical Biology* 152(4): 429–53. https://doi.org/10.1016/S0022-5193(05)80391-1.

Montoya, José M., Stuart L. Pimm, and Ricard V. Solé. 2006. "Ecological Networks and Their Fragility." *Nature* 442(7100): 259–64. https://doi.org/10.1038/nature04927.

Newman, M. E. J. 2003. "The Structure and Function of Complex Networks." *SIAM Review* 45(2): 167–256. https://doi.org/10.1137/S003614450342480.

Perkins, Theodore J., Eric Foxall, Leon Glass, and Roderick Edwards. 2014. "A Scaling Law for Random Walks on Networks." *Nature Communications* 5(1): 5121. https://doi.org/10.1038/ncomms6121.

Schelling, Thomas C. 1971. "Dynamic Models of Segregation." *The Journal of Mathematical Sociology* 1(2): 143–86. https://doi.org/10.1080/0022250X.1971.9989794.

Schelling, Thomas C. 1978. *Micromotives and Macrobehavior.* New York, NY: Norton.

Schelling, Thomas C. 2006. *Micromotives and Macrobehavior.* Revised ed. New York, NY: W. W. Norton & Company.

Segel, Lee A., and Marshall Slemrod. 1989. "The Quasi-Steady-State Assumption. A Case Study in Perturbation." *SIAM Review* 31(3): 446–77. https://doi.org/10.1137/1031091.

Sherrington, David, and Scott Kirkpatrick. 1975. "Solvable Model of a Spin-Glass." *Physical Review Letters* 35(26): 1792–96. https://doi.org/10.1103/PhysRevLett.35.1792.

Simon, Herbert A. 1962. "The Architecture of Complexity." *Proceedings of the American Philosophical Society* 106(6): 467–82. https://doi.org/10.1007/978-3-642-27922-5_23.

Sornette, Didier. 2014. "Physics and Financial Economics (1776–2014): Puzzles, Ising and Agent-Based Models." *Reports on Progress in Physics* 77(6): 062001. https://doi.org/10.1088/0034-4885/77/6/062001.

Stauffer, D. 2008. "Social Applications of Two-Dimensional Ising Models." *American Journal of Physics* 76: 470–73. https://doi.org/10.1119/1.2779882.

Tilman, David, Robert M. May, Clarence L. Lehman, and Martin A. Nowak. 1994. "Habitat Destruction and the Extinction Debt." *Nature* 371(6492): 65–66. https://doi.org/10.1038/371065a0.

Verd, Berta, Erik Clark, Karl R Wotton, Hilde Janssens, Eva Jiménez-Guri, Anton Crombach, and Johannes Jaeger. 2018. "A Damped Oscillator Imposes Temporal Order on Posterior Gap Gene Expression in Drosophila." *PLOS Biology* 16(2): e2003174. https://doi.org/10.1371/journal.pbio.2003174.

Watts, Duncan J. 1999. *Small Worlds: The Dynamics of Networks between Order and Randomness.* Princeton, NJ: Princeton University Press. https://doi.org/10.1515/9780691188331.

Watts, Duncan J., and Steven H. Strogatz. 1998. "Collective Dynamics of 'Small-World' Networks." *Nature* 393(6684): 440–42. https://doi.org/10.1038/30918.

Whitehead, Alfred North. 1925. *Science and the Modern World.* New York, NY: The Free Press.

Wimsatt, William C. 1972. "Complexity and Organization." *PSA: Proceedings of the Biennial Meeting of the Philosophy of Science Association* 1972(January): 67–86. https://doi.org/10.1086/psaprocbienmeetp.1972.3698961.

Part III

Exploring How-Possibly in Practical Contexts

9 Three Strategies for Salvaging Explanatory Value in Deep Neural Network Modeling

Philippe Verreault-Julien

9.1 Introduction

What explanatory role that highly idealized models may play has recently received much attention. One prevalent proposal that has emerged is that these models provide *how-possibly explanations* (HPEs) (e.g., Bokulich 2014; Forber 2010; Grüne-Yanoff 2013; Reutlinger, Hangleiter, and Hartmann 2018; Rohwer and Rice 2013; Verreault-Julien 2019; Ylikoski and Aydinonat 2014). One alleged type of explanatorily valuable HPEs are those considered to be *epistemically possible* (Brainard 2020; Grüne-Yanoff 2013; Grüne-Yanoff and Verreault-Julien 2021; Sjölin Wirling and Grüne-Yanoff 2021; Verreault-Julien 2019). In a nutshell, epistemically possible HPEs are explanations that are possible in the sense that the evidence does not rule them out. Although potentially fruitful, this approach to explanatory value also has a puzzling implication—the less we know, the more HPEs are epistemically possible for us.

In this chapter, I examine a particular class of puzzling epistemically possible HPEs that I call *epistemically opaque* HPEs (EO-HPEs). EO-HPEs are HPEs obtained by an epistemically opaque process such as computational simulation or deep neural network (DNN) models. In short, a process is epistemically opaque when an agent lacks knowledge or understanding of why the process yields the results that it does (e.g., Beisbart 2021; Durán and Formanek 2018; Humphreys 2009). The notion of EO-HPEs aims to capture a particular reason why we lack justification for the HPE, viz. the very process used to establish it. Contrary to HPEs acquired via a transparent process (e.g., an analytical model), EO-HPEs seem to raise different justificatory and validation challenges. The problem EO-HPEs raise is the following: how could EO-HPEs have explanatory value if they result from a process about which we lack knowledge or understanding?

I argue that, in practice, the process's opacity is not always an obstacle to EO-HPEs' explanatory value. More specifically, I present three ways EO-HPEs may have explanatory value despite their opacity. First, some EO-HPEs result from a process that is functionally transparent: we have partial, relevant, knowledge of how the process works. Second, some EO-HPEs

DOI: 10.4324/9781003342816-13

are only opaque according to some interpretations of the modal operator. Again, this implies having some knowledge of the process's capacities. Third, some EO-HPEs are pursuit-worthy even if they result from an opaque process; they may be promising despite a lack of justification. I illustrate using cases from DNN models.

This chapter makes two chief contributions. First, it elaborates on a recent and promising account of the value of HPEs as epistemically possible explanations. In particular, it identifies an obstacle to that account, namely the epistemic opacity of some modeling processes. Second, it presents different ways to salvage the explanatory value of these models despite their opacity.

9.2 The Explanatory Value of HPEs

Some models provide (defeasible and inconclusive) evidence for explanations. Highly idealized models often appear to, or actually fall short, of faithfully representing the world. As a result, whether the evidence they provide supports how-actually explanations (HAEs) and, if not, what explanatory value those models may have is contentious. One proposal that has gained a lot of ground recently is that these models provide evidence for HPEs (e.g., Bokulich 2014; Forber 2010; Grüne-Yanoff 2013; Reutlinger et al. 2018; Rohwer and Rice 2013; Verreault-Julien 2019; Ylikoski and Aydinonat 2014). There are various accounts of HPEs (e.g., Bokulich 2014; Brainard 2020; Brandon 1990; Dray 1968; Forber 2010; Hempel 1965; Verreault-Julien 2019), but arguably all of them emphasize that HPEs have modal features that differentiate them from HAEs. In Verreault-Julien (2019), I propose the following account. First, explanations are sets of propositions (see also Strevens 2013). An explanation contains two subsets of propositions: the explanans, the propositions that do the explaining, and the explanandum, the propositions that describe what is explained. Then, explanations must satisfy internal and external conditions of adequacy. The former refers to the form or structure of the explanation, and the latter to the ontological match. For instance, a deductive-nomological (DN) explanation (Hempel 1965) must have the form of a deductive argument (internal conditions) and must have a true explanans and explanandum (external conditions).

According to that account, HPEs have the general form "$\Diamond(p$ because $q)$" where p is the explanandum, q the explanans (e.g., generalization plus initial and auxiliary conditions), and $\Diamond$ denotes a modal operator meaning "it is possible that" according to a given interpretation of the operator. HAEs are simply propositions of the form "p because q." The key difference lies in the introduction of a modal operator $\Diamond$ in front of the explanation. In a nutshell, whereas HAEs are *actual* explanations, HPEs are *possible* explanations. The modal operator takes scope over the whole "p because q" to reflect that either the explanans, the explanandum, or the explanatory relation can be possible. For instance, sometimes scientists use actual causes and initial conditions to derive a possible explanandum, other times they start with an actual explanandum and try to generate it with possible initial conditions, etc.[1]

A crucial feature of that characterization of HPEs is that they have a truth value: "It is possible that (p because q)" can be true or false. This is an important contrast with accounts that view HPEs as not satisfying any external conditions (e.g., Hempel 1965). Another key feature is that the modal operator can be interpreted in different ways. For instance, an explanation can be logically, mathematically, nomologically, causally, etc., possible. To assess whether an HPE is true or false, we thus need to know the interpretation of the modal operator.

One important interpretation of the operator is in terms of *epistemic possibility* (see Brainard 2020; Grüne-Yanoff 2013; Grüne-Yanoff and Verreault-Julien 2021; Sjölin Wirling and Grüne-Yanoff 2021; Verreault-Julien 2019). Epistemic possibility tells us ways things might be relative to a given body of evidence. In a nutshell, an HPE is epistemically possible if our evidence does not rule it out.[2] Epistemically possible HPEs are often those scientists submit when considering the set of possible explanations for a phenomenon. For instance, there might be a multitude of epistemically possible explanations for "Why p?"; p might be because q, r, or another explanans. These possible explanations can be incompatible with one another, but since our evidence cannot rule them out, they are all epistemically possible. To eliminate some epistemically possible explanations, we need additional evidence to update our knowledge and rule out some of the possible explanations.[3]

To give a more concrete example, consider the phenomenon of people developing unusual blood clots following an injection of the COVID-19 Vaxzevria (AstraZeneca) vaccine in winter and spring 2021. Initially, the scientific community considered it unlikely that the vaccine might be responsible for these clotting events. However, they could not rule it out:

> But the finding leaves researchers wrestling with a medical mystery: why would a vaccine trigger such an unusual condition? "Of course, there are hypotheses: maybe it's something with the vector, maybe it's an additive in the vaccine, maybe it's something in the production process … I don't know," says Sabine Eichinger, a haematologist at the Medical University of Vienna. "It could be any of these things."
>
> (Ledford 2021, 334)

Further data collection and analysis supported the hypothesis that the vaccine was the cause (e.g., Whiteley et al. 2022). Modeling and experimental evidence, in turn, identified the adenovirus vector as the likely suspect (Baker et al. 2021), thus contributing to ruling out other causes, such as the production process or the additives.

In winter and spring 2021, explaining the clotting by citing the vaccine was an epistemically possible HPE. One reason why epistemically possible HPEs are valuable is that knowing them is necessary for considering and comparing an appropriate range of possible explanations which, in turn, is essential in determining which one(s) might be HAEs (e.g., Khalifa 2017; Ylikoski and Aydinonat 2014).[4] According to Khalifa, for instance, sound explanatory evaluation involves considering a set of plausible HPEs and comparing how the evidence or other considerations favor some over others. Not considering an appropriate range of HPEs might entail an incomplete comparison. We might fail to consider the HPE that turns out to be the HAE or be overly confident in one HPE's plausibility.

At first glance, since epistemically possible HPEs are all consistent with our evidence, it does not seem to matter how we acquire them. However, not all epistemic possibilities seem to be equally valuable (e.g., Willer 2013). In particular, we seem to be more justified in believing some over others. In the next section, I want to draw attention to the fact that the process of justifying HPEs is sometimes *epistemically opaque*.

9.3 Epistemically Opaque HPEs

Epistemic opacity refers to the general idea that we do not always know or understand all the epistemically relevant features of a process (e.g., Beisbart 2021; Creel 2020; Durán and Formanek 2018; Humphreys 2009). What are these epistemically relevant elements? Various proposals have been made. Creel (2020), which I discuss in more detail in section 4.1, distinguishes between functional, structural, and run transparency, where transparency is understood as the opposite of opacity. Durán and Formanek (2018) identify the relevant elements with the justificatory steps; a process is opaque for an agent if they do not have access and cannot survey all the steps. Zednik (2021) argues that what these relevant elements are depends on the interests of particular stakeholders. Beisbart (2021, 11644) considers that the application of a method is opaque if it has a "disposition to resist epistemic access."

Most accounts of epistemic opacity make the following distinction: epistemic opacity can be agent-relative or process-relative. Agent-relative opacity depends on the cognitive capacities of agents or epistemic communities. An otherwise transparent model may be epistemically opaque for an agent who lacks the required skills or knowledge to grasp the epistemically relevant features. Process-relative opacity, also sometimes called "essential opacity" (Alvarado 2021; Humphreys 2009), depends on the nature of the process itself. To make an analogy, one can fail to see through a window because of one's myopia (agent-relative) or because a film is applied to it (process-relative), making it opaque. In the rest of this chapter, I will only be concerned with process-relative opacity.

Two typical instances of epistemically opaque processes are computational simulations and DNN models. They are often considered to have features that make them opaque irrespective of the agents involved and, as a result, to be "black boxes." Yet, they are sometimes used for explanatory purposes, in particular, to provide evidence for HPEs. However, their epistemic opacity is prima facie problematic because of how it may affect the justification for the evidence they provide. In short, the problem is the following: How do we know whether epistemically opaque processes provide *good* evidence for HPEs? The evidence could support the HPEs, or it could not. Because of opacity, we do not know or understand all the epistemically relevant features of the processes that could help us assess the evidence.

To illustrate the vaccine case mentioned above, suppose there are two hematologists, with a similar starting evidential base. One uses a statistical model based on available data which indicates that the vaccine may explain the blood clots. The other interrogates a large language model (LLM) that says the vaccine may be responsible for the phenomenon. Both models, statistical and LLM, provide evidence for the HPE that the vaccine caused the clotting. Crucially, the evidential base of both hematologists does not rule out the vaccine explanation. But are both hematologists equally justified? Can they both equally claim to *know* the epistemic possibility? What if the evidence rules out different possibilities, which ones should we take seriously?

From an internalist perspective, opacity is an issue because features that would allow us to assess the justification provided by the evidence are not accessible to the agents. If we do not know how DNNs work and how they transform inputs into outputs, we may lack sufficient grounds for being justified in said outputs. But even from an externalist perspective, opacity is still an issue for two reasons. First, prominent forms of externalism such as reliabilism often require having evidence that supports the beliefs (e.g., Comesaña 2010; Goldman 2011). In the context of machine learning, Durán (2023) argues that "reliability indicators" may justify our beliefs in the outputs of DNNs. However, how to assess the reliability of DNNs is contentious (e.g., Buijsman 2023; Duede 2022; Freiesleben and Grote 2023). The second, and more important reason, is that even if we are justified to believe the DNNs *outputs*, that justification does not directly extend to the *explanations* we form on the basis of the DNNs and their outputs. For example, consider the now classic case of a deep learning classifier for wolf images (Ribeiro, Singh, and Guestrin 2016). Given an image as an input, the model outputs whether it contains a wolf or a husky. The model is highly accurate in the test data and is thus, in that sense, reliable. However, when probing the model to understand on which basis it makes its outputs, we learn that the model picks out wolves from the presence of snow in the background. While the outputs may be reliable, explanations formed on the basis of a spurious correlation would have limited explanatory value.

I call HPEs justified by an epistemically opaque process EO-HPEs.[5] The problem of opacity suggests that EO-HPEs would have limited explanatory value. In the next sections, I argue that despite the opacity of the process, we may have reasons to attribute value to the resulting EO-HPEs.

9.4   Salvaging Explanatory Value

I examine three cases from DNN models and propose three different strategies for salvaging explanatory value from the resulting EO-HPEs. They consist of salvaging value from (1) functional transparency, (2) modal operator interpretation, and (3) pursuit-worthiness. These strategies are not exhaustive. For instance, perhaps a reliabilist strategy might be applicable in some cases. They are also not mutually exclusive in two senses. First, two strategies may be available for the same EO-HPE, for example, we may want to consider an EO-HPE functionally transparent *and* pursuit-worthy. Second, the strategies may not always be logically independent. For instance, one might consider functional transparency necessary for selecting which modal operator should apply.[6] Nonetheless, I believe these strategies draw attention to different ways opacity might, or might not, interfere with assessing the epistemic value of EO-HPEs.

9.4.1   *Functional Transparency*

There are many scientific questions related to animal coloration (Cuthill et al. 2017). What are the best colors to avoid detection in particular environments? Do the colors depend on the observer's visual system? Why, for instance, is the tiger's fur orange? From an evolutionary perspective, we might want to know why species have evolved the color processing they have or why they have the colors that they do.

To make progress on these questions, Fennell et al. (2019) put a DNN to work to help identify which colors optimize or minimize detectability. This is crucial to understand the fitness effects of some phenotypes which, in turn, may explain why they were selected. In theory, it is possible to test empirically every single possible color on human subjects. In practice, it is impractical because color spaces are very large. For instance, testing the whole RGB gamut of 16,777,216 different colors would be a costly and time-consuming endeavor. Fennell et al. thus proposed using a neural network to predict detection time on empirically untested colors. First, the researchers collected training data by carrying out an experiment with human subjects. They observed how much time it took humans to detect a randomly colored target in two simulated environments: a temperate forest and a semi-arid desert. They also processed images in order to simulate detection time for dichromats, that is, species that perceive color via only two channels. Humans are trichromats and perceive colors through three channels, but most non-human mammals are dichromats and are effectively red-green color blind; red appears green to them. Then, the researchers trained a DNN to interpolate between experimented inputs and predict detection time. Suppose we have experimental data on magenta and cyan objects, but not on blue ones. The neural network interpolates between magenta and cyan to create the blue color and then estimates a detection time. By doing this for every shade, the researchers obtained predicted detection times for the whole RGB color space. As a result, the DNN allowed the researchers to identify the best and worst colors for detection.

Later in the article, Fennell et al. suggest that the results may help explain why some predators, for example, tigers, are not green despite the optimal concealment it would provide. Consider the following question: "Why is the tiger's fur orange and not green?" The HPE they submit can be formulated as follows: "It is epistemically possible that the tiger's orange fur was selected for because it provides excellent concealment from dichromats." Or, put slightly differently, there is little evolutionary pressure for tigers to evolve a green coat insofar as orange appears green for their prey.

This HPE relies on the hypothesis that the shade of green dichromats are able to see in place of orange is actually hard for them to detect. Although seemingly obvious, that dichromacy enhances detection ability is also a serious hypothesis (e.g., Melin et al. 2007). More importantly, it relies on the prediction that shades close to the "dark olive" optimum are actually difficult to detect for dichromats in a temperate forest environment. However, if we don't understand why the model made the predictions that it did, how can we be sure it identified actual optima and minima? Identifying actual optima and minimal does not imply that we would have an HAE because other factors may be responsible for coloration. But it makes the HPE a more serious candidate.

Here, it is useful to differentiate ways a process can be opaque. Creel (2020) distinguishes between functional, structural, and run transparency. *Functional* transparency consists of knowing the functioning of the algorithm. By "functioning," Creel means knowing what algorithm a system instantiates and having high-level knowledge of how it transforms inputs into outputs. Knowing how the computational system instantiates the algorithm, in particular, how the code produces the algorithm, is *structural* transparency. *Run* transparency consists of knowing how a computational system was run in a particular instance, including the hardware implementation and how the program interacts with data. According to Creel, these different types of transparency are logically independent. One may not know how a program was run on a particular occasion (run transparency), yet one may know the algorithm's functioning (functional transparency).[7]

This taxonomy suggests one first line of defense for the value of EO-HPEs. We may say that an EO-HPE results from a process that lacks structural or run transparency, but that is functionally transparent. In the context of explanation, functional transparency is important since it allows us to identify the difference-makers a model captures (Räz and Beisbart 2022). If an algorithm is functionally transparent, whether the model provides a valid representation of the target may be the main remaining problem to solve (Sullivan 2022).

One important aim of the field of explainable artificial intelligence (XAI) is to increase transparency in one or the other of these senses. XAI methods (e.g., Lundberg and Lee 2017; Mordvintsev, Olah, and Tyka 2015; Ribeiro et al. 2016) can increase functional transparency by telling us why the algorithm made the decision it did on a particular or multiple inputs.[8] In turn, XAI methods can help uncover HPEs (Zednik and Boelsen 2022). Whether some methods will provide the required functional knowledge of the algorithm ultimately depends on the context (Zednik 2021). Some systems may be more difficult to interpret than others. In other cases, the amount of information we need might be minimal.

Do we have functional transparency in the case of the DNN for color detection? The algorithm, is to some extent, functionally transparent for two reasons. First, as Fennell et al. note, the problem the DNN needs to solve in that context is relatively low dimensional as only the color of the spheres changes between images of a particular environment. High-dimensional data makes systems less transparent (Domingos 2012), but this is not the case here.[9] Second, Fennell et al. did carry out a limited validation experiment in which they tested detection times of 25 "easy," "intermediate," and "hard" colors with human subjects. They found the predictions consistent with the experimental results for both the dichromat and trichromat conditions. The validation experiment plays a role akin to that of explainability techniques, viz. it helped make transparent that the DNN did pick out actual features that increase or decrease detection time.[10]

Despite the process's opacity in some respects, we do have knowledge of some of its epistemically relevant parts, viz. the functioning of the algorithm. In turn, this sort of transparency improves the justification we have in the process and indicates that the EO-HPEs we obtain are not a mere product of ignorance. This strategy might not be available for all DNNs. Despite our best XAI efforts, the model might remain functionally opaque. In this case, it may be better to justify the value of the EO-HPE differently, for instance by using the other lines of defense I propose below.

9.4.2 *Modal Operator Interpretation*

How the brain works remains, for all practical purposes, a mystery. It has been suggested that artificial neural networks (ANNs), especially deep convolutional neural networks (DCNNs), may provide candidate explanations of how the brain computes inputs into outputs (see, e.g., Hassabis et al. 2017; Kriegeskorte 2015; Richards et al. 2019; Yamins and DiCarlo 2016). DCNNs seem to replicate, among others, how the brain processes visual sensory inputs using a hierarchy of representations that lead to object recognition. In particular, they are relatively good at predicting neurological data. Empirical results tend to show that artificial computer vision systems with an architecture that resembles that of biological organisms outperform those that don't. According to Kriegeskorte (2015, 431), "[t]his observation affirms the intuition that computer vision can learn from biological vision. Conversely, biological vision science can look to engineering for candidate computational theories." In short, the idea is that if models based on the architecture of the brain perform as well as or better than biological systems, then these same models may explain how the biological systems work. However, the opacity of DCNNs is an obstacle to their explanatoriness. Indeed, since we do not understand how exactly the models build the representations and transform them, how could they provide a causal explanation of brain sensory information processing? Insofar as we do not understand all the epistemically relevant features of these DCNNs, they provide EO-HPEs of neurological phenomena.

EO-HPEs are justified by an epistemically opaque process. As the previous section showed, having an EO-HPE does not imply that we are ignorant of *all* the relevant epistemic aspects. Here, I would like to apply a similar strategy and show that there are valuable things we know despite being in the presence of an EO-HPE. In section 9.2, we have seen that HPEs have the general form "$\lozenge(p$ because q)." In this formulation, the modal operator $\lozenge$ can receive different interpretations, for example, epistemic or nomological. As a result, we may reach different conclusions of epistemic possibility depending on which interpretation we adopt. For instance, one HPE may be only logically possible whereas another may be nomologically possible. Moreover, possibilities may be nested, as in "it is epistemically possible that p is nomologically possible" (Sjölin Wirling and Grüne-Yanoff 2021). In the case of DNNs, even though we lack justification for the process's results, this lack of justification might only concern some interpretations of the modal operator. In particular, I want to suggest that we may be (objectively) justified in the *mathematical* possibility results which, in turn, may serve to justify the *epistemic, causal*, possibility.

Although we lack a complete and full understanding of the capacities of ANNs (see, e.g., Zhang et al. 2021), we do have some understanding of their mathematical properties. In particular, we know that they are so-called universal function approximators (see, e.g., Cybenko 1989; Goodfellow, Bengio, and Courville 2016, section 6.4.1; Hornik, Stinchcombe, and White 1989; Zhou 2020). Universal approximator theorems are proof that any network with at least one hidden layer and a sufficiently large number of hidden units can approximate any function between inputs and output. This mathematical result is important because "it takes off the table the question of whether any particular function is computable using a neural network. The answer to that question is always "yes." So, the right question to ask is not whether any particular function is computable, but rather what's a *good* way to compute the function" (Nielsen 2015, ch. 4, emphasis in original). Assuming that biological neural activity can be represented by a function, universal approximation theorems prove that DNNs can approximate it. In other words, if a problem can be expressed by a mathematical function, then a DNN can solve it.

What do universal approximation theorems imply for the explanatory value of derived EO-HPEs? Suppose we want to explain observed neurophysiological activity in the brain (such as functional magnetic resonance imaging [fMRI] data from the visual cortex) during object recognition tasks.

Knowing how the brain processes visual information would allow us to explain the phenomena. ANNs hold the promise of learning the target function the brain instantiates. Universal approximator theorems tell us that ANNs can *in principle* learn the function but cannot guarantee that they will learn it:

> The universality theorems assure us of the representational power of neural networks with sufficient numbers of units. However, these theorems do not tell us how to set the weights of the connections, so as to represent a particular function with a feedforward net or a particular dynamical system with a recurrent net.
>
> (Kriegeskorte 2015, 425)

In terms of this chapter, universal approximator theorems may justify the following HPE: "It is mathematically possible that neurophysiological data p because of processing function q." However, knowing that the network represents a mathematically possible target function still leaves several questions unanswered. First, it remains unclear whether a given ANN will be able to learn the target function. Second, it is often uncertain what function the ANN instantiates.[11] Third, and relatedly, if we do not know what function the ANN instantiates, assessing its accuracy as an approximation of the target function is challenging. Therefore, we do not know whether this is actually the function computed by the brain (the target function) or how the brain actually computes the function. But, crucially, we know it is *mathematically possible* for ANNs to represent and approximate this function.

Although scientists are often more interested in HPEs that are causally possible, knowing that a problem has a possible mathematical solution is a valuable, albeit initial, step, in explaining a phenomenon. All that is epistemically causally possible is also logically possible. But not all that is logically possible is also causally possible. Determining the logical possibility of an explanation does some minimal headway into its causal possibility. For instance, in the context of economics, Verreault-Julien (2017) argues that the mathematical proof of the existence of a general equilibrium changed the justification economists had for its causal possibility. So, sometimes our ignorance will concern the causally possible, and not the logically possible, as is with ANNs of neurological computation.

9.4.3 *Pursuit-worthiness*

One important aspect of the "protein folding problem" (e.g., Dill and MacCallum 2012) concerns the ability to predict the three-dimensional shape of proteins—their structure—from their amino acid sequence. AlphaFold, a neural network developed by DeepMind (Jumper et al. 2021), made a breakthrough contribution to solving that problem. It surpassed other models by a wide margin in the 14th Critical Assessment of Protein Structure Prediction (CASP14), a biennial competition pitting different prediction methods against each other. AlphaFold is trained on the Protein Data Bank (PDB), a database of experimentally verified protein structures. The structure of proteins can be determined experimentally using, for instance, X-ray crystallography or cryo-electron microscopy. However, it is a difficult and expensive process. Since we know the amino acid sequences of many more proteins than their structures, it is useful to predict structures from sequences. Theoretically, since the biological function of a protein depends on its structure, this holds the promise of improving our understanding of protein function. Practically, knowing protein structure may, among other things, significantly speed up the development of new drugs.

For all its success at predicting, we have a limited understanding of how or why AlphaFold works so well:

Last, and perhaps the more immediate problem, AlphaFold2 models cannot be explained or externally validated. From our human perspective, it's essentially 'alien' technology that is currently beyond our understanding, so 'asking' why it predicted something in a particular conformation is clearly not feasible.

(Jones and Thornton 2022, 18)

The model architecture is constrained by some scientific knowledge (Jumper et al. 2021), but the associations it establishes between known structures and sequences are opaque to users. It states the confidence it has in its predictions, which provides some information with respect to their potential reliability. Yet, although useful, this information does not make the model transparent in any of Creel's (2020) senses. We are thus in the presence of a model that is relatively good at predicting the structure of proteins, but of which we are ignorant of how it arrives at those predictions.

To illustrate, consider the case of the SARS-CoV-2 virus responsible for the COVID-19 pandemic. The virus's proteins determine how it interacts with other biological systems, like humans. The function of a protein depends on its structure. Thus, knowing the structure can contribute to answering explanation-seeking questions such as "Why does protein p have function f?" "Why are some variants more infective or virulent than others?" or "Why is drug d effective against COVID-19?" Unfortunately, we do not have experimentally validated structural models of all proteins, which limits our capacity to answer such questions (see, e.g., Yan et al. 2022).

In March 2020, DeepMind (2020) released structural models for the SARS-CoV-2 membrane protein, Nsp2, Nsp4, Nsp6, and the Papain-like proteinase, and released updated versions in April and August 2020. These structural models have then served as the basis of possible explanations of phenomena related to the virus. One notable example is due to Sadek, Zaha, and Ahmed (2021), who investigated the higher infectivity of the Omicron variant using AlphaFold without relying on further experimental results. SARS-CoV-2 enters the host via the so-called spike protein. There are experimentally validated structures of the spike protein. However, how the many mutations translated into structural changes was unknown. Sadek et al. used AlphaFold to predict how these mutations would impact the structure. They concluded the following:

> Our study suggests that the higher infectivity of the Omicron variant can be explained in part by on the significant mutations in the RBD and the postfusion enhancement of the FP. Importantly, these results require further validation by X-ray crystallography and/or cryo-EM of the Omicron variant S-protein.
>
> (Sadek et al. 2021, 5)

Another study relied on the model of protein Nsp6, which is involved in the infection process. One way it does it is by interacting with sigma receptors, which themselves are linked to the endoplasmic reticulum stress response. It was suggested that drugs that target the sigma receptors might reduce the reproduction of the virus. Two drugs, haloperidol and dextromethorphan, target the sigma receptors. Researchers remarked that haloperidol seemed to reduce viral production, while dextromethorphan increased it (Gordon et al. 2020). However, there was no explanation for this difference. Pandey et al. (2020) used AlphaFold's Nsp6 structural model to simulate how it interacts with the drugs. They concluded that Nsp6 binds differently with the drugs, which may explain their differential effect on viral reproduction.

Other researchers (Gupta et al. 2021) used AlphaFold's predicted structure for Nsp2 to guide and validate their experimental cryo-electron microscopy data. The result was a complete structure of the Nsp2 protein. Analyzes of that structure suggested various possible explanations involving the interaction between the host and Nsp2 for why some variants of the virus were more virulent.

Although epistemically opaque, AlphaFold is used to generate HPEs. One reason for this, I submit, is because AlphaFold's outputs/models are *pursuit-worthy*. A scientific hypothesis or theory deserving attention is often said to be fruitful or pursuit-worthy. What exactly pursuit-worthiness entails is contentious (e.g., Laudan 1977; Šešelja and Straßer 2014; Shaw 2022; see also Kuhn 1977), and I do not aim to settle this here. For my purposes, it suffices to note that pursuit-worthiness is relevant when non-epistemic considerations matter or when we are otherwise unable to discriminate between possible explanations. Insofar as a set of EO-HPEs can be equally justified and all consistent with our knowledge, we may need to bring in additional reasons. Pursuit-worthiness allows us to demarcate between the possible explanations that should be pursued from those that should not.

Although it is still early to assess the epistemic contribution of AlphaFold and similar models (e.g., Baek et al. 2021; Yang et al. 2020), to say that the scientific community was enthusiastic about the potential uses of the model would be an understatement (e.g., Callaway 2022; Thornton, Laskowski and Borkakoti 2021). Despite AlphaFold's opacity, as the immense interest surrounding it testifies, the scientific community clearly considers its results to be significant. Moreover, scientists would not engage in costly and time-consuming experiments if they believed the predicted models were useless.

One could reply that AlphaFold's models are epistemically valuable because of the system's accuracy. After all, it won in protein structure prediction competitions. However, precisely because of its epistemic opacity, we cannot independently validate predictions except via experimentation. And, indeed, why use AlphaFold if experiments are well understood and have a high degree of validity? In a context of "urgent science" (Shaw 2022), such as a pandemic, an uncertain protein structure may be preferable to no structure at all as carrying out experiments can take a long time. Also, AlphaFold opens new areas of investigation and suggests possibilities researchers had not and could not have contemplated before. And it does so by making, in principle, testable predictions. Of course, this does not mean that *all* its results are equally valuable. AlphaFold seems to fare better in some areas than others, although its predicting ability is also surprising in others. But some of the EO-HPEs we obtain with AlphaFold's assistance manifestly deserve our attention.

9.5   Conclusion

Many HPEs have explanatory value because they are epistemically possible, viz. we cannot rule them out on the basis of our evidence. Epistemically possible HPEs play a central role in scientific progress and reasoning since they are often the precursors to HAEs. When scientists want to explain a phenomenon, they submit a list of explanations consistent with the evidence and then try to rule them out.

Although attractive, this picture of the value of HPEs also has an undesirable feature—more ignorance leads to more epistemically possible HPEs. Surely those HPEs cannot be as valuable if they are at all valuable? In this chapter, I have examined a class of potentially problematic HPEs, namely HPEs that originate from an epistemically opaque process. Looking at different cases of such HPEs stemming from DNN models, I have proposed three different strategies to salvage value in the face of opacity, namely salvaging value from (1) functional transparency, (2) modal operator interpretation, and (3) pursuit-worthiness. All these strategies provide a rationale for attributing value to the HPE even though some ignorance is involved in how we obtain them.

Interestingly, not all strategies have an equal connection to knowledge. Salvaging value via functional transparency or the modal operator interpretation has such a connection; opacity is less of an issue because we do have *some* knowledge of epistemically relevant parts. However, salvaging value via pursuit-worthiness has a more elusive relationship with knowledge; we remain ignorant of epistemically relevant parts. This makes this strategy worthy of future attention.

Acknowledgments

I would like to thank Dunja Šešelja and Emily Sullivan for their thought-provoking discussions and exchanges that led to the development of some of the main arguments of this chapter. I am grateful for the comments I received at the *Issues on the (Im)Possible 2022* conference. Finally, I would like to thank Joe Roussos and Ylwa Sjölin Wirling for their perceptive and constructive comments on a previous version of the manuscript. Their feedback has helped me make my ideas less opaque to myself and, hopefully, to others too.

Notes

1 See Grüne-Yanoff (2013) for various concrete examples and Verreault-Julien (2023) for a discussion of a possible explanatory relation.
2 The semantics of epistemic modals is fraught with difficulties, which are outside the scope of this chapter. See Egan and Weatherson (2011) for an overview of key issues.
3 A proponent of inference to the best explanation may also add that theoretical virtues can provide a basis for elimination (see, e.g., Lipton 2004).
4 See Sjölin Wirling (Forthcoming) for a discussion of how epistemic possibilities may afford understanding in the context of metaphysical inquiry.
5 I borrow the terminology from Šešelja (2022), who uses it in a different way.
6 Räz and Beisbart (2022) argue that understanding the model is necessary for explanatory understanding of phenomena. Here I remain agnostic regarding that claim.
7 In a slightly different context, Sullivan (2022) also argues that "implementation black-boxes" may not prevent understanding the higher-level functioning of an algorithm.
8 To what extent XAI methods can make a process functionally transparent is open to debate (e.g., Babic et al. 2021; Rudin 2019). My goal is not to settle it. Instead, my aim is to point out that this is an available strategy.
9 It should be noted that the researchers believe their approach could also be useful for studying color detection in higher dimensionality spaces (see also Fennell et al. 2021; Talas et al. 2020).
10 Here, I am bracketing the issue of whether testing on human subjects is a good proxy for other, non-human species.
11 Perhaps one way of understanding this is by decomposing Creel's (2020) functional transparency into multiple components. Here, we may want to say that we know how the algorithm works at a very high mathematical level, but not at a lower representational or semantic one.

References

Alvarado, Ramón. 2021. "Explaining Epistemic Opacity." Preprint. http://philsci-archive.pitt.edu/id/eprint/19384.
Babic, Boris, Sara Gerke, Theodoros Evgeniou, and I. Glenn Cohen. 2021. "Beware Explanations from AI in Health Care." *Science* 373(6552): 284–86. https://doi.org/10.1126/science.abg1834.
Baek, Minkyung, Frank DiMaio, Ivan Anishchenko, Justas Dauparas, Sergey Ovchinnikov, Gyu Rie Lee, Jue Wang, et al. 2021. "Accurate Prediction of Protein Structures and Interactions Using a Three-Track Neural Network." *Science* 373 (6557): 871–76. https://doi.org/10.1126/science.abj8754.

Baker, Alexander T., Ryan J. Boyd, Daipayan Sarkar, Alicia Teijeira-Crespo, Chun Kit Chan, Emily Bates, Kasim Waraich, et al. 2021. "ChAdOx1 Interacts with CAR and PF4 with Implications for Thrombosis with Thrombocytopenia Syndrome." *Science Advances* 7 (49): eabl8213. https://doi.org/10.1126/sciadv.abl8213.

Beisbart, Claus. 2021. "Opacity Thought Through: On the Intransparency of Computer Simulations." *Synthese* 199(3): 11643–66. https://doi.org/10.1007/s11229-021-03305-2.

Bokulich, Alisa. 2014. "How the Tiger Bush Got Its Stripes: 'How Possibly' Vs. 'How Actually' Model Explanations." *The Monist* 97(3): 321–38. https://doi.org/10.5840/monist201497321.

Brainard, Lindsay. 2020. "How to Explain How-Possibly." *Philosopher's Imprint* 20(13): 1–23.

Brandon, Robert N. 1990. *Adaptation and Environment*. Princeton: Princeton University Press.

Buijsman, Stefan. 2023. "Over What Range Should Reliabilists Measure Reliability?" *Erkenntnis* 89: 2641–61 https://doi.org/10.1007/s10670-022-00645-4.

Callaway, Ewen. 2022. "What's Next for AlphaFold and the AI Protein-Folding Revolution." *Nature* 604(7905): 234–38. https://doi.org/10.1038/d41586-022-00997-5.

Comesaña, Juan. 2010. "Evidentialist Reliabilism." *Noûs* 44(4): 571–600. https://www.jstor.org/stable/40959693.

Creel, Kathleen A. 2020. "Transparency in Complex Computational Systems." *Philosophy of Science* 87(4): 568–89. https://doi.org/10.1086/709729.

Cuthill, Innes C., William L. Allen, Kevin Arbuckle, Barbara Caspers, George Chaplin, Mark E. Hauber, Geoffrey E. Hill, et al. 2017. "The Biology of Color." *Science* 357 (6350): eaan0221. https://doi.org/10.1126/science.aan0221.

Cybenko, G. 1989. "Approximation by Superpositions of a Sigmoidal Function." *Mathematics of Control, Signals and Systems* 2(4): 303–14. https://doi.org/10.1007/BF02551274.

DeepMind. 2020. "Computational Predictions of Protein Structures Associated with COVID-19." https://www.deepmind.com/open-source/computational-predictions-of-protein-structures-associated-with-covid-19.

Dill, Ken A., and Justin L. MacCallum. 2012. "The Protein-Folding Problem, 50 Years On." *Science* 338(6110): 1042–46. https://doi.org/10.1126/science.1219021.

Domingos, Pedro. 2012. "A Few Useful Things to Know About Machine Learning." *Communications of the ACM* 55(10): 78–87. https://doi.org/10.1145/2347736.2347755.

Dray, William H. 1968. "On Explaining How-Possibly." *The Monist* 52(3): 390–407.

Duede, Eamon. 2022. "Instruments, Agents, and Artificial Intelligence: Novel Epistemic Categories of Reliability." *Synthese* 200(6): 491. https://doi.org/10.1007/s11229-022-03975-6.

Durán, Juan Manuel. 2023. "Machine Learning, Justification, and Computational Reliabilism."

Durán, Juan M., and Nico Formanek. 2018. "Grounds for Trust: Essential Epistemic Opacity and Computational Reliabilism." *Minds and Machines* 28(4): 645–66. https://doi.org/10.1007/s11023-018-9481-6.

Egan, Andy, and Brian Weatherson. 2011. *Epistemic Modality*. Oxford: Oxford University Press.

Fennell, John G., Laszlo Talas, Roland J. Baddeley, Innes C. Cuthill, and Nicholas E. Scott-Samuel. 2019. "Optimizing Colour for Camouflage and Visibility Using Deep Learning: The Effects of the Environment and the Observer's Visual System." *Journal of The Royal Society Interface* 16(154): 20190183. https://doi.org/10.1098/rsif.2019.0183.

Fennell, John G., Laszlo Talas, Roland J. Baddeley, Innes C. Cuthill, and Nicholas E. Scott-Samuel. 2021. "The Camouflage Machine: Optimizing Protective Coloration Using Deep Learning With Genetic Algorithms." *Evolution* 75(3): 614–24. https://doi.org/10.1111/evo.14162.

Forber, Patrick. 2010. "Confirmation and Explaining How Possible." *Studies in History and Philosophy of Science Part C: Studies in History and Philosophy of Biological and Biomedical Sciences* 41(1): 32–40. https://doi.org/10.1016/j.shpsc.2009.12.006.

Freiesleben, Timo, and Thomas Grote. 2023. "Beyond Generalization: A Theory of Robustness in Machine Learning." *Synthese* 202(4): 109. https://doi.org/10.1007/s11229-023-04334-9.

Goldman, Alvin I. 2011. "Toward a Synthesis of Reliabilism and Evidentialism? Or: Evidentialism's Troubles, Reliabilism's Rescue Package." In *Evidentialism and Its Discontents*, edited by Trent Dougherty, 254–80. Oxford: Oxford University Press.

Goodfellow, Ian, Yoshua Bengio, and Aaron Courville. 2016. *Deep Learning.* Cambridge, MA: MIT Press.

Gordon, David E., Gwendolyn M. Jang, Mehdi Bouhaddou, Jiewei Xu, Kirsten Obernier, Kris M. White, Matthew J. O'Meara, et al. 2020. "A SARS-CoV-2 Protein Interaction Map Reveals Targets for Drug Repurposing." *Nature* 583 (7816): 459–68. https://doi.org/10.1038/s41586-020-2286-9.

Grüne-Yanoff, Till. 2013. "Appraising Models Nonrepresentationally." *Philosophy of Science* 80(5): 850–61. https://doi.org/10.1086/673893.

Grüne-Yanoff, Till, and Philippe Verreault-Julien. 2021. "How-Possibly Explanations in Economics: Anything Goes?" *Journal of Economic Methodology* 28(1): 114–23. https://doi.org/10.1080/1350178X.2020.1868779.

Gupta, Meghna, Caleigh M. Azumaya, Michelle Moritz, Sergei Pourmal, Amy Diallo, Gregory E. Merz, Gwendolyn Jang, et al. 2021. "CryoEM and AI Reveal a Structure of SARS-CoV-2 Nsp2, a Multifunctional Protein Involved in Key Host Processes." bioRxiv. https://doi.org/10.1101/2021.05.10.443524.

Hassabis, Demis, Dharshan Kumaran, Christopher Summerfield, and Matthew Botvinick. 2017. "Neuroscience-Inspired Artificial Intelligence." *Neuron* 95(2): 245–58. https://doi.org/10.1016/j.neuron.2017.06.011.

Hempel, Carl G. 1965. "Aspects of Scientific Explanation." In *Aspects of Scientific Explanation: And Other Essays in the Philosophy of Science*, 331–496. New York: Free Press.

Hornik, Kurt, Maxwell Stinchcombe, and Halbert White. 1989. "Multilayer Feedforward Networks Are Universal Approximators." *Neural Networks* 2(5): 359–66. https://doi.org/10.1016/0893-6080(89)90020-8.

Humphreys, Paul. 2009. "The Philosophical Novelty of Computer Simulation Methods." *Synthese* 169(3): 615–26.

Jones, David T., and Janet M. Thornton. 2022. "The Impact of AlphaFold2 One Year On." *Nature Methods* 19(1): 15–20. https://doi.org/10.1038/s41592-021-01365-3.

Jumper, John, Richard Evans, Alexander Pritzel, Tim Green, Michael Figurnov, Olaf Ronneberger, Kathryn Tunyasuvunakool, et al. 2021. "Highly Accurate Protein Structure Prediction with AlphaFold." *Nature* 596 (7873): 583–89. https://doi.org/10.1038/s41586-021-03819-2.

Khalifa, Kareem. 2017. *Understanding, Explanation, and Scientific Knowledge.* Cambridge: Cambridge University Press. https://doi.org/10.1017/9781108164276.

Kriegeskorte, Nikolaus. 2015. "Deep Neural Networks: A New Framework for Modeling Biological Vision and Brain Information Processing." *Annual Review of Vision Science* 1(1): 417–46. https://doi.org/10.1146/annurev-vision-082114-035447.

Kuhn, T. 1977. "Objectivity, Value Judgment, and Theory Choice." In *The Essential Tension: Selected Studies in Scientific Tradition and Change*, 320–39. Chicago: University of Chicago Press.

Laudan, Larry. 1977. *Progress and Its Problems: Towards a Theory of Scientific Growth*. Berkeley; Los Angeles: University of California Press.

Ledford, Heidi. 2021. "How Could a COVID Vaccine Cause Blood Clots? Scientists Race to Investigate." *Nature* 592(7854): 334–5. https://doi.org/10.1038/d41586-021-00940-0.

Lipton, Peter. 2004. *Inference to the Best Explanation*. 2nd ed. London: Routledge.

Lundberg, Scott M., and Su-In Lee. 2017. "A Unified Approach to Interpreting Model Predictions." Proceedings of the 31st International Conference on Neural Information Processing Systems. Vol. 30. Red Hook, NY : Curran Associates, Inc. https://dl.acm.org/doi/10.5555/3295222.3295230.

Melin, Amanda D., Linda M. Fedigan, Chihiro Hiramatsu, Courtney L. Sendall, and Shoji Kawamura. 2007. "Effects of Colour Vision Phenotype on Insect Capture by a Free-Ranging Population of White-Faced Capuchins, Cebus Capucinus." *Animal Behaviour* 73(1): 205–14. https://doi.org/10.1016/j.anbehav.2006.07.003.

Mordvintsev, Alexander, Christopher Olah, and Mike Tyka. 2015. "Inceptionism: Going Deeper into Neural Networks." *Google AI Blog*.

Nielsen, Michael A. 2015. *Neural Networks and Deep Learning*. Determination Press.

Pandey, Preeti, Kartikay Prasad, Amresh Prakash, and Vijay Kumar. 2020. "Insights into the Biased Activity of Dextromethorphan and Haloperidol towards SARS-CoV-2 NSP6: In Silico Binding Mechanistic Analysis." *Journal of Molecular Medicine* 98(12): 1659–73. https://doi.org/10.1007/s00109-020-01980-1.

Räz, Tim, and Claus Beisbart. 2022. "The Importance of Understanding Deep Learning." *Erkenntnis* 89: 1823–40. https://doi.org/10.1007/s10670-022-00605-y.

Reutlinger, Alexander, Dominik Hangleiter, and Stephan Hartmann. 2018. "Understanding (With) Toy Models." *The British Journal for the Philosophy of Science* 69(4): 1069–99. https://doi.org/10.1093/bjps/axx005.

Ribeiro, Marco Tulio, Sameer Singh, and Carlos Guestrin. 2016. "Why Should I Trust You?: Explaining the Predictions of Any Classifier." In *Proceedings of the 22nd ACM SIGKDD International Conference on Knowledge Discovery and Data Mining*, 1135–44. KDD '16. New York, NY, USA: Association for Computing Machinery. https://doi.org/10.1145/2939672.2939778.

Richards, Blake A., Timothy P. Lillicrap, Philippe Beaudoin, Yoshua Bengio, Rafal Bogacz, Amelia Christensen, Claudia Clopath, et al. 2019. "A Deep Learning Framework for Neuroscience." *Nature Neuroscience* 22(11): 1761–70. https://doi.org/10.1038/s41593-019-0520-2.

Rohwer, Yasha, and Collin Rice. 2013. "Hypothetical Pattern Idealization and Explanatory Models." *Philosophy of Science* 80(3): 334–55. https://doi.org/10.1086/671399.

Rudin, Cynthia. 2019. "Stop Explaining Black Box Machine Learning Models for High Stakes Decisions and Use Interpretable Models Instead." *Nature Machine Intelligence* 1(5): 206–15. https://doi.org/10.1038/s42256-019-0048-x.

Sadek, Ali, David Zaha, and Mahmoud Salama Ahmed. 2021. "Structural Insights of SARS-CoV-2 Spike Protein from Delta and Omicron Variants." bioRxiv. https://doi.org/10.1101/2021.12.08.471777.

Šešelja, Dunja. 2022. "What Kind of Explanations Do We Get from Agent-Based Models of Scientific Inquiry?" Preprint. http://philsci-archive.pitt.edu/20532/.

Šešelja, Dunja, and Christian Straßer. 2014. "Epistemic Justification in the Context of Pursuit: A Coherentist Approach." *Synthese* 191(13): 3111–41. https://doi.org/10.1007/s11229-014-0476-4.

Shaw, Jamie. 2022. "On the Very Idea of Pursuitworthiness." *Studies in History and Philosophy of Science* 91: 103–12. https://doi.org/10.1016/j.shpsa.2021.11.016.

Sjölin Wirling, Ylwa. Forthcoming. "Understanding with Epistemic Possibilities: The Epistemic Aim and Value of Metaphysics." *Argumenta*, Forthcoming.

Sjölin Wirling, Ylwa, and Till Grüne-Yanoff. 2021. "Epistemic and Objective Possibility in Science." *The British Journal for the Philosophy of Science.* https://doi.org/10.1086/716925.

Strevens, Michael. 2013. "No Understanding Without Explanation." *Studies in History and Philosophy of Science Part A* 44(3): 510–15. https://doi.org/10.1016/j.shpsa.2012.12.005.

Sullivan, Emily. 2022. "Understanding from Machine Learning Models." *The British Journal for the Philosophy of Science* 73(1): 109–33. https://doi.org/10.1093/bjps/axz035.

Talas, Laszlo, John G. Fennell, Karin Kjernsmo, Innes C. Cuthill, Nicholas E. Scott-Samuel, and Roland J. Baddeley. 2020. "CamoGAN: Evolving Optimum Camouflage with Generative Adversarial Networks." *Methods in Ecology and Evolution* 11(2): 240–47. https://doi.org/10.1111/2041-210X.13334.

Thornton, Janet M., Roman A. Laskowski, and Neera Borkakoti. 2021. "AlphaFold Heralds a Data-Driven Revolution in Biology and Medicine." *Nature Medicine* 27(10): 1666–69. https://doi.org/10.1038/s41591-021-01533-0.

Verreault-Julien, Philippe. 2017. "Non-Causal Understanding with Economic Models: The Case of General Equilibrium." *Journal of Economic Methodology* 24(3): 297–317. https://doi.org/10.1080/1350178X.2017.1335424.

Verreault-Julien, Philippe. 2019. "How Could Models Possibly Provide How-Possibly Explanations?" *Studies in History and Philosophy of Science Part A* 73: 22–33. https://doi.org/10.1016/j.shpsa.2018.06.008.

Verreault-Julien, Philippe. 2023. "Toy Models, Dispositions, and the Power to Explain." *Synthese* 201(5): 171. https://doi.org/10.1007/s11229-023-04084-8.

Whiteley, William N., Samantha Ip, Jennifer A. Cooper, Thomas Bolton, Spencer Keene, Venexia Walker, Rachel Denholm, et al. 2022. "Association of COVID-19 Vaccines ChAdOx1 and BNT162b2 with Major Venous, Arterial, or Thrombocytopenic Events: A Population-Based Cohort Study of 46 Million Adults in England." *PLOS Medicine* 19 (2): e1003926. https://doi.org/10.1371/journal.pmed.1003926.

Willer, Malte. 2013. "Dynamics of Epistemic Modality." *Philosophical Review* 122(1): 45–92. https://doi.org/10.1215/00318108-1728714.

Yamins, Daniel L. K., and James J. DiCarlo. 2016. "Using Goal-Driven Deep Learning Models to Understand Sensory Cortex." *Nature Neuroscience* 19(3): 356–65. https://doi.org/10.1038/nn.4244.

Yan, Weizhu, Yanhui Zheng, Xiaotao Zeng, Bin He, and Wei Cheng. 2022. "Structural Biology of SARS-CoV-2: Open the Door for Novel Therapies." *Signal Transduction and Targeted Therapy* 7(1): 1–28. https://doi.org/10.1038/s41392-022-00884-5.

Yang, Jianyi, Ivan Anishchenko, Hahnbeom Park, Zhenling Peng, Sergey Ovchinnikov, and David Baker. 2020. "Improved Protein Structure Prediction Using Predicted Interresidue Orientations." *Proceedings of the National Academy of Sciences* 117(3): 1496–503. https://doi.org/10.1073/pnas.1914677117.

Ylikoski, Petri, and N. Emrah Aydinonat. 2014. "Understanding with Theoretical Models." *Journal of Economic Methodology* 21(1): 19–36. https://doi.org/10.1080/1350178X.2014.886470.

Zednik, Carlos. 2021. "Solving the Black Box Problem: A Normative Framework for Explainable Artificial Intelligence." *Philosophy & Technology* 34(2): 265–88. https://doi.org/10.1007/s13347-019-00382-7.

Zednik, Carlos, and Hannes Boelsen. 2022. "Scientific Exploration and Explainable Artificial Intelligence." *Minds and Machines* 32(1): 219–39. https://doi.org/10.1007/s11023-021-09583-6.

Zhang, Chiyuan, Samy Bengio, Moritz Hardt, Benjamin Recht, and Oriol Vinyals. 2021. "Understanding Deep Learning (Still) Requires Rethinking Generalization." *Communications of the ACM* 64(3): 107–15. https://doi.org/10.1145/3446776.

Zhou, Ding-Xuan. 2020. "Universality of Deep Convolutional Neural Networks." *Applied and Computational Harmonic Analysis* 48(2): 787–94. https://doi.org/10.1016/j.acha.2019.06.004.

10 Modeling Climate Possibilities

Joe Roussos

10.1 Introduction

This chapter examines modal modelling in climate science. It considers two related topics. The first is the use of climate models to attribute extreme weather events (EWEs) to climate change. Here the debate is between advocates of different approaches, called the risk and storyline approach. The way each method explores and makes use of possibilities is discussed, with particular attention given to the storyline method, as the non-probabilistic, possibility-focused method. The second topic is the interpretation and use of collections of climate models. Here, the debate is between those who think that collections of models can be used to produce conditional probabilities for future climate possibilities and those who think that they can merely identify such possibilities.

Each topic is the subject of a current debate within climate science and philosophy of science, and each has an important modal component. The debates are similar in that each involves a contrast between probabilistic and non-probabilistic methods. Each debate turns crucially on details about these models and their uncertainties, details which cannot be fully explored in a chapter like this. (No familiarity with climate science is assumed, and the chapter begins with an introduction to climate modelling in section 10.2. Readers familiar with climate science and modelling can skip to section 10.3.) The focus is instead on the modal claims and methods employed by the disputants and on their connections to other philosophical questions. As we will see, the nature of climate change pushes scientists to consider questions about what is possible and how one might defend modal claims.

A brief introductory note on terminology. "Possibilities" are ways the world might be, and they are often specified relatively coarsely. An "epistemic possibility" is a way the world might be, as far as we know.[1] Since our topic is climate science, what "we know" in this context is the well-established results of climate science, and the description of the possible situation will be in terms of relevant climate variables. What the well-established results are and what it means for something to be possible "as far as" that knowledge goes, is discussed below. By contrast, an "objective possibility" is a way the world

DOI: 10.4324/9781003342816-14

might be which is not indexed to any body of knowledge but is instead determined by some set of facts, including the natural laws governing the world. Since our topic is climate science, that set includes relevant facts about the Earth's climate and the laws of nature governing its evolution.

10.2   The Project of Climate Modelling

Climate science studies how the Earth's atmosphere, oceans, land surface, ice sheets, and biosphere interact in creating its climate. Roughly speaking, an area's climate is its prevailing weather conditions over a suitably long period (often 30 years), making "climate" an aggregate concept. Some important climate variables are air temperature, sea temperature, and precipitation; an area's climate is often summarised by referring to the average and extreme values for these variables, seasonally and annually. The core subject of climate science at present is, of course, human-caused climate change.

Simulation models are among the most important tools of climate science. Climate science uses a spectrum of models, with different degrees of complexity and resolution. The most complex are called global climate models (GCMs), and they model the circulation of the atmosphere and oceans, and their interactions with the Earth's land surface and ice sheets. Such models play an indispensable role in the practice of climate science; scientists would not be able to make the claims they currently make about climate change without using these models (Goodwin 2015).

At a simulation model's heart is a description of how the system works and how it evolves from one state to another. For GCMs, this is a representation of the dynamics of the atmosphere and oceans and of how this system is driven by the energy input from solar radiation. These models draw on well-understood theories from thermodynamics, fluid mechanics, nonlinear dynamics, and so on. As is common with models, scientists use their theoretical understanding of the climate system to identify the most important aspects for inclusion in GCMs, and they make simplifications and idealisations elsewhere. There are two important sources of uncertainty here. The first is our limited understanding of the physical processes of the climate, such as the impact of aerosols on the Earth's energy balance. The second is the effect of the idealisations in the model.

The model's structural-dynamical core is essentially a set of partial differential equations involving variables which are continuous functions of space and time. These are very difficult, if not impossible, to solve exactly. To make progress, scientists use numerical approximation methods (Pättiniemi and Koskinen, forthcoming, 3–5; Winsberg 2018, 45–46). They begin by discretising space and time and then create an algorithm for approximating the solution to the continuous equations. Computers can then produce approximate solutions, using step-by-step methods (Parker 2009). These approximation methods introduce more uncertainty into climate modelling, as does disagreement or uncertainty about the best numerical techniques to employ.

This process, of designing a model structure and realising it in a computer program using numerical approximations, leaves various important things out. This is in part due to the resource-intensive nature of the computing required—simplifications are introduced to ensure that the simulations generate results in a reasonable amount of time. Cost is an important factor in determining the size of the grid, which in turn determines which processes in the climate are modelled explicitly: those which happen at greater-than-grid scales. Anything that happens strictly within a grid box cannot be resolved by a discretised model and so needs to be added into the model separately. A separate model, called a "parameterization," is used for this—it typically has a simple form, depending on a few parameters. Turbulence and clouds are important examples of features treated via parameterizations in GCMs (Petersen 2012; Winsberg 2018, 47–50).

To understand the effects of human activity on the climate, the models are initialised with "forcing" scenarios, describing a profile of emissions over some period. The models then simulate the response of the climate and produce the kinds of output now familiar from the Intergovernmental Panel on Climate Change (IPCC) reports, for example, estimates of the change in global mean surface temperature by 2100 under a particular scenario. These results are called "projections," and they are presented as conditional possibilities: states of the climate that might result if the forcing scenario obtains.

Climate science makes regular use of collections of such models, called ensembles. The results of such collections are often studied statistically and used to produce probability distributions for climate variables of interest, such as changes in temperature or precipitation (under a specific scenario). An important use of climate models is for model experiments, which alter a single factor while holding other things equal, to test the influence of that factor on the climate. This is, roughly, how climate fingerprinting works: models simulate the recent history of the climate, with and without human emissions of greenhouse gases. The simulations with human emissions closely match the observed warming, which provides evidence of anthropogenic global warming.[2]

10.3   Extreme Event Attribution

One important question for climate scientists is how climate change affects EWEs. Heavy rains, bad storms, heatwaves, and droughts would all occur without anthropogenic climate change, but it is widely held that climate change will alter their frequency and severity. So, when an EWE occurs, people naturally want to know whether or to what degree climate change caused it. This question is particularly important for adaptation as "based on the occurrence of a particularly damaging extreme event, plans could be made to adapt to an increasing frequency of such events in future" (Stott et al. 2016, 24).

Two methods are currently popular for attributing EWEs to climate change: a risk-based approach and a storyline-based approach. There is a significant debate between proponents of these approaches about the efficacy and relevance of the attribution information they provide (Otto et al. 2016; Shepherd 2016; Stott et al. 2016; Trenberth, Fasullo, and Shepherd 2015), which has attracted recent philosophical discussion (Lloyd and Oreskes 2019; Winsberg, Oreskes, and Lloyd 2020). We need not interest ourselves in the details of the debate between these methods; what is interesting for our purposes is that each method is in an important sense modal. The risk approach asks a counterfactual question about a class of events. The storyline approach presents a possible causal history for a specific event.

10.3.1   *The Risk Approach*

By the "risk approach," I mean the set of methods employed by the World Weather Attribution (WWA) collaboration, as defined in their protocol paper (Philip et al. 2020), and elaborated and discussed in various other places (Oldenborgh et al. 2021; Otto et al. 2016; Stott et al. 2016). Here is a basic description of the approach. Consider an event like the heatwave in western Europe in July 2022, during which the UK saw temperatures of 40°C for the first time. The risk approach takes this event to be a token of a type and asks a question like: what is the probability of a heatwave of this type, given anthropogenic climate change? It then compares this to the probability of such an event in a counterfactual climate without anthropogenic warming. The focus on a type of event, rather than the specific 2022 heatwave, is because "any climate event under consideration, for example, a heat wave, drought, or flood, has evolved in its own unique way and is therefore, in principle, attributable to a unique set of causes that is not applicable to any other event" (Stott et al. 2016, 25). The method seeks to evaluate the *change* in the probability of a kind of event, due to global warming; the results are expressed as ratios of probability (e.g., human influence doubled the probability of a record warm summer) or fractions of attributable risk (e.g., 0.5 of the risk of a record warm summer is attributable to human influence).

To define a type of event, scientists seek to identify in what sense the event was extreme and the features that typify it. Attribution studies occur immediately after the event, and the definition is formed with explicit reference to questions that stakeholders (typically the public) are asking and how the event affected them (Otto et al. 2016). The type-definition is then realised in a model in terms of modelled variables. For the 2022 UK heatwave, WWA used "two event definitions, i.e., … the annual maximum of 2-day average temperatures over this region as well as the annual maximum of the daily maximum temperature" (Zachariah et al. 2022, 4). By contrast, for the 2019 European heatwave, affecting mostly France and Spain, they used the definition of "the highest 3-day averaged daily mean temperature for each year" (Vautard et al. 2019, 6). There is a similar variation in the type-definitions for other EWEs, such as droughts, heavy rainfalls, and windstorms.

The probabilities are calculated from the frequencies of this type of event in simulation model runs. The fraction of attributable risk requires two such frequencies. The first comes from the "actual" world, which is simulated by GCMs and includes anthropogenic climate forcing in the form of humanity's history of emissions since the start of the industrial era. The models are run many times to generate many data points for weather consistent with this climate state. A frequency is calculated for extreme events of the same type as, say, the 2022 UK heatwave. The second frequency comes from a counterfactual "natural" world, which is simulated without anthropogenic climate forcings. Roughly speaking, the models simulate how the climate would have developed from pre-industrial times to 2022, without human emissions. They are then run many times, and the frequency of extreme events of the 2022 heatwave type is measured. These frequencies are then compared, to calculate the fraction of the risk of this event type due to climate change.

The risk approach defines the probability of an event type, in a given climate state, as the fraction of simulated weather states containing that type of event. This implies a kind of frequentist notion of probabilities, but instead of counting the number of events occurring in the real world, they count the number of simulation runs with the event type. There are two such frequencies, coming from what I called above "actual" and "counterfactual" simulations. The "actual" world simulations are meant to represent possible weather states consistent with the climate state that the world is actually in, that is, one with anthropogenic climate change and the history represented by our historical data. The counterfactual simulation generates weather states compatible with a different climate state: that of a world without climate change. Any climate state is compatible with lots of different weather events; being in our current climate state doesn't determine any particular extreme event. This is why the risk approach calculates two frequencies and compares them.

So, both the "actual" and counterfactual simulations generate possibilities. What kind of possibilities are these? I argue that they are best interpreted as a species of objective possibility. The "actual" simulation should be read as saying: given a set of facts about the world's actual climate state, these weather states are possible. This is a better fit than epistemic possibility, since we know that one such state actually occurred—the one with the 2022 UK heatwave. So, the generated possibilities cannot be the straightforward kind of epistemic possibilities, which are "states compatible with what we know."

Less straightforwardly, these possibilities are generated by models that are based on imperfect knowledge of the actual climate state, and so they represent epistemically limited attempts to identify objective possibilities. We do not know all the details of the 2022 UK heatwave—we have only so many measurements (from the extant meteorological stations) and for only certain climate variables (e.g., perhaps we have so-called "near surface" temperature measurements but not any higher up in the atmosphere). Additionally, as we have seen, climate scientists characterise the event of interest in the model via a relatively crude proxy definition, such as a 2-day average maximum temperature (near surface). However, each model run containing a heatwave of this type will include as much detail about that modelled heatwave as the model resolves. This might be *more detailed* than the knowledge we have of the actual heatwave—for example, by including facts about the atmospheric temperature at several points in the atmospheric column—or less detailed, e.g., because the model represents a region like England using a series of discrete blocks, which may be coarser than the actual set of measurements from meteorological stations. So, it is possible that some modelled heatwaves might, for all we know, correspond to the actual heatwave and indeed be realisations of how it actually happened in more detail than we possess.

Do these nuances make the modelled events epistemic possibilities? I think not, because of the aims of the modellers. An EWE attribution study does not aim to examine micro-realisers of the macro-specified actual token event. It aims to study the frequency of this *type* of event, in a given climate state, for the purpose of assessing the probability of the actual event. So, in the context of the study, the modelled simulations should be understood as (attempting to) identify objective possibilities. They are *conditional* possibilities in the sense that they are specified relative to a background climate state, and they are *climatological* possibilities, as the relevant kind of possibility is compatibility with laws of nature governing the climate—laws of physics, chemistry, biology, and so on. Another way to see that these are objective possibilities is that they are justified by defending the model's ability to describe the actual world (van Oldenborgh et al. 2021): the model is evaluated on how well the pattern of extreme events matches the observed distribution of extreme events. This is taken to provide positive evidence that model runs correspond to ways the world might have been.

The same goes for the possibilities identified by the counterfactual simulation of the world without anthropogenic climate change. There, our epistemic state is much worse, since we don't have observations of such a world in 2022. But, nevertheless, the scientists' aim is to use their models to describe objective climate possibilities, conditional on a set of background climatological facts. One account of objective climate possibilities that can play this role, due to Joel Katzav, is discussed below in section 10.4.

One reason it matters that these are objective possibilities is because of the debate about the storyline method. That alternative attempts to answer questions about the token event in the actual world, which storyline advocates claim is more relevant to decision-makers than what might have happened counterfactually.

10.3.2 *The Storyline Approach*

The storyline approach to EWE attribution is quite different. Here, the investigation takes the occurrence of the event as given and asks questions about the contribution of climate change. Example questions offered by Trenberth et al. (2015) include: "Given the weather pattern, how were the temperatures, precipitation and associated impacts influenced by climate change?"

and "Given a flood, where did the moisture come from? Was it enhanced by high ocean temperatures that might have had a climate change component?" So, instead of asking how much more likely this *kind* of event is due to climate change, a storyline attribution study asks how climate change might have contributed to *this particular event*. Ted Shepherd, a major developer of the method, described it as "analogous to accident investigation (where multiple contributing factors are generally involved and their roles are assessed in a conditional manner)" (Shepherd 2016, 32).

It is also an accident investigation that focuses on particular *aspects* of the EWE. Storyline advocates distinguish between the thermodynamic and dynamic factors influencing an event. In this context, thermodynamic factors include the temperatures of the seas and the atmosphere, while dynamic factors include atmospheric circulation patterns, such as blocking high-pressure systems that push precipitation away. Speaking roughly, dynamic factors influence the nature of each EWE and its placement in space and time, while thermodynamic factors influence the severity of EWEs, for example, by changing how much energy is available to drive a storm. Importantly, "thermodynamic aspects of climate change are generally robust in theory, observations and models" (Shepherd et al. 2018, 563), while dynamical aspects are significantly less certain and less well-grounded in theory. The circulation of the atmosphere is subject to significant natural variation and changes due to climate change are small relative to this variability. Thus, "forced circulation changes are not well established and it is difficult to detect changes in circulation-related extremes in observations because of small signal-to-noise" (Trenberth et al. 2015). As Lloyd and Oreskes put it, proponents of the storyline approach "suggest that for a given severe weather event where we do not have a physically credible model that includes the dynamics, then 'under such conditions,' it is better for event attribution to focus on thermodynamics of the event (Trenberth et al. 2015, 729; Shepherd 2016, 703). Their proposal is that in such cases, we should take the extreme event as a given constraint and ask if thermodynamic factors are involved in such a way as to worsen it" (Lloyd and Oreskes 2018, 311).

The storyline method thus involves an accident investigation for a particular event, focused on thermodynamic factors. At a high level, the aim is an assessment of the possible contributions of climate change to the EWE. The main method, as well as intermediate output, is the "storyline." A storyline is defined as "a physically self-consistent unfolding of past events" (Shepherd et al. 2018, 557). An attribution study typically investigates several storylines, which stipulate particular scenarios for investigation. For example, Chan et al. look at eight storylines in their investigation of the 2010–2012 UK drought (Chan et al. 2022, table 1). One is "drier preconditions," which tests the sensitivity of the drought to drier periods in the run-up to the drought; this was done by varying the precipitation in the 3 and 6 months prior to the drought, in a model (Chan et al. 2022, 1757). This is the general method for EWEs: climate models are used to probe relationships between thermodynamic factors and atmospheric conditions.[3] Such relationships are probed using counterfactual analysis, generated by comparing climate simulations with and without a test factor.

So, the storyline approach has in common with the risk approach the use of models to probe possibilities and also to compare several counterfactual simulations. The differences lie in the overall goals of the investigation and in the intended output. No probability is assigned to a storyline, which is described as a physics-based, plausible causal narrative. Notably, each storyline is just one such causal narrative, with no qualifiers other than plausibility attached. Once these causal factors have been studied, they are assembled into a broad assessment of the event. This can be understood as an epistemically possible history of the EWE. The purpose of identifying such a possibility is to seek to understand the driving factors involved in the event, and it is in the choice and role of these factors that special emphasis is given to establishing plausibility.

For our purposes, it is interesting to dwell on one way that scientists have defended the value of storylines, despite their seeming limitations. The focus on the narrative possibility's basis in physical theory, and in well-understood climate processes linked to climate change, is critical in defending its relevance and usefulness (Shepherd 2016; Shepherd et al. 2018; Trenberth et al. 2015). While the risk approach offers quantified probabilities for its possibilities (expressed in terms of dynamical conditions), these may lack *credibility*, according to Shepherd (2016, 32), due to the model uncertainties plaguing dynamical simulations. This is an organic example of scientists arguing that the possibilities they have modelled are not "mere" possibilities, but are "real" or "serious" possibilities (cf. section 4.2).

As we saw above, storyline advocates argue that the changes in circulation modelled by the risk approach have low signal-to-noise ratios, especially for EWEs (Hoerling et al. 2013). This is in part due to natural variation and in part because, for contingent historical reasons, there has been insufficient exploration of model uncertainties, specifically those associated with model structure (Knutti, Masson, and Gettelman 2013). This is offered as a reason why the possibilities relied upon in the risk approach are *not* credible. As the same models are used in each approach, this requires some subtlety. Proponents of the storyline method here distinguish between two uses of models. The first is an investigation of the *processes* involved in climate change. Confidence in these processes is built by noting the robustness of model behaviour and its agreement with observations from multiple sources (Lloyd 2015). The claim is that this use is less sensitive to model uncertainties, and it is for this purpose that models are used in storyline studies. The second use is the generation of *quantitative climate predictions*, with associated probabilities. Here the uncertainty in modelled circulation changes becomes critical. Critics claim that such predictions are overly reliant on models as a single source of climate information and are, therefore, sensitive to "missing physics" not currently in the models, the under-sampling of extreme climate possibilities, and several other challenges (Shepherd et al. 2018, 564–566). The storyline method relies on thermodynamics, which has well-understood links to climate change, and avoids quantitative predictions. Thus, it is claimed that it more credibly identifies possibilities. (It is not important here whether this argument succeeds, as my interest is in tracing the role and nature of the modal claims made.)

Advocates want the possibilities represented by storylines to be more than just credible, however. They must be suitable for motivating the right kind of response in the audience of the attribution study. Storyline advocates aim to support adaptation efforts that guard against future climate risks. They note that "it may be difficult to convince people to invest in defences against a hypothetical risk, but easier to do so if an event has previously occurred so clearly could occur again, but potentially with more impact" (Shepherd 2016, 33). The storyline approach produces a possible causal history for a concrete historical event, the effects of which are easily observable and recent in the memory of the presumed audience for the attribution study.

Shepherd et al. (2018, 560) draw explicitly on behavioural psychology in describing how and why their approach works to support adaptation efforts. First, they argue that "availability bias is apparent in the history of physical infrastructure measures to mitigate natural disaster risk"—that is, policymakers overestimate the risk of events they have seen happen and underestimate those they have not. What is needed to counteract this is an "episodic understanding" of possible hazards; a term inspired by the notion of "episodic memory" from Tulving (1972): a form of memory where one relives an event rather than merely recalling facts. Surveying a range of failures to invest in risk mitigation measures, Shepherd et al. argue that the failure to act occurs "despite good [factual] understanding of the case for such investments, because the necessary episodic understanding is not present" (Shepherd et al. 2018, 560). The storyline approach aims to develop this episodic understanding by providing a narrative of an EWE in which the role of climate change is made vivid and the predicted future is thus made more tangible.

Given all this, the storyline can be understood as a how-possibly explanation of the EWE. Not in Hempel's sense, of a potential how-actually explanation, nor in the sense of establishing *that* the EWE itself is possible since this is trivial for a past event. It is also not a how-possibly explanation in the sense of dispelling a (presumably prior) belief that it could not have happened, by showing how it was possible. Instead, this is a species of how-possibly explanation in which the storyline shows how known thermodynamic processes of climate change could have influenced the (known, past) EWE. It aims at providing a plausible, psychologically compelling link between climate change and a particular past event, in order to help non-specialists reason about climate actions.

It is perhaps similar to the how-possibly explanations in evolutionary biology discussed by Brandon (quoted in Verreault-Julien 2019): "What good is a speculative how-possibly explanation? The short answer is: it shows how known evolutionary mechanisms *could produce* known phenomena" (Brandon 1990, 180, emphasis added). Such how-possibly explanations often focus on possible causes; for example, Verrault-Julien characterises them as establishing the proposition that "possibly, p because q." In the case of the storyline approach, there is also a causal aim, though perhaps a weaker one: it seeks to establish the possible causal *contributions* of climate change, as well as their nature and valence—that is, increasing or decreasing the severity of the EWE.

Notably, storyline how-possibly explanations are not produced in service of developing a how-actually explanation. Attribution studies tend to occur fairly soon after the event in question, and they aim to contribute to the heightened discussion of that event taking place in its aftermath. The emphasis on behavioural psychology, and the references to "risk," show the scientists' interest in influencing and supporting policy discussion. While some EWEs of course garner greater interest and may, therefore, receive more comprehensive study, this is not typically the case. The output of the storyline approach is explicitly an *epistemically possible* causal history and does not aspire to more than possibility. Indeed, significant attention is devoted to discussing the value and use of the storyline *qua* possibility—for example, its potential role in "robust decision-making" methods which take as inputs unquantified possibilities (Shepherd et al. 2018). These are discussed more in section 10.5.

10.4   Ensembles of Climate Models

This section considers the debate over how to make use of collections, or ensembles, of climate models. This is a more general topic than the previous one, and attribution studies regularly make use of ensembles, so the considerations discussed here apply there also.

Let's start with a basic question: why do climate scientists trust models? There are numerous reasons, including the following (Knutti et al. 2010, 2757 and references therein). First, these models are based on accepted physical principles that we have independent reason to believe govern the relevant domains, such as the conservation of mass, momentum, and energy, and the laws of thermodynamics. Second, these models can reproduce complex and significant features of the past and current climate, including its natural variability and its response to anthropogenic influences during the period for which we have data. Third, there is remarkable agreement in the predictions of climate change produced by these models and by various non-model-based methods, such as paleoclimate data. This convergence between different sources of evidence is taken to be significant and confirmatory when it comes to the core hypotheses of climate change (Lloyd 2015; Oreskes 2004, 2018), and it is taken to provide additional support for the claim that the models correctly capture some core mechanisms of climate change.

What, then, are we to make of the diversity of and disagreement between such models? Between 50 and 100 GCMs exist globally, created and maintained by different research groups and national meteorological centres. They were developed for various reasons, many contingent and historical,[4] and disagree to a greater or lesser degree across a wide range of variables and kinds of output. They all purport to describe the same system but have numerous differences including at the structural-dynamical level. Their diversity therefore partially represents the uncertainties described above. An important response to this puzzle is the Coupled Model Intercomparison Project (CMIP). CMIP gathers these models and studies their relations and the distributions of their predictions. CMIP6, the iteration of the project supporting the IPCC's sixth assessment report, contains around 100 models from 49 modelling groups worldwide.[5] It aims to evaluate these models, summarise their output, and use facts about the distribution of model results to produce improved output.

The IPCC assessment reports use CMIP results to construct probability distributions for key climate variables. Put very roughly, the distribution of model results is used to fit a statistical model, which then generates probabilities for, say, global mean temperature change at 2100. It is this practice which is at issue in this section. Some climate scientists and philosophers argue that the use of probabilities is unjustified and misleading. Amongst them, a subset proposes instead that these models should be seen as identifying possible future climate states but no more. So, the two positions are, roughly,

1 **Probabilist:** climate models can provide probabilities for future climate states in the form of probability distributions over climate variables; and
2 **Possibilist:** climate models can identify possible future climate states, where these might be "real" or "serious" possibilities but cannot be assigned even qualitative likelihoods.

The probabilist position is strictly stronger than the possibilist one: if collections of models can tell us how probable various outcomes are, they can also tell us whether these outcomes are possible.

What kind of possibilities are these? Broadly speaking, divergence between models appears to be regarded as something to be reduced—a reduction in the range of estimates for a variable in successive CMIP versions is presented as progress. This indicates that the uncertainty represented by the ensemble is conceived of as epistemic, and thus that the results represent epistemic possibilities. Thus, a reduction in the scope of possibilities corresponds to an increase in knowledge. If the divergence were thought to represent the natural variation of the climate system, so that each model result picked out an objective possibility, there would be no cause to celebrate the reduction of divergence. And, indeed, it is acknowledged that some degree of divergence is desirable: there are multiple, objectively possible, future states of the climate consistent with any forcing scenario, due to the internal variability of the system. In other words, a perfect climate model should generate a range of results representing objective possibilities compatible with the modelled scenario. So, a crucial challenge is identifying how much ensemble spread represents epistemic uncertainty and how much represents objective variability.

Part of the debate is therefore about how to understand and represent the uncertainty in climate ensemble results. Uncertainty (and its representation) has an important connection to decision: in addition to being an epistemic lack, uncertainty is a barrier to successful action. Climate change is an enormous practical challenge, and so part of the debate here is about which representation of uncertainty best supports successful decision-making.

10.4.1 *Probabilism and Its Discontents*

A full discussion of the statistical methods of climate science is beyond this chapter, but here are some important basics. First, at a basic level, these methods involve fitting a statistical model to the set of model results and using this to infer probability distributions. Second, the probability distributions are often produced from relatively small numbers of runs of each model, which are assumed to be representative of that model's output (Katzav et al. 2021).[6] Third, CMIP ensembles are what are called "ensembles of opportunity"— they were gathered from the models which happened to exist, rather than being designed or selected to have specific properties (Tebaldi and Knutti 2007). Each is intended by its authors to be as good as possible, where "goodness" is determined by overlapping sets of purposes and standards. Finally, modellers acknowledge that there is significant uncertainty that is not captured by these methods, and so the derived probability distributions are sometimes adjusted "by hand." For example, in the IPCC's fifth assessment report, the 5–95% range of model results for global mean temperature change in 2100, under scenario RCP8.5, is 2.6–4.8°C. If model frequencies were used directly to calculate probabilities, this would correspond to a 90% probability interval. Instead, the IPCC reported this range as "likely," that is, greater than 66% probability, due to unmodelled but acknowledged uncertainties (Thompson, Frigg, and Helgeson 2016).

Criticisms of probabilistic methods have addressed both particular statistical methods and general justificatory strategies. One initially popular method was to treat the models as independent samples from some distribution that is centred on the truth. If this were the case, it would legitimate the use of standard statistical tools for handling samples: the sample mean would be an estimator for the population mean (i.e., the true value for the variable in question); adding more models would lead to the ensemble mean to converge on the truth, and the variance of the sample could be used to produce confidence intervals for the true value (Tebaldi and Knutti 2007). However, this treatment of models as samples is not justified, as has been argued both conceptually (Katzav 2014; Parker 2010, 2011) and empirically (Knutti et al. 2010; Stainforth et al. 2007a). The models are not independent: climate modellers have common training, make use of similar techniques for similar reasons, and share both experience and computer code, with the result that the models share many components (Winsberg 2018). For this reason, the errors in these models are not random. Additionally, no attempt was made to ensure that this sample is representative—as noted, this ensemble was collected rather than constructed.

This is not the only method, however. Annan and Hargreaves (2011, 4529) describe an alternative that acts as though "the truth and the models are … drawn from the same distribution," rather than assuming that the models centre on the truth. This has performed significantly better in some studies and has received a recent philosophical defence by Dethier (2022). Nonetheless, the conditions under which it has been shown to succeed are limited, and its general applicability relies on the quality of the underlying models, which is a topic of significant concern to possibilists, as discussed below.

Katzav et al. (2021) note that probabilities could in principle receive support from theory. (Recall that support from theory is a reason that climate models are trusted at all by scientists.) They argue, however, that "theory provides limited guidance in interpreting model and ensemble output, including what model biases imply for output accuracy. Thus, theory tends to leave open the extent to which output spans the range of possibilities, or whether these possibilities are weighted in a way that reflects reality" (Katzav et al. 2021, 9).

10.4.2 The Possibilist Challenge

There are several versions of what is here called the possibilist interpretation,[7] but they share a common critique of the adequacy of climate models. Much of the discussion takes at its starting point a pair of influential articles led by climate scientist David Stainforth, which criticise probabilistic methods and propose alternative ways of using these outputs (Stainforth et al. 2007a; 2007b). Their critique of probabilistic methods centres on an exploration of the uncertainties of climate models themselves. It is complex, but four points are worth noting.

The first is a problem inherent in climate modelling: the models are "simulating a never before experienced state of the system" and therefore "cannot be meaningfully calibrated" against historical data (Stainforth et al. 2007a, 2145). The second is that the models are multiply interconnected and share biases. They are, therefore, neither independent nor representative of the scientific uncertainty about their domains. The third point is that the models display extreme uncertainty, as evidenced by their wide divergence in important variables. They report an ensemble whose estimates for the change in Mediterranean average precipitation range from −28% to +20%. This wide range is despite the limited diversity of models, highlighted in the last point. The fourth issue is that this uncertainty is poorly understood and underexplored, with the expected result that further investigation of uncertainty, especially structural model uncertainty, should be expected to significantly *widen* the range of predicted values for important variables.

Their conclusions are stark:

> Can we rule-out or weight the models in such ensembles? Much effort is currently focussed on assigning models' weights based, to some extent, on their ability to reproduce current climate… As long as all of our current models are far from being empirically adequate, we consider this to be futile. Relative to the real world, all models have effectively zero weight… The lack of any ability to produce useful model weights, and to even define the space of possible models, rules out the possibility of producing meaningful PDFs for future climate based simply on combining the results from multi-model or perturbed physics ensembles; or emulators thereof. Models can, however, provide insight without being able to provide probabilities.
>
> (Stainforth et al. 2007a, 2155)

Under certain conditions, they argue, we can rule specific models out of consideration—such as they are outside of their domains of applicability. Other than that, we should regard these ensembles as providing a "lower bound on the maximum range of uncertainty" (Stainforth et al. 2007a, 2156). It is a maximum range of uncertainty since it considers all applicable ensemble results, but merely a lower bound because of the unexamined model uncertainty.[8] They conclude that the "range of possibilities highlighted for future climate at all scales clearly demonstrates the urgency for climate change mitigation measures and provides non-discountable ranges which can be used" by decision-makers (Stainforth et al. 2007a, 2159). So, the function of these models is to *highlight possibilities that cannot be ignored* for decision-making purposes.

Though well-respected, this work clearly lies at an extreme when it comes to climate science's view of climate models and probabilistic methods. (Witness the continued use of probabilities in the IPCC's sixth assessment report.) Other climate scientists who are critical of the sampling approach to generate probabilities nevertheless assert that models are impressively reliable on *particular aspects* of the climate and that groups of models often do better than individual models, even than the *best* individual models (Annan and Hargreaves 2011; Knutti et al. 2010). This motivates their continued investigation of probabilistic methods for the treatment of model ensembles and perhaps explains the wider continued reliance on these methods.

The most pessimistic of the possibilists is Gregor Betz, who takes inspiration from Stainforth et al. but draws markedly more negative conclusions. Betz introduced his possibilistic view in the context of the provision of policy advice (Betz 2007), where it was referred to as the "scenario approach." Betz's motivation was the view that in a democracy, scientific input to policy should be value-free (cf. Betz 2013). He argues that the probabilistic method relied on unmotivated and subjective judgements, either in the form of Bayesian treatments of model uncertainty or more directly via expert elicitation of probabilities.

While this value-free ideal might be unpalatable to some, there is much value in Betz's work which is independent of it. In particular, his discussion of which possibilities are worthy of attention and his demarcation of categories of modal hypotheses. Betz advocates for reporting as possible all scenarios which cannot be ruled out given the current body of knowledge (Betz 2007, 6), and he calls these "serious possibilities" following Levi (1980). Betz (2015) also considers relativising this to a body of knowledge, to generate kinds of serious possibility such as logical, physical, etc. In later work, Betz (2010) shifts to a more substantively specified notion. He distinguishes between three categories of possibilities: verified possibilities, verified impossibilities, and possibilistic hypotheses. A verified possibility has been demonstrated to be consistent with background knowledge, a verified impossibility has been demonstrated to be inconsistent with background knowledge, and possibilistic hypotheses have been articulated but not verified (Betz 2010, 93). (Verifying impossibilities is also referred to as falsifying possibilistic hypotheses.)

In later works (Betz 2010, 2015), Betz's critique of probabilism is more squarely focused on the inadequacy of climate models. The core question for Betz is whether climate simulations can verify and falsify possibilities. He is motivated by the more extreme position presented in the second Stainforth-led paper: while Stainforth et al. (2007a) seem to suggest that models can positively identify relevant possibilities, Stainforth et al. (2007b, 2168) state that "we cannot confirm the relevance of any climate forecast." Betz concurs: models cannot verify possibilities. One reason for this is quite straightforward: since models contain contrary-to-fact assumptions, they do not represent serious possibilities. This is a simple consequence of his definition of serious epistemic possibility, which perhaps opens room for a version of possibilism on which climate models do verify possibilities: as Sjölin Wirling and Grüne-Yanoff (2021) note, *objective* possibilities regularly involve contrary-to-fact specifications. Katzav's variety of possibilism, discussed below, may be an example of this strategy.

Betz also argues that, due to underdetermination, models cannot falsify hypotheses. This leaves relatively limited room for their use in science: models can play a role in formulating possibilistic hypotheses since they can identify emergent phenomena and draw out the implications of large, complex datasets (Betz 2010, 97–98). To support this purpose, Betz advocates for a move away from focusing on the convergence of simulation results and towards encouraging a proliferation of models and an exploration of the scenarios they generate.

Joel Katzav criticises Betz for not basing his possibilism in the actual practice of climate modelling (Katzav 2023) and offers a somewhat more optimistic possibilism. Katzav develops a notion called "real possibility," which is intended as a species of objective possibility (Katzav 2023, 2). In the original formulation, to say that something is a "real possibility at some time t is, roughly, to say that it is consistent with the overall way things have been up until t and that nothing known excludes it" (Katzav, Dijkstra, and de

Laat 2012, 270). A state is compatible with the way things are at a time, if "in something like the circumstances obtaining over the period, the domain's laws and/or mechanisms, or a similar set of such laws and/or mechanisms, would bring the state of affairs about" (Katzav 2014, 236). So real possibilities are objective possibilities, with two qualifiers. The first is that they haven't been excluded by our current knowledge, and the second is that they have "the potential to be realised in the actual world" (Katzav 2023, 8).

Models can provide us with real possibilities by capturing (perhaps roughly) the central processes, laws, and mechanisms of the system. Where Betz's definition of serious possibility means that contrary-to-fact assumptions prevent climate models from representing serious possibilities, Katzav's objective conditions and use of qualifiers such as "the *basic* way things are" and "*something like* the circumstances" are meant to avoid this problem. "Representing the basic way the climate system is over a period of time is compatible with being false to a substantial degree" (Katzav 2014, 236). Katzav also argues that real possibilities are what we need. While epistemic possibilities may be too weak to ground climate action, the close link of real possibilities with the way things are in the actual world motivates taking them seriously. Thus, "when we take outputs to be epistemically possible, we ought also to be able to suppose that they represent real possibilities" (Katzav 2023, 2; cf. Katzav et al. 2021).

For Katzav, "realness" comes in degrees. Two slightly contrary takes on degrees of realness or remoteness are offered across Katzav's work. Katzav et al. (2012, 273) employ Popper's notion of corroboration and suggest that "a claim's degree of corroboration is … correlated with the extent to which the claim is consistent with the overall way things are and, therefore, with the extent to which the claim is a real possibility." This allows for "realness" to depend on the extent to which a claim has been tested, in addition to the results of particular statistical tests. This degreed notion is intended to allow a pure possibilistic method to rank possibilities by how remote they are—where remoteness is inverse to realness. In later work (Katzav 2014, 237), however, we are cautioned that model errors threaten such judgements of remoteness. Instead, we are directed towards counterfactual assessments of remoteness: if models indicate it is a real possibility that human emissions have such-and-such an effect on the climate, then we can conclude that the occurrence of this effect is less remote (alternatively, more real) than it would have been without anthropogenic climate change.

10.5 Uncertainty, Decision, and Possibility

This final section links the previous discussions of possibility to a common background issue: how the results of climate models can be used to support decisions. This can be framed as a debate about the appropriate representation of uncertainty for climate model results, where "appropriateness" involves making use of the information contained in those results while not going beyond that information by offering false precision. The link to decision-making is crucial: more informative representations of uncertainty allow for more discerning decisions, and so we seek a representation of uncertainty that is as informative as is justifiable.

Possibility is at the heart of uncertainty: uncertain agents entertain multiple possibilities, without knowing which is actual. Modern representations of uncertainty therefore begin with a notion of possibility, formally represented with a set of possible worlds. These represent the elementary outcomes or ways the world might be, about which we are uncertain. Sets of worlds represent coarser descriptions of what might happen, called events or propositions. When modelling an agent's uncertainty, it is usual to include a world for every outcome the agent can distinguish, specified in a way they find relevant. Using these, we can represent the agent's beliefs about what might happen or what might be the case. So, for a climate scientist thinking about climate change, these might be possible future states of the climate, resolved at the level of detail that fits their understanding, and specified using climate variables used in their work. These worlds then represent different ways the climate might be.

Uncertainty can be represented crudely using only sets of possible worlds: we can start with a large set of worlds W, and represent an agent's uncertainty with the subset of these worlds that they consider possible, $U \subseteq W$. By introducing propositions, we can also represent differences between the fineness of the agent's understanding and the true (or modeller's) picture, by representing the agent's uncertainty with sets of worlds rather than worlds. The set U can also represent worlds that are compatible with a body of knowledge, impersonally specified, or the worlds considered possible by a group. Either way, this is a crude but serviceable representation of uncertainty: it merely tells us that a set of states is possible, without giving any ranking between them or offering quantitative information about their relative or absolute likelihoods.

The starting point of the possibilistic interpretation of climate models is the question: do climate simulations represent (or assist us in identifying) possibilities? An answer to this question might be: yes, an ensemble of climate models can produce a subset of worlds C, specified using climate variables contained in those models, which are to be considered possible. Betz is sceptical of this answer, while the other possibilists discussed endorse it.

Despite its crudeness, this representation of uncertainty can support decision-making. Indeed, there are many decision rules one can specify that rely only on properties that this representation provides. Here are two examples. Suppose that a decision-maker is choosing between two actions and that they are aware of two possibilities so that they have four outcomes to consider. The decision rule *Caution*$_{SP}$ recommends the action with the best worst-case outcome. So, the decision-maker considers the two outcomes of each action and notes the worst-case. They then compare these worst-cases and pick the action which results in the better one. As the name suggests, this rule displays great caution, as the decision is made solely on the basis of the worst possible outcome. An alternative is *Robustness*$_{SP}$, which recommends the action that performs tolerably well over the widest set of possibilities, for some threshold of toleration. Lempert and Schlesinger (2000) suggest a rule like this for climate decisions.

This link to the decision highlights one way in which modal modelling matters. With their focus on influencing decision-making, scientists like Stainforth and Shepherd take their modal models to be relevant to what we do in the actual world. The fact that climate modelling is, in an important sense, modal does not at all diminish its importance for decision-making. This should allay the worry of authors like Dethier (2023), who worry that the possibilistic interpretation competes with the idea that models are relevant to hypotheses about the actual world.

Returning to the discussion of representations of uncertainty, the next question faced by possibilists is whether climate ensembles can support a *more informative* representation than mere sets of possibilities.

Probability offers a much more informative representation and correspondingly supports more sophisticated decision rules. A brief bit of formalism: a probability measure is a function, Pr, which assigns a number between 0 and 1 to each subset of W.[9] Such a function is called a probability measure if, and only if, it obeys these three properties: (Pr1) $Pr(\varnothing) = 0$, (Pr2) $Pr(W) = 1$, and (Pr3) if $X \cap Y = \varnothing$ then $Pr(X \cup Y) = Pr(X) + Pr(Y)$.[10] As we saw above, there are significant doubts about whether climate ensembles can support probabilities. These worries are related to general limitations that probabilistic representations of uncertainty face (Halpern 2017, 16–23).

Chief among these is the worry that a probabilistic representation is very demanding: every possibility must be comparable to every other, not just ordinally but cardinally, and enough information needs to be available to generate precise numbers for each alternative while obeying the additivity condition. If the knowledge state being represented is incomplete, with certain possibilities not being comparable, these conditions will not be met. Similarly, if ignorance about what is possible is significant, probabilities may not be suitable, as Pr1 assumes that the set W contains all possibilities. Probabilities also cannot represent (or have difficulty representing) the weight of evidence underlying comparisons, as $Pr(X) = 0.5$ may be used to represent both a situation in which there is plenty of evidence for each of X and $\bar{X}$, and to represent a situation in which there is no evidence at all on the matter.

A range of other representations of uncertainty have been developed to address these limitations. These include imprecise probabilities, Dempster-Shafer functions, plausibility measures, and possibility measures (Genin and Huber 2021; Halpern 2017). The latter are of interest to us, because of their association with the possibilist interpretation and, in particular, Katzav's degreed notion of realness/remoteness—Katzav et al. (2021, §4.2) explicitly suggest them as a tool for working with ensemble outputs. There is a prima facie attractiveness to this, as the principle underlying formal possibility theory is sometimes glossed as "any hypothesis that is not known to be impossible cannot be ruled out," which accords with some of Betz's statements seen above and with Katzav's Popperian gloss on real possibility. The two questions facing the possibilist are again: can ensembles support possibility measures and is this the most informative representation justifiable?

A possibility measure is a function, Po, which assigns a number between 0 and 1 to each subset of W. Such a function is called a possibility measure if, and only if, it obeys these three properties: (Po1) $Po(\varnothing) = 0$, (Po2) $Po(W) = 1$, and (Po3) if $X \cap Y = \varnothing$ then $Po(X \cup Y) = \max\{Po(X), Po(Y)\}$. Po1 and Po2 are identical to Pr1 and Pr2, the difference is in Po3. Since $Po(X \cup Y)$ is equal to the largest of $Po(X)$ and $Po(Y)$, $Po(X \cup Y) \leq Po(X) + Po(Y)$, and so possibility measures are sub-additive. That is to say that a proposition is at least as possible$_m$ as each of the propositions it includes, and no more possible$_m$ than the most possible$_m$ of them. The subscript "m" is added to "possible" to draw attention to its use as a technical term, one which has an unclear connection with the notions of possibility discussed above, and with those used in modal epistemology and metaphysics.

Possibilities$_m$ are variously interpreted. Dubois and Prade, important theorists of possibility measures, report that their originator L. A. Zadeh used them to represent feasibility, "as in the example of *how many eggs can Hans eat for his breakfast*" (Dubois and Prade 2015a, 33), which is a modal notion but not an epistemic or metaphysical one. Dubois and Prade offer different interpretations of their own: the function Po represents "the state of knowledge of an agent (about the actual state of affairs)... distinguishing what is plausible from what is less plausible, what is the normal course of things from what is not, what is surprising from what is expected." Later they say that $Po(X) = 0$ "means that state [X] is rejected as impossible," while $Po(X) = 1$ "means that state [X] is totally possible (=plausible)" (Dubois and Prade 2015a, 34). These are potential interpretations depending on the use case, rather than synonyms: plausibility, typicality, and surprise are distinct notions. None corresponds obviously to "degree of possibility," for either an epistemic or objective notion of possibility, and it is unclear why total (presumably full, unqualified) possibly should be equated with *plausibility*. The most useful interpretation for our purposes begins with the observation that possibilities$_m$ are often relativised to a body of knowledge K. So, possibility$_m$ measures consistency with K (in a fuzzy sense) and thus could be used to articulate a kind of epistemic possibility.

Possibilities$_m$ are significantly less constrained than probabilities. Importantly, it is not the case that, in general, $Po(X) = 1 - Po(\bar{X})$. Instead, $1 - Po(\bar{X})$, called the "necessity$_m$" of X or $Ne(X)$, is thought of as representing a distinct part of the uncertainty about X. Where $Po(X)$ is sometimes described as measuring the extent to which X is consistent with the knowledge on which Po is based, $Ne(X)$ measures the extent to which that knowledge implies X (Dubois and Prade 2001). Zero necessity$_m$ does not put any constraints on possibility$_m$; indeed, $Ne(X) = 0$ is compatible with $Po(X) = 1$, their conjunction is glossed as "X is consistent with the knowledge base but not implied by it"—leaving room for ay belief attitude.

Now that we know what possibility measures are, we can ask whether climate model ensembles can support the assignment of numerical degrees of possibility$_m$, although they cannot support probabilities (according to possibilists). One difficulty in establishing this is that "there are several representations of epistemic states that are in agreement with the above setting ... [b]ut all these representations of epistemic states do not have the same expressive power. They range from purely qualitative to quantitative possibility distributions, using weak orders, qualitative scales, integers, and reals" (Dubois and Prade 2015b). So, the proposal to use possibility measures to represent climate model ensemble results is radically underspecified until advocates describe the degree of expressive power they take climate models to support. If these possibility measures merely capture an incomplete comparative ranking of possibilities, then this should be relatively easy to construct. Katzav's example of a counterfactual assessment of the remoteness of an effect with and without climate change may be an example of a straightforward comparison. As more comparisons are added and the ranking approaches a complete ordinal scale, more information will need to be supplied by the possibilist, including information on how to construct rankings from sets of apparently undifferentiated model results. This is by no means impossible: possibility measures have mathematical connections to imprecise probabilities, which are relatively easy to use to generate rankings—see, for example, Roussos, Bradley, and Frigg (2021). The worry is that this will soon contradict the assessment of Stainforth et al. (2007a) that we cannot weigh or differentiate between the results of ensemble members in a principled way. Recalling that this assessment was the source of much of the possibilist programme, this presents a serious complication.

Possibilists thus have a difficult line to walk if they wish to go beyond mere sets of possibilities, without ascribing more power to climate model ensembles than their own critique of probabilism allows. A more detailed case needs to be made for possibility measures, one which specifies precisely how expressive a representation is intended and which argues in detail that climate ensembles can support it. Ideally, this case would also argue that we can do no better than the proposed possibilistic account. Attention could then turn to the development of statistical methods for constructing these possibility$_m$ distributions.

10.6 Conclusion

Climate science is a rich source of examples for the philosopher interested in modal modelling. The nature of climate change as a central problem for science, and the desire to attribute weather phenomena to human-caused climate change, creates conditions in which scientists are naturally led to engage in modal reasoning. The debates discussed in sections 10.3 and 10.4 are only a small sample of such reasoning and the role that possibility plays in it.

The discussion of EWE attribution illustrates how scientists are in need of sophisticated modal concepts. Both the risk and storyline side of the debate aim to articulate a species of possibility which has desirable properties for their purposes. Philosophers can play a crucial role here. In the case of the storyline method, I argued that the form of possibility identified has important similarities with that identified independently by Katzav, in his discussion of model ensembles. Helping scientists to articulate their different goals and the required modal notions to support those goals can help to advance scientific practice.

The discussion of ensemble results in section 10.4 highlights another reason to engage with scientific practice. Detailed engagement with the scientific practices of modal modelling can furnish us with new modal concepts—such as Katzav's "real possibility." Real possibilities, with their dual status as nearby objective possibilities that have not been epistemically excluded, are compelling candidates for decision-making attention. Their significance is therefore not limited to the philosophy of climate science or discussions of climate policy.

Section 10.5 of this chapter highlighted how paying attention to the role of possibilities in decision-making can guide both our modal theorising and our scientific practice. It can guide theoretical work in modal epistemology because considerations about decision-making determine the kind of possibilities we want to have access to for practical purposes. It can inform scientific practice by identifying the conditions that need to be met to utilise an uncertainty representation such as possibility measures, and the kind and quality of information required to generate such measures from climate model outputs.

Notes

1 It is common to talk in terms of "possible worlds," as I will do in section 10.5. There is no issue with doing so here, so long as one accepts that a "world" can be specified coarsely and that the epistemically possible worlds do not stand in a simple relationship to the objectively possible worlds.

2 Readers interested in more details about climate models and the philosophical issues they raise should read the excellent collected volume edited by Lloyd and Winsberg (2018).

3 In some overview papers, contributions are assessed directly; for example, Trenberth et al. use observations of high sea-surface temperatures, and the known process in which these raised temperatures provide moisture to storms, to make a qualitative statement about the impact of climate change-related increased sea temperatures on the huge amount of snow in the 2010 "Snowmageddon" storm (Trenberth et al. 2015, 727).

4 For example, having a climate modelling centre is a way to signal that a country takes climate change seriously. Less politically, since the purposes and interests of the modellers influence the idealisations and simplifications employed in the model's development, there is a natural desire to have "our model" reflect "our priorities." It is difficult to disentangle the contributions these factors have. In this chapter, the focus is on the uncertainty and purpose-diversity reasons for the range of models, but political and social reasons should not be forgotten.

5 https://www.carbonbrief.org/cmip6-the-next-generation-of-climate-models-explained/.
6 Recall: climate is an aggregate notion and the climate system has significant natural variability, so even a perfect model is expected to generate different outputs for repeated runs.
7 This section reviews and discusses some important entries in the debate in climate science and philosophy but does not pretend to be comprehensive. Important philosophical contributions are neglected from other parts of the philosophy of science (Hansson 2011), as are important papers from other disciplines such as economics (Sugden 2009), which have influenced this debate.
8 These articles, though seminal, are over 15 years old. So it is worth noting that Stainforth remains sceptical of the use of probabilities in the analysis of climate ensembles—see Katzav et al. (2021) and Stainforth and Calel (2020).
9 Alternatively, if one wished to distinguish impossibility from zero probability, and represent both, one could define P on U.
10 Property 1 is redundant but included here for comparison with possibility measures.

References

Annan, James D., and Julia C. Hargreaves. 2011. "Understanding the CMIP3 Multi-model Ensemble." *Journal of Climate* 42: 4529–38.
Betz, Gregor. 2007. "Probabilities in Climate Policy Advice: A Critical Comment." *Climatic Change* 85(1): 1–9. https://doi.org/10.1007/s10584-007-9313-9.
Betz, Gregor. 2010. "What's the Worst Case? The Methodology of Possibilistic Prediction." *Analyse & Kritik* 1: 87–106.
Betz, Gregor. 2013. "In Defence of the Value Free Ideal." *European Journal for Philosophy of Science* 3: 207–20.
Betz, Gregor. 2015. "Are Climate Models Credible Worlds? Prospects and Limitations of Possibilistic Climate Prediction." *European Journal for Philosophy of Science* 5(2): 191–215. https://doi.org/10.1007/s13194-015-0108-y.
Brandon, R. N. 1990. *Adaptation and Environment*. Princeton: Princeton University Press.
Chan, Wilson C. H., Theodore G. Shepherd, Katie Facer-Childs, Geoff Darch, and Nigel W. Arnell. 2022. "Storylines of UK Drought Based on the 2010–2012 Event." *Hydrology and Earth System Sciences* 26(7): 1755–77. https://doi.org/10.5194/hess-26-1755-2022.
Dethier, Corey. 2022. "When Is an Ensemble Like a Sample? 'Model-Based' Inferences in Climate Modelling." *Synthese* 200(52): 1–20.
Dethier, Corey. 2023. "Against 'Possibilist' Interpretations of Climate Models." *Philosophy of Science* 90(5): 1417–26. https://doi.org/10.1017/psa.2023.6.
Dubois, Didier, and Henri Prade. 2001. "Possibility Theory, Probability Theory and Multiple-Valued Logics: A Clarification." *Annals of Mathematics and Artificial Intelligence* 32: 35–66.
Dubois, Didier, and Henri Prade. 2015a. "Possibility Theory and Its Applications: Where Do We Stand?" In *Springer Handbook of Computational Intelligence*, edited by Kacprzyk, Janusz, and Witold Pedrycz, 31–60. Springer Handbooks. Berlin, Heidelberg: Springer. https://doi.org/10.1007/978-3-662-43505-2_3.
Dubois, Didier, and Henri Prade. 2015b. "Practical Methods for Constructing Possibility Distributions." *International Journal of Intelligent Systems* 31(3): 215–39.
Genin, Konstantin, and Franz Huber. 2021. "Formal Representations of Belief." *The Stanford Encyclopedia of Philosophy*, edited by Edward N. Zalta and Uri Nodelman . Metaphysics Research Lab, Stanford University. https://plato.stanford.edu/archives/fall2022/entries/formal-belief.

Goodwin, William M. 2015. "Global Climate Modelling as Applied Science." *Journal for General Philosophy of Science* 46(2): 339–50.

Halpern, Joseph Y. 2017. *Reasoning about Uncertainty*. Cambridge, MA: MIT Press. https://doi.org/10.7551/mitpress/10951.001.0001.

Hansson, Sven Ove. 2011. "Coping with the Unpredictable Effects of Future Technologies." *Philosophy & Technology* 24(2): 137–49.

Hoerling, Martin, Arun Kumar, Randall Dole, John W. Nielsen-Gammon, Jon Eischeid, Judith Perlwitz, Xiao-Wei Quan, Tao Zhang, Philip Pegion, and Mingyue Chen. 2013. "Anatomy of an Extreme Event." *Journal of Climate* 26(9): 2811–32. https://doi.org/10.1175/JCLI-D-12-00270.1.

Katzav, Joel. 2014. "The Epistemology of Climate Models and Some of Its Implications for Climate Science and the Philosophy of Science." *Studies in History and Philosophy of Science Part B: Studies in History and Philosophy of Modern Physics* 46(B): 228–38. https://doi.org/10.1016/j.shpsb.2014.03.001.

Katzav, Joel. 2023. "Epistemic Possibilities in Climate Science: Lessons from Some Recent Research in the Context of Discovery." *European Journal for Philosophy of Science* 13(4): 57. https://doi.org/10.1007/s13194-023-00560-7.

Katzav, Joel, Henk A. Dijkstra, and A. T. J. (Jos) de Laat. 2012. "Assessing Climate Model Projections: State of the Art and Philosophical Reflections." *Studies in History and Philosophy of Science Part B: Studies in History and Philosophy of Modern Physics* 43(4): 258–76. https://doi.org/10.1016/j.shpsb.2012.07.002.

Katzav, Joel, Erica L. Thompson, James Risbey, David A. Stainforth, Seamus Bradley, and Mathias Frisch. 2021. "On the Appropriate and Inappropriate Uses of Probability Distributions in Climate Projections and Some Alternatives." *Climatic Change* 169(1): 15. https://doi.org/10.1007/s10584-021-03267-x.

Knutti, Reto, C. Furrer, C. Tebaldi, J. Cermak, and G. Meehl. 2010. "Challenges in Combining Projections from Multiple Climate Models." *Journal of Climate* 23(10): 2739–58.

Knutti, Reto, David Masson, and Andrew Gettelman. 2013. "Climate Model Genealogy: Generation CMIP5 and How We Got There." *Geophysical Research Letters* 40(6): 1194–99. https://doi.org/10.1002/grl.50256.

Lempert, Robert J., and Michael E. Schlesinger. 2000. "Robust Strategies for Abating Climate Change." *Climatic Change* 45(3): 387–401. https://doi.org/10.1023/A:1005698407365.

Levi, Isaac. 1980. *The Enterprise of Knowledge: An Essay on Knowledge, Credal Probability, and Chance*. Cambridge, MA: MIT Press.

Lloyd, Elisabeth A. 2015. "Model Robustness as a Confirmatory Virtue: The Case of Climate Science." *Studies in History and Philosophy of Science Part A* 49(February): 58–68. https://doi.org/10.1016/j.shpsa.2014.12.002.

Lloyd, Elisabeth A., and Naomi Oreskes. 2018. "Climate Change Attribution: When Is It Appropriate to Accept New Methods?" *Earth's Future* 6(3): 311–25. https://doi.org/10.1002/2017EF000665.

Lloyd, Elisabeth A., and Naomi Oreskes. 2019. "Climate Change Attribution: When Does It Make Sense to Add Methods?" *Epistemology & Philosophy of Science* 56(1): 185–201.

Lloyd, Elisabeth A., and Eric Winsberg, eds. 2018. *Climate Modelling: Philosophical and Conceptual Issues*. Cham, Switzerland: Palgrave Macmillan. https://doi.org/10.1007/978-3-319-65058-6.

Oldenborgh, Geert Jan van, Karin van der Wiel, Sarah Kew, Sjoukje Philip, Friederike Otto, Robert Vautard, Andrew King, et al. 2021. "Pathways and Pitfalls in Extreme Event Attribution." *Climatic Change* 166(1): 13. https://doi.org/10.1007/s10584-021-03071-7.

Oreskes, Naomi. 2004. "The Scientific Consensus on Climate Change." *Science* 306(5702): 1686. https://doi.org/10.1126/science.1103618.

Oreskes, Naomi. 2018. "The Scientific Consensus on Climate Change: How Do We Know We're Not Wrong?" In *Climate Modelling: Philosophical and Conceptual Issues*, edited by Elisabeth A. Lloyd and Eric Winsberg, 31–64. Cham, Switzerland: Palgrave Macmillan. https://link.springer.com/book/10.1007/978-3-319-65058-6

Otto, Friederike E. L., Geert Jan van Oldenborgh, Jonathan Eden, Peter A. Stott, David J. Karoly, and Myles R. Allen. 2016. "The Attribution Question." *Nature Climate Change* 6(9): 813–16. https://doi.org/10.1038/nclimate3089.

Parker, Wendy S. 2009. "Does Matter Really Matter? Computer Simulations, Experiments and Materiality." *Synthese* 169(3): 483–96.

Parker, Wendy S. 2010. "Whose Probabilities? Predicting Climate Change with Ensembles of Models." *Philosophy of Science* 77(5): 985–97. https://doi.org/10.1086/656815.

Parker, Wendy S. 2011. "When Climate Models Agree: The Significance of Robust Model Predictions." *Philosophy of Science* 78(4): 579–600. https://doi.org/10.1086/661566.

Pättiniemi, Ilkka, and Rami Koskinen. Forthcoming. "Climate Models." In *The Routledge Handbook of Philosophy of Scientific Modeling*, edited by Tarja Knuuttila, Natalia Carrillio, and Rami Koskinen. London, UK: Routledge.

Petersen, Arthur. 2012. *Simulating Nature*. 2nd ed. Florida: CRC Press.

Philip, Sjoukje, Sarah Kew, Geert Jan van Oldenborgh, Friederike Otto, Robert Vautard, Karin van der Wiel, Andrew King, et al. 2020. "A Protocol for Probabilistic Extreme Event Attribution Analyses." *Advances in Statistical Climatology, Meteorology and Oceanography* 6(2): 177–203. https://doi.org/10.5194/ascmo-6-177-2020.

Roussos, Joe, Richard Bradley, and Roman Frigg. 2021. "Making Confident Decisions with Model Ensembles." *Philosophy of Science* 88(3): 439–60.

Shepherd, Theodore G. 2016. "A Common Framework for Approaches to Extreme Event Attribution." *Current Climate Change Reports* 2(1): 28–38. https://doi.org/10.1007/s40641-016-0033-y.

Shepherd, Theodore G., Emily Boyd, Raphael A. Calel, Sandra C. Chapman, Suraje Dessai, Ioana M. Dima-West, Hayley J. Fowler, et al. 2018. "Storylines: An Alternative Approach to Representing Uncertainty in Physical Aspects of Climate Change." *Climatic Change* 151(3): 555–71. https://doi.org/10.1007/s10584-018-2317-9.

Stainforth, David A., M. R. Allen, E. R. Tredger, and Leonard Smith. 2007a. "Confidence, Uncertainty and Decision-Support Relevance in Climate Predictions." *Philosophical Transactions of the Royal Society A* 365: 2145–61.

Stainforth, David A., and Raphael Calel. 2020. "New Priorities for Climate Science and Climate Economics in the 2020s." *Nature Communications* 11(1): 3864. https://doi.org/10.1038/s41467-020-16624-8.

Stainforth, David A., T. E. Downing, R. Washington, A Lopez, and M. New. 2007b. "Issues in the Interpretation of Climate Model Ensembles to Inform Decisions." *Philosophical Transactions of the Royal Society A* 365(1857): 2163–77.

Stott, Peter A., Nikolaos Christidis, Friederike E. L. Otto, Ying Sun, Jean-Paul Vanderlinden, Geert Jan van Oldenborgh, Robert Vautard, et al. 2016. "Attribution of Extreme Weather and Climate-related Events." *Wiley Interdisciplinary Reviews. Climate Change* 7 (1): 23–41. https://doi.org/10.1002/wcc.380.

Sugden, Robert. 2009. "Credible Worlds, Capacities and Mechanisms." *Erkenntnis (1975-)* 70(1): 3–27.

Tebaldi, C., and R. Knutti. 2007. "The Use of the Multi-Model Ensemble in Probabilistic Climate Projections." *Philosophical Transactions of the Royal Society A: Mathematical, Physical and Engineering Sciences* 365(1857): 2053–75. https://doi.org/10.1098/rsta.2007.2076.

Thompson, Erica, Roman Frigg, and Casey Helgeson. 2016. "Expert Judgment for Climate Change Adaptation." *Philosophy of Science* 83(5): 1110–21. https://doi.org/10.1086/687942.

Trenberth, Kevin E., John T. Fasullo, and Theodore G. Shepherd. 2015. "Attribution of Climate Extreme Events." *Nature Climate Change* 5(8): 725–30. https://doi.org/10.1038/nclimate2657.

Tulving, E. 1972. "Episodic and Semantic Memory." In *Organization of Memory*, edited by E. Tulving, and W. Donaldson, 381–402. New York, NY: Academic Press.

Vautard, Robert, Olivier Boucher, Geert Jan van Oldenborgh, Friederike Otto, Karsten Haustein, Martha M. Vogel, and Sonia I. Seneviratne, et al. 2019. "Human Contribution to the Record-Breaking July 2019 Heat Wave in Western Europe." *World Weather Attribution*, August, 32.

Verreault-Julien, Philippe. 2019. "How Could Models Possibly Provide How-Possibly Explanations?" *Studies in History and Philosophy of Science Part A* 73: 22–33.

Winsberg, Eric. 2018. *Philosophy and Climate Science*. Cambridge: Cambridge University Press.

Winsberg, Eric, Naomi Oreskes, and Elisabeth A. Lloyd. 2020. "Severe Weather Event Attribution: Why Values Won't Go Away." *Studies in History and Philosophy of Science Part A* 84: 142–49.

Wirling, Ylwa Sjölin, and Till Grüne-Yanoff. 2021. "Epistemic and Objective Possibility in Science." *The British Journal for the Philosophy of Science*. https://doi.org/10.1086/716925.

Zachariah, Mariam, Robert Vautard, Dominik L. Schumacher, Maja Vahlberg, Dorothy Heinrich, Emmanuel Raju, Lisa Thalheimer, et al. 2022. "Without Human-Caused Climate Change Temperatures of 40°C in the UK Would Have Been Extremely Unlikely." *World Weather Attribution*, July, 26.

11 Prospective Modeling

Alfred Nordmann

11.1 Opening

What might be the case and *What can be done*—these are interrelated but categorically different propositions. Both concern the possible, and both involve modeling practices in science, engineering, and daily life. After a quick review of modeling possible and actual states—what is and what might be—this paper moves to the other side of the categorical divide. Modeling what can be done is "prospective modeling" and consists in the demonstration of technical power. This form of modeling the possible does not implicate the ontological questions of reality and modality which arise within a representational framework. Also, it does not entangle us in the predicament of referring to the future. Prospective models establish a claim to what can be done, and they do so in terms of what has already been achieved in the model. This proposal joins forces with other attempts to fill a gap in extant accounts of modeling. It goes further only by breaking explicitly with the idea that models instantiate theory or represent reality. Prospective models do not relate to a given target system by way of picture and pictured but can give rise to a targeted or intended system by way of rehearsal and reenactment. They provide a template for mimetic action in another setting than that of the model itself.

11.2 Possibilitas et potentia

Different conceptions of possibility provide the backdrop for the comparison of two ways of modeling the possible. In the terms proposed by Peter Galison, one of these conceptions is ontologically concerned, and the other ontologically indifferent (Galison 2017). Ontology pursues the question of being and everything that really exists. This question is not particularly troubling for all that is right in front of our eyes. Ontological concern, if not anxiety, comes with "neutrinos, angels, devils, and classes" (Goodman 1983, 33), with counterfactuals and dispositions, with the physical structures behind or below the appearances, with remembered and reported events, and the modal qualification of what is not actually and not necessarily, but possibly the case.

DOI: 10.4324/9781003342816-15

In the classical modal sense of *possibilitas*, a possible state of affairs has the property or quality that it might actually be the case—somewhere, when the time is right, or when certain conditions are met that do not presently obtain. Here, ontological concern takes the guise of an epistemological quandary: how to ascertain that a non-actual state of affairs has the quality of possibly being the case? Involving counterfactual claims about unrealized but realizable dispositional properties, these are questions of truth or falsity, they revolve around veridical statements which posit something that is not actually now but latently now the case.

In contrast to all of this, there are ways of modeling the possible that do not induce ontological anxiety in that they do not posit the reality of the non-actual. Aside from modal possibility as a property that qualifies reality, there is potency (*potentia*) and what is effectively possible for me or you or something to do. The latter refers to a capability or power that, once acquired and demonstrated, can be exercised again and again. To be sure, the exercise of power involves many questions of knowledge, but not the problem of induction or other questions concerning an elusive, perhaps unknowable reality. Take a carpenter, for example, who acquired and demonstrated the power to build a cabinet with drawers and doors that open and close neatly and effortlessly. With the demonstration of this power comes the possibility of doing it again, building one or many more such cabinets. To be sure, the carpenter might be lazy or sloppy or cheap, from now on delivering ill-fitted cabinets. We do not know, in other words, that the carpenter will produce first-class cabinets, but still we do not doubt the carpenter's power to do so, which was manifested once and for all in the prototype. We might wonder whether the carpenter's powers transfer from this situation to that, from oak to pine, from showroom to living room, but these are practical or technical concerns regarding problems of performance and the exercise of a power which has been acquired and demonstrated.[1]

Moving from the carpenter to the modeler, the shift from possibility to potency corresponds to a shift from modeling what might be the case to "prospective modeling" as the acquisition and demonstration of power in a model. These two types of modeling and of conducting research are complementary: They are different registers or language games of "possibility" and aim for different kinds of knowledge, veridical knowledge and working knowledge. For example, anxious to know the truth, one can always ask, "how can I be certain that the carpenter or experimentalist has the power to do something once and again; how do I know that the accomplishment of the cabinet or the observed effect in the experiment was not a matter of cheating or of luck?" By asking this perfectly legitimate skeptical question, one is not undermining the categorical difference between "what is possibly the case" and "what is possible to do"—one is only changing register, shifting the language game from the probing of powers to ascertaining the reality of possibility. In other words, talk of actuality can always lead to talk about modal possibility or *possibilitas*.

11.2.1   *What Might Be the Case—But Is Not*

Models of what is actually the case are always also models of what might possibly be the case. Indeed, from the modeling point of view, the actual is always just a special instance of the possible. Modeling the possible is therefore part and parcel of science. Any description of what is contingently the case specifies possibilities of how it could be otherwise. A potted history of conceptions of modeling serves to illustrate how modal possibility (*possibilitas*) comes cheap in all kinds of ways.

From the point of view of theory, the models of a theory form a class of possible situations that instantiate the theory. They thus circumscribe the space of what is physically and not just logically possible—what might be the case if the requisite initial conditions obtain, irrespective of whether these are quickly and easily realized in an experimental setting, or realizable at all.

From the point of observation and experiment, models of a particular phenomenon or event provide a reconstruction of that situation by formulating in the language of mathematics and physical theory a structural or dynamic analogue. When we informally say, for example, that a virus is "spreading like wildfire," there are propagation models that take this analogy quite literally, modeling the spread of disease with tools developed to dynamically represent wildfires—or *vice versa*. If models on the first view are actual or possible instantiations of a theory, models on this second view provide representations of an actually observed phenomenon. These open a space for possibility in that any claim of a causal dependency supports counterfactual considerations of parameter variations and their effects. By specifying the conditions and constraints under which the modeled phenomenon can be said to arise, by probing the sensitivity and specificity of such models, hypothetical alternatives or other worlds come into view.

From the point of view, thirdly, of models as mediators between theory and reality, the emphasis is on fitting, tuning, calibrating, and thus on the countless parameters that determine a situation and the selection by the model of salient inter-dependencies (Morgan and Morrison 1999, Nordmann 2008). Here, a free play of "what ifs" configures a specific relation of model, theory, and reality. By varying parameters one performs a causal analysis as one seamlessly moves between the actual and the possible—tuning the system by finally setting the dial to the representation of an actual phenomenon or to the design specification of an experiment that might realize a possible outcome.

So, if the business of science is that of representation or theoretical description of the world, the actual and the possible are always in play, and there is nothing very special about "modeling the possible." To be sure, theoretical entities, counterfactuals, disposition terms, and questions of projectibility pose the ontological difficulties which occupied analytic philosophers throughout much of the 20th century (e.g., Goodman 1983). In recent decades, these quandaries were overshadowed by an epistemological provocation. The problem shifted from the reality of the non-actual to the question of how one can tell the truth by saying something false. Beginning with discussions of similarity, idealization and abstraction, fictionalism and the as if, philosophers became more interested in hypothetical entities than in theoretical entities, more interested in the conceptual function of "siligen atoms" (Winsberg 2006) than in the reality of electrons. If for the construction of a model one willfully invents a silicon atom with only two bonds like hydrogen, this seems perverse in light of a commitment to represent what is and what might be the case. Coming to terms with this seeming perversity, the debate has now shifted to toy models and the ways in which scientific claims to truth are like those and yet different from those of fictional writing (Toon 2012, Knuuttila 2021a)—and thus the debates are coming around full circle to the empiricist conviction by Heinrich Hertz and others that scientific models can be constructed from hidden masses and other hypothetical entities as long as they generate true predictions (Hertz 1899, see van Fraassen 2008).

If fictionalism lost sight of the actual and the possible, this is because it takes for granted that models are malleable tools for representing what is and what might be the case. It is thus that scenario modeling in climate science and elsewhere has come into view, and thereby—firmly within the framework of *possibilitas*—the construction of whole batteries of models that describe alternative possible worlds. Here, to be sure, it makes a difference whether scenarios are mere extrapolations or, instead, visions of a discontinuous reality in which our current models might not even apply. By way of extrapolation one asks what might be the case if present trends continue, or what might be the case if we change the value of a parameter in successful models of what has been observed. Alternatively and without extrapolation, there is a form of modeling the possible that considers social interactions in a society of humans and robots, or of ecosystems on Mars after the introduction of genetically modified insects and plants. Here, it is especially interesting to see how descriptions of known reality are adapted to model these speculative possibilities. But despite all these different modeling practices, we remain within the sphere of the veridical, of picturing real and hypothetical states, of modeling the possible in terms of *possibilitas*, seeking to ascertain for non-actual states whether they have the property that they might be the case.

11.2.2   *What Can Be Done—Now and Forever*

There are ways of modeling the possible that do not induce ontological anxiety or epistemological puzzlement in that they do not posit the reality of the non-actual and do not employ idealization, abstraction, or fictional entities. They model what can be done in the very mundane terms of what has been done in the model. These are here called "prospective models." The various reasons for this choice of term will be given further on, in the meantime here a first explication of what these are:

> In prospective modeling the technical achievement of building the model opens to view a field of action.

The technical achievement of building the model consists in the acquisition and demonstration of the power to do something—and the model shows what can be done in virtue of its having been done in the construction of the model. This opens to view a field of action simply because what has been done, can be done again, and might be done in another medium, on another scale, under different conditions (compare Knuuttila and Koskinen 2021).

The most familiar example for prospective modeling can be found in architectural modeling. The model submitted for an architectural design competition does not represent a building but *is* a building at a certain scale, with a certain level of detail and execution. As discussed by Sabine Ammon, the architectural model intends a full-scale building that may or may not be built in the future. As such, it is a physical model of something non-existent. If it wins the competition and is reenacted in brick and mortar, the original model can serve as a representation of certain features of the realized edifice. Even then, however, it continues to present the physical features of the model as a site for experimental and discursive learning (Ammon 2015, Ammon and Hinterwaldner 2017). The model edifice is considered from many points of view and invites modeling of the model in the sense of exploring features of the model: how people move through and inhabit the building, the intake and output of energy, the electrical wiring or emergency evacuation procedures. Through these kinds of models, many facts are established. These inform not only the decision to build but also the modifications throughout the planning and building process. For design theory, for the history of art and architecture, and even for the careers and incomes of many architects, these models are achievements in their own right. Some of the most famous buildings in the history of architecture exist as models only because there was never a commission to reenact them in brick and mortar. And in many interesting cases, they exist as models only because there are no brick-and-mortar technologies to reproduce them, for example, because the material properties of steel and glass are just too different from those of cardboard and paper.

There is a great affinity in this description to accounts of models as artefactss and boundary objects, suggesting also that prospective models might be thought of as "models for" rather than "models of" (Fox Keller 2000; Knuuttila 2021b; Star and Griesemer 1989). Indeed, all prospective models are material achievements or built artefacts but not *vice versa*. Some artefactual models serve to represent what is the case, others accomplish a performance of what can be done. While both kinds of artefactual models can serve as boundary objects and thus as a platform on which theorists and practitioners from many disciplines come together for experimental exploration and learning, this is not the primary function or feature of prospective models. To be sure, prospective models as artefacts can be passed around and shared, they are models for intervention and parameter variation, but first of all they are accomplished constructions. Prospective modeling foregrounds the successful creation of the model as an artefact and of the artefact as a model. It makes a claim not to depiction or truth but to a way of building and effecting things.

Architectural and other prospective models show what can be done and invite us to do it again, to recapitulate, imitate, reconstruct, or reenact the achievement in the model. In contrast to the representational models of what might be the case, the prospective model does not refer to other times and places where a merely latent reality becomes manifest. Instead, like pulling a curtain, these models instantaneously open to view a space of action. In and through them, powers have been acquired and this achievement is akin to a proof of concept or proof of principle. As models of effective possibility (*potentia*) and of what can be done they provide a productive anticipation of reality-in-the-making. In order to fulfill their promise, they need to be taken up, developed, extended, expanded, or enacted. Drawing on Nelson Goodman's notion of exemplification, prospective models might thus be defined:

> A composition of symbols or material elements is a prospective model just in case that it exemplifies the actions by which this composition was achieved.

If first and foremost prospective models are accomplished constructions, the model exemplifies this accomplishment somewhat as a swatch of fabric exemplifies the fabric or a Rolls-Royce exemplifies the property of being a Rolls-Royce.[2] The prospective model exemplifies the features which it is constructed to possess—as does the swatch of fabric and as does the Rolls-Royce. Once these features are taken to signify something, and once the prospective model is taken to be salient for purposes of reenactment, variation, or development, its affordances are used to construct the target of a mimetic relation: The model of an edifice then becomes a guide of sorts for the building of some brick-and-mortar edifice.

11.3 Prospects

In 1947, the Norwegian ethnographer and explorer Thor Heyerdahl showed that the precolumbian inhabitants of Peru had the technical skill and material means to build boats by which they could travel to and settle Polynesia. He did so by building a boat from material that would have been available and then taking that boat to travel to Polynesia. The famous Kon-Tiki expedition did not prove that Polynesia was, in fact, settled by South Americans—on the contrary, the theory behind it was deeply flawed and was never generally accepted (Holton 2004; Melander 2019). The expedition did not represent a historical event or the design of precolumbian boats. All it did was show what could have been done by precolumbian South Americans. Most scientists then and now believe that the indigenous population of Polynesia stems from Asia. But if the settlers of Polynesia were to have come from South America, after all, they could have accomplished the feat in some such manner. Heyerdahl engaged in modeling effective possibility (*potentia*).[3] He claimed for himself a way of doing things which actual historical people might have adopted as well, establishing quite generically that other people can do what he has managed to do. Whether a specific people, so to speak, took him up on it, is quite a different matter. But if they chose to do so, theory-building ethnographers could now situate historical actors within a field of action that was opened to view by Heyerdahl's prospective model.

The examples from ethnography and architecture provide a clue that "prospective" here is not opposed to "retrospective" and does not refer to "future prospects."[4] As it is discussed here, prospective modeling has nothing necessarily to do with anticipating, predicting, or shaping the future. Instead, it belongs to a family of practices that open to view what can be done in a field of action.

11.3.1 *Staking a Claim*

Models of what is or might be the case inhabit the sphere of veridical knowledge, they are claims to truth or empirical adequacy, predictive accuracy, or proper fit of theory and reality. Models of what can be done lay claim on another kind of achievement in the sphere of working knowledge. They stake claims on ways of doing or effecting something, proofs of principle, prototypes, and demonstrators.[5]

Prospective modeling is thus a form of prospecting as a technical practice that seeks out gold deposits, investment opportunities, ideally a goldmine of riches. It refers to the activity of prospectors, geological surveyors, real estate agents, or data-mining algorithms that identify promising sites for development. Prospecting is the search for deposits of gold, oil, or other precious things like, nowadays, internet users who might become voters or customers. Prospecting in this sense is often associated with "staking a claim" to a piece of land or to a novel practice—laying the ground for further construction.

Prospective models claim the ground on which a certain kind of work can be done or a project pursued. Most prominently in current science and technoscience, it claims the ground on which researchers can do the work of making and building by reenacting what prospectively has been made and built *in silico*.

In a pair of papers, Johannes Lenhard quotes Uzi Landman from Georgia Tech's Computational Materials Science Center (Lenhard 2004, 2006). To do materials research computationally is different from using computers to represent the materials and their properties. Instead, it consists in the creation of a computational simulacrum, a kind of substitute reality, allowing for the manipulation and exploration of the computationally rendered materials.

The simulated materials are analogous to a human disease process that is implemented in a genetically prepared cancer mouse, inviting a conflation of ontologies. The cancer mouse is not a tool for representing human disease but a physical medium for its instantiation, and in computational materials science or synthetic biology the computer is also not a tool for representing data about gold molecules but a symbolic device for the physical implementation of dynamic processes. And accordingly, just like a medical researcher who diagnoses and treats human cancer in the model organism, Uzi Landman speaks of simulated materials as if they were the real thing: "To our amazement, we found the gold atoms jumping to contact the nickel probe at short distances. Then, we did simulations in which we withdrew the tip after contact and found that a nanometer-sized wire of gold was created. That gold would deform in this manner amazed us, because gold is not supposed to do this" (Lenhard 2004, 93). A medical researcher might learn to interrupt the expression of some gene and to treat the disease in an animal model, Uzi Landman here deformed *in silico* gold so as to create a nanosized gold-wire. His demonstration *in silico* is paradigmatically a prospective model, showing what can be done. In this case, some years later the gold nanowire was experimentally obtained in the lab—a reenactment in a different physical medium than that of the computer.

11.3.2   *Hopeful Monsters*

Uzi Landman's creation of the unlikely nanowire was prospective in that it laid the ground for open-ended and varied reenactments of what was once and for all accomplished in the simulation model. It is also prospective in the sense of Arie Rip who speaks of "Technology as Prospective Ontology," taking as prospective the relation between a technological promise and its realization. This relation is not one of original and copy. Rip likens it rather to the writing of a text and how it is taken up by a reader, and thus to an evolving process. Like any other technical artefact, the prospective model establishes an option, a way of doing things. It does not fully determine how these options will be taken up. The architectural model, the Kon-Tiki voyage, the simulated nanowire rehearse a way of doing or effecting things. Particular performances, if they take place at all, will evolve and adapt it. To make this evolutionary point, Rip adopts from economic historian Joel Mokyr a notion that originated with the evolutionary biologist Richard Goldschmidt: The "hopeful monster" is a significant mutation which may give rise to new lineages and whole new ways of doing things – gold behaving amazingly as it is not supposed to, precolumbian technology venturing out further than anyone would have thought:

> Engineers add to the furniture of the world, and thus shift its ontology – if we use the term "ontology" in a simplistic way. This "adding" is not a simple, linear activity of first making something, and making

it available, which is then added to the world. There is a strong prospective element. Artefacts start as technological options, a promise of functionalities, in other words "hopeful monstrosities." This is visible, sometimes literally, in the prototypes: these embody a prospective. When they are developed further, introduced and taken up on location, they remain unfinished.

(Rip 2009, 405–6)

Building on this notion, albeit implicitly, Wolfgang Liebert and Jan C. Schmidt articulated a program of "prospective technology assessment" (Liebert and Schmidt 2010). Technology assessment is often in the hands of social scientists who try to imagine socio-cultural consequences and implications of new technologies. They are trying to look towards the future. However, in order to assess ecological impacts, dual-use potential, and disruptive features of new technologies, what is needed is a kind of prospecting that scours specific design features, developing prospective models that exhibit what some envisioned technical system or device can do, beyond what it is designed to do. Adapting Rip's terminology, a fusion reactor, for example, embodies a prospective which can be articulated by a model which shows what these reactors can do in the service of military rather than civilian interests (Franceschini, Englert and Liebert 2013). This is a kind of technology assessment that needs to be performed by scientists and engineers, it is modeling the possible in terms of what one can do, and not in terms of what might be the case.

11.3.3   Setting the Scene

Rip's use of "prospective" is familiar in another context, namely that of writing or reading a prospectus for a book to be written or a trip to be taken or an investment to be made. Here, the prospectus is an advertisement of what can be done coupled with exemplary demonstrations of the capability to do it. While offering a promising prospect of happy results, everyone knows that the final realization will deviate from the original promise, necessarily so because there is work involved, thus further learning, which often necessitates modifications.

This openness to what can happen is highlighted by yet another, perhaps most elementary use of "prospect" and "prospective," the one that exhibits most clearly the opening to view of a field of action. It is known from traditional stage-designs in the theatre where a field of action is opened and delimited at once by the so-called backdrop or prospect—the backdrop referring to the setting against which everything is seen, and the prospect being all that is seen from the spectator's point of view. This might be a perspectival painting of a temple or a landscape, situating a particular action within the horizon of a world. Prospective models establish such a horizon within which particular research activities can unfold.

When Don Eigler and Erhard Schweizer individually positioned 35 xenon atoms to spell the letters "IBM," they did not refer to their achievement as a model. Yet it was immediately taken to be a template for further action, a horizon for all nanotechnological research to come: the willful manipulation and placement of individual atoms held the promise of "shaping the world atom by atom," of building and making all kinds of things at an atomic scale (Nordmann 2021a). The first design of a genetic circuit was of similar significance for Synthetic Biology. The repressilator is an artificial device that corresponds to a naturally occurring feedback mechanism. In contrast to Eigler and Schweizer's "IBM," Elowitz and Leibler's repressilator is often referred to as a model synthetic regulatory circuit which enters many computer-aided research-designs in Synthetic Biology (Elowitz and Leibler 2000).[6]

11.4   Epistemic Matters

It is one thing to characterize prospective modeling, another to clarify its epistemic role in science and technoscience. How are prospective models generated and validated? And how is knowledge gained and codified through prospective models? Do prospective models fortify or explode the distinctions of basic and applied research, science and engineering, and the like?

11.4.1   *Systematicity—Principles of Composition*

On the one hand, the IBM-logo and the repressilator open to view or raise the curtain on a field of action and further research: what can be done in the model, can be done again and again in different contexts, more and less varied and for various purposes. On the other hand, together with other such models, they circumscribe a disciplinary or technical practice.

During their training at art academies, aspiring artists used to gather around models of great art such as the Laocoon sculpture group. This artwork, for example, is a prospective model in more ways than one. It is significant for the very achievement of its making as an exemplary rendition of the agony of death. By scrutinizing the model one can see that it consists in the selection of a pertinent moment in the causal trajectory from a snake-bite to the death of Laocoon and his sons. Its status as prospective model was explicitly discussed by numerous theorists of art. Most prominently, according to Gotthold Ephraim Lessing the model shows what can be done by way of choosing a single pregnant moment that narrates a course of events.[7] Implicitly and explicitly, books like Lessing's *Laocoon* propose that one can learn to reenact this choice in the visual arts more generally. Moreover, the Laocoon sculpture does not stand alone. Along with the Venus de Milo, Belvedere Torso, Boy with Thorn, and many other exemplars, it defines a canon within which a new generation of artists was formed.

Similarly, students of architecture study many prospective models of buildings that have and have not been built, and students of synthetic biology learn about repressilators, toggle switches, and other techniques for working with genetic circuits (Knuuttila and Loettgers 2013; Kohl and Falk 2020). For mechanical engineering, Föttinger-type torque converters and hydro-dynamic coupling may play a role analogous to the Laocoon group in the history of art (Förster 1960).[8] Designed for the propulsion of ships, its technical reenactments in other fields like automatic transmission proved far more consequential. In the patent applications, claims are made, not to theories and ideas, but to accomplished apparatus. Föttinger explicitly stakes this claim:

> Having now particularly described and ascertained the nature of my said invention and in what manner the same is to be performed, I declare that what I claim is: –
>
> 1. An improved hydraulic power transmitting apparatus [...] said apparatus being characterised by the fact that the same turbine wheel or wheels is or are adapted to operate as the driven wheel or wheels for two or more methods of running.
>
> [...]
>
> 18. The improved apparatus constructed, arranged or adapted to operate substantially as described with reference to the accompanying drawings for the purposes specified.
>
> (Föttinger 1910)

The new arrangements and adaptations of wheels for Föttinger couplings continue until today with reenactments at different scales for different purposes. But the accompanying drawings and their symbolic conventions place Föttinger's invention within a much larger family of transmission systems, which define a technical field of action for mechanical engineers who are socialized by learning what can be done through combining and recombining established machine elements (Nordmann 2020; Reuleaux 1876).

11.4.2 Working Knowledge—Affordances

Prospective models of the possible are models of what can be done. Models of what might be the case are just like models of what is the case—only conditionalized to indicate that the modeled state has the property of being possible and not actual, that is, of being actual only when certain conditions are fulfilled. Models of what can be done model a power that needs to be acknowledged by way of reenactment: To be sure, what can be done once, can be done again and invites reenactment in another setting—but whether it actually affords such reenactment can be determined only when people actually embark on this reenactment.

As we saw above, it is a difficult task to ascertain whether something might be the case—whether the property of being possible can be ascribed. This is a matter of verification in an impossibly difficult situation. Modeling what can be done is much easier, and more complicated only insofar as it takes two steps to ascertain whether an achievement affords reenactability – it is a matter of verification and validation.[9]

The verification of the prospective model consists in the achievement of the model itself. Does Föttinger's hydraulic power transmitting apparatus really transmit power in a closed system without physical contact of solid parts? Can the repressilator really disrupt or generate circadian rhythms? Can the model really stake the claim to have successfully rehearsed an achievement of power or control?

Validation concerns fitness for purpose. Does the prospective model really allow for mimetic reenactment? Is the technical achievement in the model reproducible in another setting – adaptable to different scales, materials, purposes? It would seem that all these questions can be answered ostensively, that is, by the mere one-off production of the facts, an immediate proof of existence. And yet, they are subject to technical negotiations regarding salience, simplicity, efficiency, robustness, and the like. This ambivalence is captured in Paul Teller's discussion of actual possibilities (Teller 2025). An architectural model establishes an actual possibility without referring to past, present, or future buildings. And yet there is a difference between El Lissitzky's Γ-shaped "skyhooks" and the executable blueprint of a commissioned building. The former establishes what can be done and thus done again in the medium of drawing, balsawood, cardboard, and paper, providing an aspirational form for reenactment in other settings. The latter is specified and qualified in many ways, it incorporates and makes explicit many, though still not all of the conditions that need to be fulfilled in order to enact the blueprint in a material building. All prospective models show what can be done but they constrain in different ways and with different degrees of precision what counts as "doing it again."

Veridical models of what is or might be the case are representational in that they involve a picturing relation of aboutness in the general sense of Hertz, Boltzmann, and the "picture theory of science [*Bildtheorie*]" (van Fraassen 2008). To be sure, there are very different views about this relation and what is implied by claims that this is a model of that. And yet, whether we talk about pictures in terms of correspondence or coordination, in terms of likeness or structural mapping, it involves an aboutness-relation as representations provide a symbolic stand-in for something that is to be described, depicted, or pictured.[10] At issue in these debates are matters of ontological anxiety or concern, such as the question whether there has to be real similarity or at least structural isomorphism, and whether that which is to be represented offers a standard of correctness or truth against which the representation can be measured.

The prospective model offers a *Vorbild* (archetype or paradigm, as such like a role model), and its reenactment performs a *Nachbildung* (emulation or imitation, as such like an apprentice or a disciple).[11] This picturing relation is mimetic—not in the sense of making a true copy of an original, but in the sense of imitation, participation, or joining in with a way of doing things, falling in step, taking up, adapting, even parodying the model (Nordmann 2021b). *Vorbild* and *Nachbildung* thus do not stand in the veridical relation of representation but in the mimetic relation of anticipation and recapitulation, of production and reproduction, of rehearsal and performance, or mutual assimilation. They do not refer to an independent standard of correctness or truth, with one having priority and serving as the measure of the other.

11.4.3 Prospective Explanation—Material Abductions

It would seem that the discussion of prospective models leads us away from scientific or technoscientific research practice. But it also leads back to central concerns of the philosophy of science.

Uzi Landman generated a prospective computer model in which he observed the amazing behavior of gold. His achievement implied repeatability. To the extent that Landman was showing what gold atoms could do *in silico*, he would expect that their behavior could be reproduced in the lab—though it would still take some work to make this happen. One can tell that Landman himself is working with an implicit notion of prospective modeling for the simple reason that it never occurs to him to consider his computer model as a hypothetical construct which generates a testable prediction. He never speaks about the confirmation or corroboration of a hypothesis. Instead, on this and other occasions he speaks about the facts established in the construction of the model and how the model can thus "guide" laboratory experiments.

> A most interesting result of the simulations pertains to the "universal" nature of the mechanical and electrical properties of junctions. [...] The similarity between the force curves for the two wires [observed in the model] confirms formation of nanowires of similar nature, irrespective of the previous history of the junction. Indeed, such considerations guided recent experiments [...]
>
> (Landman, Barne, Luedtke 1997)

Before the amazing behavior of gold was demonstrated "for real" by material experimentation in the laboratory, Landman did not consider his achievement a mere hypothesis that still awaits confirmation. Instead, his finding *in silico* would guide researchers to its experimental reenactment. And after that reenactment succeeded, he does not offer the computer model as an explanation of the physical phenomenon. Instead, it is from now on a question of correlating what is done in the model and what can be done in the laboratory: "Structural transformations, electronic spectra and ballistic transport in pulled gold nanowires are investigated with *ab initio* simulations, and correlated with recent measurements" (Häkkinen et al. 2000)—the two performances remain independent achievements which show what gold can do in *silico,* suggesting that it can do this also in the lab. The production of effects in the model and then in the laboratory become calibrated, fine-tuned, and perhaps assimilated to one another. Accordingly, Landman received in 2005 the Aneesur Rahman Prize for Computational Physics and, more specifically, for prospective modeling, namely "[f]or pioneering computations that have generated unique insights into the physics of materials at the nanometer length scale, thereby fostering new theoretical and experimental research."[12]

Landman's significant achievement in the model stands on its own. It exhibits in a novel manner an interesting dynamic process (*Vorbild*). And what has been done once, can be done again: A laboratory experiment might serve to reenact the achievement of the model (*Nachbildung*). Of course, as one knows from architecture, such reenactments do not always succeed—it might turn out that the material conditions are just too different. But whether they succeed or not, the relation between the calculated and the experimental phenomena is not, representationally, one of projection, mapping, or denotation. It is instead akin to a technical construction where a prototype and its subsequent instantiations are built in the same way and therefore, evidently, are alike, possessing shared properties of design and constructive procedure.

It has become perfectly normal to describe model-construction as involving a building into the model of certain physical features or experimentally observed behaviors. Accordingly, models of what is or might be the case can be viewed as technical artefacts that do the work of representation. Prospective modeling offers a complementary perspective according to which the production of an experimentally observed phenomenon is guided by a prospective model and thus becomes a technical reproduction of, for example, a computer simulation. This technical reproduction or reenactment is an independent achievement that adds to what is achieved in the model. With one of these independent achievements guided, perhaps inspired by the other, they do not stand in the representational relation of one being the picture of the other. Instead they stand in the mimetic relation of *Vorbild* and *Nachbildung*, of an achieved performance to its imitations, reproductions, or reenactments in different settings at different occasions.

In contemporary research practice, this relation holds not only for Uzi Landman's computational physics and simulation modeling. It also holds for many theoretical reconstructions, indeed: "explanations," of laboratory phenomena. These so-called explanations consist in the construction of a prospective model that functions like the construction of an abductive inference.

The paradigmatic Peircean-type abduction creatively invents or conjures up a set of premises such that, if the premises were true some otherwise surprising phenomenon would appear to be a matter of course. In other words, abductive inference affords an explanation, irrespective of it being a true or best or plausible explanation. And therefore, on the standard view of abduction, it has to be followed by deduction and induction in order to validate or disqualify the proposed explanation or explanatory mechanism. But even if the explanation turns out to be false, abduction achieves the sudden transformation of a strange happening into an *explanandum* that exists in the sphere of intelligibility or explainability. The construction even of a seemingly absurd explanation opens to view the challenge of triangulation as described by Mary Morgan (2025). Before, we were wondering about a fact and whether it even exists (such as the fact of biological evolution), but now, after an abductive inference to an explanatory mechanism (natural selection), we suddenly find ourselves discussing whether natural selection provides an adequate explanation of the fact of evolution.

This constructive achievement of an abduction is even more evident in the case not of the symbolic but of the material invention of an explanatory pathway. One example of this was Thor Heyerdahl's Kon-Tiki expedition. Its *explanandum* was the settlement of the Polynesian islands. In a spectacular way it opened to view the possibility that precolumbian inhabitants of Peru came to Polynesia by boat: It could have happened like this, and thus this still highly unlikely event is definitely in the sphere of technical action.

Similarly, seeking to improve structural steel, materials researchers and designers identify embrittlement phenomena, showing, for example, that a certain kind of thermal conditioning improves the steel. For purposes of explanation and understanding, they turn to theoretical chemistry or computational metallurgy where a prospective model is produced. It could have happened like this, the phenomenon is definitely in the sphere of theoretical understanding: "These studies convincingly demonstrated that modern quantum mechanical techniques are sufficiently accurate to determine the values of the interfacial energies that are important to a complete understanding of embrittlement" (Eberhart 1994, 333). In other words, these studies convincingly demonstrated that modern quantum mechanics offers theoretical resources sufficient to reproduce observed phenomena of embrittlement, thus showing that initially surprising and unexpected phenomena can be rendered intelligible in the sphere of theory. In this case, building a prospective model followed the now-common practice of first producing a calculated structure and generating from it a simulated high-resolution image, finally inserting that image within an actual micrograph, showing that it seamlessly blends in. The theoretical resources thus enabled the modeler to construct the visual likeness between a symbolically constituted calculated phenomenon and the materially constituted experimental phenomenon. Is this an explanation of the phenomenon? "With cautious optimism, the answer is yes" (Eberhart 1994, 333). But whether or not one calls this an explanation, the prospective model provides a material abduction that opens to view a way of doing things guided by the model. Irrespective of whether the algorithms that produce an effect in the model provide a true or best explanation of what was done in the laboratory, the prospective model shows one way of doing it and thus affords opportunities to do it again under varied conditions.[13]

11.5   Conclusion

Finally, at the risk of repeating it too often, here is a terse summary of the argument:

Modal possibility—what might be the case—comes cheap. It shadows or precedes all modeling the actual: Of necessity, what is actual is also possible, and the parameters of any model of real situations or events can be varied to suggest a non-actual possibility. Modal possibility (*possibilitas*) also shadows prospective models of what can be done: Evidently, whatever one actually does or achieves in and through the construction of a model, must have been possible to do.

In the sphere of the veridical and when models are judged for their correctness, the question of what is actually, what might possibly, and what cannot be the case is interesting, even important. This is different in the sphere of working knowledge where models and other constructions exhibit through the artefact at hand how things work together to perform work or produce an effect. Here, it is just trivial that what was really done is evidently possible. Instead, what is of interest and significance is that something has been done in the creation of the construction, and that it can be done again, perhaps in the same way or some other way.

A composition of symbols or material elements is a prospective model just in case that it exemplifies the actions by which this composition was achieved. In other words, we do and make, compose and configure things all the time, in the realm of symbols and material things. Whenever something becomes salient for having been made, composed, or configured, this exemplification of a way of doing things can become productive: prospective models afford specific reenactments.

What has been done, if only just once, in the construction of a prospective model, signifies the possibility or power (*potentia*) of doing it again, perhaps under different conditions, at a different scale, with different materials, by way of different routines: it is an invitation to consider a field of action that has opened up to view (*a prospect*).

Prospective models do not relate to a target in the way of copy and original, representation and reality, blueprint and execution. They are not true of or signify something other than themselves. The relation to a target only comes to view when the achieved construction becomes salient—when it is offered up for extraction, exploitation, adaptation. The relation of the model to the target is then one of rehearsal and reenactment (*Vorbild* and *Nachbild*), that is, features of the model become subject to emulation, repetition, parody, or selective reproduction. (To be sure, many prospective models are intentionally constructed to become salient in such a way, e.g., architectural models, which then succeed as models in unpredictable ways.)

In the sphere of veridical knowledge and of classical modern science, the predominant epistemic interest is to provide a theoretical description of reality and to represent what is or might be the case, what and how things are, their properties. Here, models of theories and models of reality play a prominent role. In the sphere of working knowledge, the predominant epistemic interest is to exhibit what things can do, their affordances in the context of a work, i.e., when they are composed, constructed, or configured in certain ways.

To the extent that contemporary research incorporates both forms of knowledge-production, "prospective modelling" can help elucidate practices and procedures that do not easily fit the canon of classical methodology. For example, what is a proof of concept or proof of principle, and in which sense does it provide proof? Is this perhaps a species of prospective modeling— offering an exemplification or self-vindicating demonstration of what can be done? More challenging even is the practice of "explaining" experimental findings by rebuilding them from bits and pieces of theory in a simulation model. As in Peircean abduction, the very fact that some explanation can be constructed opens to view a prospect for action—extracting information, designing further experiments, mining not data but all the elements that render the construction. The epistemic value of these "explanations" is not that they offer a true or relatively best account, but that they offer a digital twin that is composed of bits and pieces of well-established theory: tried and true algorithms, modules, and routines.

It is this activity of prospecting that gives prospective models their name. What matters here is not the contrast to "retrospective" but, instead, the field of action opened up to view by what is accomplished in the building of a model: Prospective models afford productive anticipations of reality-in-the-making.

Notes

1 According to Rom Harré, when we attribute powers, dispositions, tendencies to materials or things, this always has a conditional structure: "If properly prompted, the material or thing will then exhibit the power or dispositional property" (Harré 2000). The question of whether things actually have this power is a question of veridical knowledge: can we know the general conditional statement to be true? However, we do not attribute to the carpenter the power or disposition, if prompted, to produce cabinets. Having built a cabinet for the carpentry examination, the carpenter demonstrates working knowledge, that is, the acquired capability or power to build cabinets—this is something that carpenters can do, perhaps each in their own way, whether prompted to do so or not. This is not a claim about the existence or not of specific causal powers, but a claim of what can be done, whether or not the causal processes are fully known. Francis Bacon's philosophy of science recommends that one proceed from the carpenter's achievement towards the knowledge of causal powers (Bacon 2000). Yet another account has been proposed by Nancy Cartwright when she speaks of nature as an artful modeller (Cartwright 2019)—a proposal that deserves a separate discussion.

2 To be sure, these are elliptical formulations which Goodman takes pains to elaborate in a representational idiom: "Likewise, to say that a car on the showroom floor exemplifies a Rolls-Royce is to say elliptically that the car exemplifies the property of being a Rolls-Royce. But such ellipsis can be dangerous in technical discourse where to say that x exemplifies a B means that x refers to and is denoted by a B" (Goodman 1968, p. 53). – This paper adopts a rather wide notion of representation which includes all kinds of picturing, referring, denoting, signifying,

or standing in – all of which in the sphere of the veridical, raising questions of aboutness and truth. *Pace* Goodman, however, exemplification is here not considered a species of representation.

3   For a more mainstream example, consider archaeologists who gather together in order to shoot stone projectiles at animal hides to plausibilize their collection of weapon-points (Gero 2001, 312).

4   The term "prospective model" rarely appears in actual research practice. Not surprisingly, in these rare cases it is used interchangeably with "predictive model" and in opposition to "retrospective model," e.g., in respect to climate change or the economy.

5   "Working knowledge" is knowledge of how things can work together to perform work within the setting of a work, paradigmatically a clock-work and by extension any socio-technical system (Baird and Nordmann 1994, Nordmann 2015). The contrast between veridical knowledge and working knowledge is a matter of discussion in the philosophy of technology, also regarding the difference of science and technoscience.

6   To be sure, it is interesting and deserves further study how the word "model" is actually used in regard to the repressilator. Some understand the repressilator as a model of a cellular mechanism, others investigate the repressilator as a model for naturally occurring and synthetic gene regulatory networks, still others learn from the repressilator how to build networks with at least three proteins. These views are not strictly held apart and indeed, they are not mutually exclusive in that they correspond to different conceptions of „model": model of, model for, prospective model.

7   Theorists of modeling know Lessing from Nancy Cartwright's discussion of fables (1999, 37–44, see Nordmann 2008).

8   Föttinger himself likened engineering to the fine arts. He did not just declare but argued for the notion that a "good designer [*Konstrukteur*] is an artist" (Föttinger 1916, 18).

9   Every artefact and thus every model affords something. A classical model of a physical system affords understanding, prediction, calculation, and the like. As for modeling the possible, however, "being possible" is a modal property and not an affordance in veridical models of what might be the case. In prospective models of what can be done, however, the possible appears as a relational term or affordance: The invitation to reenact what has been achieved is not a mere property but needs to be taken up.

10   See note 2 and Nordmann 2006.

11   Model citizens, role models, fashion models are not usually included in reflections on models and modeling. There is not a something that they are models of. They also do not typically serve as models for investigation. But they are prospective models, par excellence. The model on the runway is a creative achievement and opens to view a new fashion as a way of doing things that is reenacted by very different people: it becomes appropriated into various economic, physical, or cultural settings. I thank Sadegh Mirzaee for bringing this to my attention.

12   See www.aps.org/programs/honors/prizes/prizerecipient.cfm?last_nm=Landman& first_nm=Uzi&year=2005 (accessed June 17, 2024).

13   It cannot be shown here that this is a typical story about the division of labor between theorists and experimentalist in surface science, materials research, and many other fields: Theorists provide "explanations" by showing that they can reconstitute the phenomenon by means of theory. This includes optimization in a design cycle where bits and pieces of theory, tried and true modules become incorporated in the creation and fine-tuning of a simulacrum of the empirically obtained target phenomenon. To ever more closely approximate a perfect imitation is a mimetic practice, and so is that of researchers further down the road who emulate in their experimental practice what was done in the model.

References

Ammon, Sabine. 2015. "Perspektiven architekturphilosophischer Entwurfsforschung. "In *Architektur und Philosophie: Grundlagen. Standpunkte, Perspektiven*, edited by Jörg H. Gleiter, and Schwarte Ludger, 185–95. Bielefeld: transcript.

Ammon, Sabine, and Inge Hinterwaldner, eds. 2017. *Bildlichkeit im Zeitalter der Modellierung. Operative Artefakte in Entwurfsprozessen der Architektur und des Ingenieurwesens*. München: Fink 2017.

Bacon, Francis. 2000. *The New Organon*. Cambridge: Cambridge University Press.

Baird, Davis, and Alfred Nordmann. 1994. "Facts-Well-Put." *British Journal for the Philosophy of Science* 45: 37–77.

Cartwright, Nancy. 1999. *The Dappled World: A Study of the Boundaries of Science*. Cambridge: Cambridge University Press.

Cartwright, Nancy. 2019. *Nature, The Artful Modeler: Lectures on Laws, Science, How Nature Arranges the World and How We Can Arrange It Better*. Chicago: Open Court.

Eberhart, Mark. 1994. "Computational Metallurgy." *Science* 265: 332–3.

Elowitz, Michael, and Stanislas Leibler. 2000. "A Synthetic Oscillatory Network of Transcriptional Regulators." *Nature* 403: 335–8.

Förster, Hans Joachim. 1960. Föttinger-Wandler und -Kupplungen für Kraftfahrzeuge. Automobil-Industrie, no. 8, 53-83.

Föttinger, Hermann. 1910. "Improvements in Devices for the Transmission of Power between Adjacent Turbine Wheel Shafts (Patent GB 19090686)." http://worldwide. espacenet.com/publicationDetails/originalDocument?CC=GB&NR=190906 861A&KC=A&FT=D&ND=3&date=19091202&DB=worldwide.espacenet. com&locale=en_EP

Föttinger, Hermann. 1916. *Technik und Weltanschauung*. Danzig: Königlich Technische Hochschule.

Fox Keller, Evelyn. 2000. "Models of and Models for: Theory and Practice in Contemporary Biology." *Philosophy of Science* 67 (Proceedings of the 1998 Biennial Meeting of The Philosophy of Science Association Part II: Symposia Papers), S72–S86.

Franceschini, Giorgio, Matthias Englert, and Wolfgang Liebert. 2013. "Nuclear Fusion Power for Weapons Purposes." *The Nonproliferation Review* 20(3): 525–544.

Galison, Peter. 2017. "The Pyramid and the Ring: A Physics Indifferent to Ontology." In *Research Objects in Their Technological Setting*, edited by Bernadette Bensaude-Vincent, Sacha Loeve, Alfred Nordmann, and Astrid Schwarz, 15–26. Abingdon: Routledge.

Gero, Joan. 2001. "The Social World of Prehistoric Facts: Gender and Power in Paleoindian Research." In *Interpretive Archaeology: A Reader*, edited by Julian Thomas, 304–16. London and New York: Leicester University Press.

Goodman, Nelson. 1968. *Languages of Art. An Approach to a Theory of Symbols*. Indianapolis: Bobbs-Merrill.

Goodman, Nelson. 1983. *Fact, Fiction, and Forecast*. Cambridge: Harvard University Press.

Harré, Rom. 2000. "Dispositions and Powers." In *A Companion to the Philosophy of Science*, edited by William Newton-Smith, 97–101. Malden: Blackwell.

Hertz, Heinrich. 1899. *The Principles of Mechanics: Presented in a New Form*. London: Macmillan.

Knuuttila, Tarja. 2021a. "Imagination Extended and Embedded: Artifactual Versus Fictional Accounts of Models." *Synthese* 198 (Suppl 21), 5077–5097. doi.org/ 10.1007/s11229-017-1545-2.

Knuuttila, Tarja. 2021b. "Models, Fictions and Artifacts." In *Language and Scientific Research*, edited by J. Gonzalez Wenceslao, 199–222. Cham: Palgrave Macmillan.

Knuuttila, Tarja, and Rami Koskinen. 2021. "Synthetic Fictions: Turning Imagined Biological Systems into Concrete Ones." *Synthese* 198, 8233–8250. doi.org/10.1007/s11229-020-02567-6.

Knuuttila, Tarja, and Andrea Loettgers. 2013. "Basic Science Through Engineering? Synthetic Modeling and the Idea of Biology-Inspired Engineering," *Studies in History and Philosophy of Biological and Biomedical Sciences* 44(2), 158–169. doi: 10.1016/j.shpsc.2013.03.011.

Kohl, Thorsten, and Johannes Falk. 2020. "Knowledge Objects of Synthetic Biology: From Phase Transitions to the Biological Switch." *Journal for General Philosophy of Science* 51: 1–17.

Häkkinen, Hannu, Robert N Barnett, Andrew G Scherbakov, and Uzi Landman. 2000. "Nanowire Gold Chains: Formation Mechanisms and Conductance." *The Journal of Physical Chemistry B* 104: 9063–9066.

Holton, Graham E.L. 2004. "Heyerdahl's Kon Tiki Theory and the Denial of the Indigenous Past." *Anthropological Forum* 14(2): 163–181.

Landman, Uzi, Robert N Barnett, and W.D Luedtke. 1997. "Nanowires: Size Evolution, Reversibility, and One-Atom Contacts." *Zeitschrift für Physik D* 40: 282–287.

Lenhard, Johannes. 2004. "Nanoscience and the Janus-Faced Character of Simulations." In *Discovering the Nanoscale*, edited by Davis Baird, Alfred Nordmann, and Joachim Schummer, 93–100. Amsterdam: IOS Press.

Lenhard, Johannes. 2006. "Surprised by a Nanowire: Simulation, Control, and Understanding." *Philosophy of Science* 73: 5 (Proceedings of the 2004 Biennial Meeting of The Philosophy of Science Association Part II: Symposia Papers), S605–S616.

Lessing, Gotthold Ephraim. 1962. *Laocoon: An Essay about the Limits of Painting and Poetry*. Indianapolis: Bobbs-Merrill.

Liebert, Wolfgang, and Jan C Schmidt. 2010. "Towards a Prospective Technology Assessment: Challenges and Requirements for Technology Assessment in the Age of Technoscience." *Poiesis and Praxis* 7(1–2): 99–116.

Melander, Victor. 2019. "David's Weapon of Mass Destruction: The Reception of Thor Heyerdahl's 'Kon-Tiki Theory." *Bulletin of the History of Archaeology* 29(1): 1–11.

Morgan, Mary. 2025. "Alternative Worlds: Reasonable Worlds? Plausible Worlds?" In *Modeling the Possible: Perspectives from Philosophy of Science*. New York: Routledge.

Morgan, Mary, and Margaret Morrison, eds. 1999. *Models as Mediators*. Cambridge: Cambridge University Press.

Nordmann, Alfred. 2006. "Collapse of Distance: Epistemic Strategies of Science and Technoscience." *Danish Yearbook of Philosophy* 41: 7–34.

Nordmann, Alfred. 2008. ",Getting the Causal Story Right': Hermeneutic Moments in Nancy Cartwright's Philosophy of Science." In *Nancy Cartwright's Philosophy of Science*, edited by Stephan Hartmann, Carl Hoefer, and Luc Bovens, 369–88. New York: Routledge.

Nordmann, Alfred. 2015. "Werkwissen Oder How to Express Things in Works." *Jahrbuch Technikphilosophie* 1: 81–9.

Nordmann, Alfred. 2020. "The Grammar of Things." *Technology and Language* 1(1): 85–90.

Nordmann, Alfred. 2021a. "First and Last Things: The Signatures of Visualization-Artists." *Technology and Language* 2(2): 96–105.

Nordmann, Alfred. 2021b. "Biotechnology as Bioparody: Strategies of Salience." *Perspectives on Science* 29(5): 568–82.

Reuleaux, Franz. 1876. *The Kinematics of Machinery: Outlines of a Theory of Machines*. London: Macmillan.

Rip, Arie. 2009. "Technology as Prospective Ontology." *Synthese* 168: 405–22.

Star, Susan, and James Griesemer. 1989. "Institutional Ecology, 'Translations' and Boundary Objects: Amateurs and Professionals in Berkeley's Museum of Vertebrate Zoology, 1907-39." *Social Studies of Science* 19(3): 387–420.
Teller, Paul. 2025. "Actual Possibility." In *Modeling the Possible: Perspectives from Philosophy of Science*. New York: Routledge.
Toon, Adam. 2012. *Models as Make-Believe: Imagination, Fiction, and Scientific Representation*. New York: Palgrave-Macmillan.
van Fraassen, Bas. 2008. *Scientific Representation: Paradoxes of Perspective*. Oxford: Clarendon Press.
Winsberg, Eric. 2006. "Handshaking Your Way to the Top: Simulation at the Nanoscale." *Philosophy of Science* 73(5): 582–94.

12 Alternative Worlds

Reasonable Worlds? Plausible Worlds?

Mary S. Morgan[1]

12.1 Introduction

Scientists and historians are practically adept at reasoning about alternative possible worlds relevant for their field. The question here is: how do they judge that an alternative world is possible and how do they limit their investigation of 'the possible' so that it does not fall over into 'the impossible'? Given that the gap between possible and impossible worlds is not well defined, and there are no clear lines of demarcation, scientists rely not just on specific, subject-matter, knowledge and more general field constraints, but on developing systematic ways of exploring alternative accounts about how the world works in order to police the boundaries between possible- and impossible-world accounts.

There are two elements in the analysis here: first, to examine how scientists explore 'alternative worlds' – that is, their alternative accounts of the world; then to gain insight into their ways of judging the validity of these alternative worlds in terms that take into account both their theoretical considerations and empirical ones.

The term: 'alternative worlds' is chosen both to deepen and widen the focus on how scientists frame their thinking about possible worlds, and the practices they use to explore and validate them. Sjölin Wirling and Grüne-Yanoff have suggested (this volume) three themes in the literature characterising the essentials of such thinking about alternative worlds and judgements upon them[2]. The first is that the imagination is central to the mode of thinking, the second is that 'background knowledge' plays a part in judging the validity of outcomes, and thirdly, that similarity may lie at the root of such judgements. The argument, and examples, here find that both imagination and background knowledge are surely involved, but questions, and analyses, how they fit together in scientific practices. Imagination comes when scientists thinking out not only alternative theoretical accounts for their phenomena of interest but also alternative descriptions of their actual worlds. Background information is important on both sides – in developing theoretical and empirical accounts. Similarity issues are also critically relevant in the account and examples here, but by modes of direct comparisons (rather than

DOI: 10.4324/9781003342816-16

via analogical reasoning as Sjölin Wirling and Grüne-Yanoff suggest) in judging the validity of the imagined world accounts not just in terms of possible or impossible worlds but in enabling finer-grained judgements.

One of the usual sites for systemic exploration into alternative worlds is by working with models. Another medium for exploration is, as we shall see, in the use of counterfactuals; and sometimes they work together. In both modes, scientific imagination is needed to think through how the world works in conjunction with background information. Both modes of work provide reasoning tools, and they do so in part by limiting both the imagination and the terrain of investigation to make such reasoning practicable. That combination may produce sensible accounts, but it may also produce non-sensical or impossible world accounts. This raises important questions about how such reasoning and outcomes can be judged – what are the quality criteria that scientists use to judge the reasonableness and plausibility of their model or counterfactual reasoning and outcomes? As we shall see, these characteristics are matters of scientific community practices, rather than laid down as philosophical rules or precepts. The community cases used here come from economics, but their modes of investigation are generic across many fields, and so the analysis is more widely relevant.

The term 'alternative worlds', as we will see, covers a range of things in scientific work, from the world according to scientists' theories and models of it, to the actual world (including its historical development and empirical descriptions), and in both frames, judgements are based not just on outcomes but in terms of pathways to outcomes. Scientists using other modes of investigation of possible worlds may well also use other forms of alternative worlds and use them in different ways. For example, Koskinen (2017) is concerned with 'how-possibly' alternatives to the existing world, alternatives which might be made differently, or fulfil the same function but be constituted differently. Nordmann's account of 'prospective models and modelling' (this volume) is concerned with a more open notion of alternative worlds that speak to the design and build approaches of some of the arts and sciences. These two approaches have resonance with the work of chemists in figuring out the pathways and recipes to synthesise new things in their world.

12.2   Sketching Out Alternative Worlds in Models

Let me start with models to see how they are used to sketch out alternative worlds (and leave counterfactuals till later). Scientists' models come in all sorts, from the model organisms of biology to the mathematical, statistical or diagrammatic models of economics and physics. In these latter formats, models can be considered 'artefacts', a label due to Knuuttila (2011) meaning that they are made-up, constructed objects that help their scientists to think about their worlds. In the 'models as mediators' view (Morrison and Morgan 1999), this generic kind of model is typically constructed from a mixture of all sorts of elements, some theoretical notions, some empirical elements, each with more or less validity as descriptions of, or claims about, the world. And in this 'models as mediators' account, these constructed/artefactual objects are used to mediate between the theoretical and the empirical knowledge of a science, a role they exercise by virtue of being made up of elements of both and being partially of them both (but not being in any direct sandwich-line between them, nor being shorthand versions of either). Because of this make-up, models cannot simply be judged as purely deductive machines or purely inductive ones. Rather, this mediating function is fulfilled by scientists using their models to explore the implications of their theory in relation to the world it is about, even where that world is only loosely and partially described in their models. It is not just their construction, but their usage that is of interest here. It is in usage that scientists manipulate their models as tools of reasoning to outline their alternative world accounts. It is in usage that those alternative worlds created by their models provide the materials for scientists to judge the validity and usefulness of their models.

The 'models as mediators' account did not offer a general recipe into how that mediating role worked, nor how it should, or could, be characterised in creating alternative world accounts. Various papers in that Morgan and Morrison (1999) collection explored particular examples of such mediation. Marcel Boumans (1999) perceptively observed that bringing in empirical characteristics into a largely theory-driven model created a certain 'built in justification'. Another paper by Ursula Klein (1999) discussed the way paper-tool models enabled chemists to move back and forward between particulars and general levels. R.I.G. Hughes (1999) discussed the Ising model's role across various empirical domains in physics based on his earlier 'DDI' account of the usage of models in science: a three-step process: 'denote' (construct the model), 'demonstrate' (by manipulating the model) and 'interpret' (those outcomes). This was extended in Morgan 2012, to insert 'questions' as the second step (needed to prompt the specific demonstration process and so focus the inferences to follow) and 'narrative' – a broader sense than 'interpret' (intended to link the questions to both the demonstrations and the outcomes, and so mediate between theories and empirics). Mari and Giordani (2014), in a slightly different but helpful generalising move, suggested that the 'models as mediators account' offered scientists two 'tools': "a model is used both as a theoretical tool for interpreting our concepts and an operational tool for studying the corresponding portion of the world" (p 83). Nordmann (this volume, p4) notices the importance of 'what if' questions, as suggested in Morgan 2012, to return the focus onto the model usage as one tool, characterising the mediating activity of using models as "...a free play of 'what ifs' [that] configures a specific relation of model, theory and reality... as one seamlessly moves between the actual and possible [worlds]".

The argument here continues with this characterisation, namely on how the mediating quality of model usage serves both ends at the same time, namely that scientists use models both to explore their theories and their relevance to the world jointly. The point of such model building and using is to find models that are useful in helping to characterise the world and to answer their questions about the world. They judge the quality of their models in terms of the alternative accounts of the world that the model offers: are those worlds plausible and reasonable or impossible or non-sensical? These qualities are relevant to both the theory and the empirical domains: models need to make sense in both domains. And they are not separate domains, but related ones. Asking questions with the model focusses on how the world might work in theory terms, and so on the explanatory power of model work in opening up alternative world possibilities. But equally, if the model explorations have no contact with empirical domains, they might be quite suspect. Models in usage need to produce alternative world accounts that are theoretically reasonable and empirically plausible[3].

The question addressed here then asks: How does a scientist in a field distinguish between the alternative worlds generated by such model usage? This paper explores two different dimensions of judgement of these alternative worlds found through modelling. The first involves reasoning with a model (formal or informal) to see how the elements of the model knit together along a possible path to an outcome: this looks for and judges the *reasonable*, or perhaps a better label, the *well-reasoned* path and outcome. The second comes from judging the congruence of both the pathways implied by model reasoning and their outcomes with what is already known about those aspects of the world, asking if this is a *plausible world* account, that is: could it plausibly happen, or have happened, like this?

These two characteristics, well-reasoned and plausible, are rather loose criteria. And implying that lines can be easily drawn between the plausible and implausible, and between the reasonable and unreasonable, are equally problematic, because of the difficulty of characterising both the 'reasonable' and the 'plausible'. The distinction I want to create is:

Reasonable – focusses more on the theoretical aspects of the model and its usefulness in providing a well-reasoned account of the possible world-in-the-model in terms that are in line with both existing ideas and well-attested knowledge about how the world works; or in developing theories beyond these boundaries but still ones that can be accepted by the community of the time. Judging this reasonableness involves not only judging the outcome of model reasoning but, more importantly, judging the paths of model reasoning on the way. Think of it this way: models are not black boxes, but reasoning devices in which the scientist can see into and learn from the model manipulations on the way. However convincing are the inputs and outcomes (the assumptions and predictions), if the path of reasoning in the model world produces completely non-sensical points in terms of its subject implications, then that model world might well be judged an impossible world. And even if the inputs and outcomes of the model seem possible, it might be that explorations with the model show some pretty surprising paths.

Plausible – focusses more on using subject matter evidence to judge the model materials. The aim is not only to avoid the impossible world pathways and outcomes that come from models that might be very little constrained by evidence ('common knowledge', or formally obtained by scientific methods); but also to strengthen judgements of possibility from the theoretical side into judgements of plausibility using evidence from the actual world. This is where we find overlap with counterfactual questions and modes of reasoning.

Scientists judge the validity of their models in terms of the alternative world accounts they create: are they reasonable and plausible, or are they impossible, unbelievable, and non-sensical? How does this work in practice? The arguments here are based on the analysis of cases and practices in economics, cases that are often regarded as paradigm cases in their own scientific field, and some have excited the interests of commentaries in philosophy of science.[4] These cases are used to parse out the kinds of criteria that scientists use (or maybe misuse) in judging the usefulness of models in their field. By approaching these cases as model-based explorations of alternative worlds, the aim is to see how economic scientists who use artefactual models judge the plausibility and reasonableness of how these worlds work in the models they create. I stress again, it is the use of models that is key to these judgements, not the model judged as a constructed, but largely passive, object. The arguments begin with a discussion of reasonable or well-reasoned alternative worlds, then of plausible such worlds, and finish with a discussion of practices where reasonable and plausible worlds are found together. This prompts some further reflections on the role of the actual world in defining and making judgements in relation to alternative world accounts, for the mode of investigations may well go along with other notions of alternative worlds.

12.3   Reasoning about the Possibilities of the Model World: Well-Reasoned and Reasonable Alternative Worlds

Let me start with a compressed example, one which has a strangely mixed fictional/factual framing. The Edgworth Box[5] – an innovative diagrammatic model developed in the late 19th century that became paradigmatic to economists - uses the example of Robinson Crusoe as the basis for reasoning about the situation and behaviour of two individuals, isolated on an island, in exchanging labour for goods. An implausible, fictional model-based account? Perhaps. The book of that title was regarded as one of the first novels and so a fictional account, but its prose dressed it up as a news account: a kind of realistic storytelling. In fact, the story was probably based on a real account of a ship's captain abandoned on an island following his crew's mutiny, and who did not make it back to Scotland for many years. Did that fictional/factual aspect of the original situation matter to the evolution of the diagrammatic and mathematical model of that situation, and its reasoning usage, amongst economists? Surely not. For them, its innovation was to capture the important heart of a knotty problem in economics, namely how two people haggle or reason to get to a point of exchange. The realism of assumptions or accuracy of outcomes (predictions) was much less important to that community than the insight from the reasoning process the diagrammatic model enabled and demonstrated. This was reasoning about how one got from the starting assumptions to the possible outcomes; the path to the best outcome perceived from both sides was not self-evident or direct and depended on the details of the representational model, which enabled economists to see the reasoned pathway to that best outcome. That pathway, and outcome, were then taken to be relevant for the general exchange problem, yet the outcome was only agreed upon because of the conviction given by the economic reasoning middle in using the model. To label this as either a possible or impossible world makes little sense. It was a highly imaginative pathway and solution, translated through an artefactual model into a well-reasoned world in economic terms. And because it was well reasoned, it seemed reasonable to the scientific community of the day.

Of course, it is possible to argue about lots of elements of the practical reasoning processes found in the sciences. And it is difficult to characterise the qualities of accounts and outcomes that make them seem reasonable or possible, though it may be much easier to see them as impossible. In these terms, all such judgements of what counts as reasonable are those tested against and accepted by the community of scientists in the field at the time. There is no outside judgement of what is reasonable except in subject contexts. This is not an attempt to revitalise the notion of 'normal science'. On the contrary. In the Edgworth Box example above, the assumptions, elements, accounts and outcomes were all reasonable for that community – it was the way the assumptions about behaviour and starting points were put together into a new diagrammatic model, and the reasoning with that model, that were seen as novel and unusual and led to new ways to think about the problem. Contemporaries took it as a starting point and developed it because it was considered an original contribution that generated a new direction in terms of both representational device and reasoning mode within its field (evidenced by the fact that historians have documented its continuing innovation over the following decades; see Humphrey 1996).

Another key example of model reasoning in economics, one that starts with an impossible world outcome and is developed in such a way that can suggest a reasonable world outcome, is found in Malthus's account of the population problem of his day.[6] His starting point was to attack the writers of his time (the end of the 18th century) who assumed that mankind and its society were set irreversibly on a path towards a socio-economic utopia. He complained bitterly, and with brilliant rhetoric, against both their rose-tinted visions and their lack of evidence of such an evolutionary path that would convince him, or anyone of sense, about this process and its outcome. Malthus thought they were imagining an impossible future and challenged them in their own rhetorical terms to show any evidence of the path by which, for example, "a man becomes an ostrich" (an example he chose to match their outlandish claims about social evolution). Malthus eschewed their rhetoric in his own riposte, beginning with sober arguments that enabled him to separate out a possible from an impossible world in the context of the viable future of a population. This distinction was a key element in his argument against the impossibility of their promised future utopia.

Malthus was working before modelling became the standard way of doing economics, yet he had a set of assumptions which he combined to think through the future path of mankind in quite a formal way. He started with two postulates that surely did not sound unreasonable: "That food is necessary to the existence of man" and "That the passion between the sexes…" will continue "nearly in its present state" (Malthus 1798/1976, pp. 19–20). Then, to generate his 'model' reasoning about the future, he argued that population growth would grow geometrically and food supply arithmetically. These modelling assumptions might seem as if they were picked out of thin air, but they rested upon two generic pieces of evidence of his day. One was that the food supply had recently grown rapidly in Britain because of the contemporaneous 'agricultural revolution'; his assumption formed an upper bound, for such further growth possibilities seemed more limited. The population growth evidence was based on the experience of immigrant communities (such as the Amish in the USA) whose population growth was not restricted by limited land and so food supply – another upper bound. These input assumptions were poorly attested by modern standards, but for his period passed as fairly good pieces of evidence of the most optimistic possibilities for both growth rates. His reasoning with these resources (that is, his postulates and assumptions about growth rates), all of which seemed reasonable in themselves yet, when combined, quickly opened up an impossible world outcome when tried out on the British situation. Thus, the population of his time (around 1800) was thought to be 7 million; and his 'model account' of arithmetic versus geometric growth told him that:

- in the first 25 years, population would grow to 14 million, and food output support 14 million;
- in the next 25 years, population would grow to 28 million, but food growth supports only 21 million;
- and, after 100 years, population would grow to 112 million, but food only supports 35 million. (Figures abstracted from Malthus text, 1798/1976, pp. 22–23.)

Thus, while the model world seemed based on a set of reasonable assumptions, its usage in tracing out the future of the population quickly revealed an impossible world.

What was impossible about the future world that Malthus outlined? Note that there was no logical or deductive failure in the reasoning: the arithmetical calculations told Malthus how many people could be fed and how many would die from starvation (or not exist) in the future world if his assumptions held. But while it is possible to imagine these numbers in the arithmetical model, not so in the actual world that his model was used to explore. Since this implied mass death of the arithmetic world was surely an impossibility in the actual economic world, Malthus asked himself what would happen instead. Using these same starting assumptions of behaviour and evidence but adding in some additional economic arguments about how people behave when there is pressure on food supplies (delaying marriage, etc.), he created an alternative, and now possible, world account. This was a narratively reasoned account, weaving together some other elements of economic consideration into a story of repeating periods of relative well-being and then hungry poverty as the population growth cycled above and below the food supply growth.

Here was an alternative account of how the world worked, and in the process provided a way to save the original assumptions (both theoretical and empirical) and to understand why neither the envisaged utopia nor mass starvation was a likely outcome given the constraints of the actual world. As he pointed out, there was no observable evidence to match his additional reasoning categories; they were the conceptual (or theoretical) categories of economics, and so this example fits well with Mari and Giordani's account of model's mediating work. Yet it was a well-reasoned account, and might be considered entirely reasonable for its period (and reads so since the same conceptual categories are used by economists today). Narrative plays an important role here; it is where the elements of the account are joined up and made sense of together; it is not a description, but rather a causal account, in which the reasoning is made evident. This facility of narrative to make sense of a set of elements in a model and embed reasoning with the model is ubiquitous not just in economics (as we shall see), but in other sciences, particularly the natural sciences, where time is an essential element in explanations[7].

We can find other similar examples in economics where the model's mathematics implies a particular outcome or set of outcomes that are judged impossible in the actual world by the economist. For example, solving an algebraic model of how different generations overlap and manage their economic responsibilities to each other (called the 'overlapping generations model') revealed a negative root, which implied that people lived negative years of life! This is like the Malthus example, where a rational mathematical model implied an impossibility in the economic world envisaged in the model, and in this case, that root was rejected in favour of working with the positive root solution of the model.

Another more complex example leads us into the next section of the argument. This example is less obviously an impossible world example and comes with an algebraic model of the aggregate economy (the so-called 'macroeconomy') by Samuelson in 1939.[8] His little 3-equation model was simulated by hand (this was the first simulation of a model, before computers) by plugging in alternative parameter values and running the model for nine periods.

The question explored was: 'What would happen if the government spent an extra unit in this little model economy?' The economic context was the 1930s, and the model was built to represent Keynes's breakthrough in theorising and policy work. The values of the parameters on this model of his theory were chosen to explore the theory in the model (not because they had been established by any prior statistical work). Different parameter values produced different narrative paths and outcomes for the little economy: cycles, a small increase then stability, or a huge take-off. This final simulation implying a rocket-trajectory would be regarded as an impossible world to any economists of that day, and indeed since. So again, this was a mathematical model possibility, but was just *implausible* for anyone with background knowledge of how far and fast economies might grow, whereas Malthus' model world was more genuinely an *impossible* actual world: so many more mouths than food must mean mass starvation, but it would never have gotten to that outcome as far as economists were concerned, and thus Malthus reasoned out an alternative account.

The point here is not that Samuelson's algebraic model portrayed an impossible world, nor that it was an idealised toy model, nor that the reasoning in the model was at fault. Rather, the issue was that certain posited parameter values in the equations created extremely implausible paths in his explorations with the model; other values created plausible paths. Just because of this finding of variety, Samuelson speculated as to the further possible world results that his model might generate: he solved the algebraic model so as to chart the full range of behaviours, and concluded that almost anything could happen in his model world. His model might, of course, be saved by bringing in more that was known about the world into the model world, just as Malthus had done. For Samuelson, this would have meant by using a parameter value on the equations that were more plausible and so create narrative reasoning (pathways and outcomes) with the model that could be judged plausible. Exploring such plausibility comes next.

12.4   Exploring the Limits of the Actual World: Plausible Alternative Worlds

One of the main foci for judging plausibility is paying attention to the context of the actual world and using data about the actual world in formulating the alternative model world. One place where this is dealt with in quite well-understood ways, according to some well-known recipes, is in the statistical work of economics, labelled econometrics.

In this field, there are standard modes by which the statistical qualities of the model are assessed (goodness of fit, etc.). The recipes for gaining statistical satisfaction have been developed over the past century, and rely in economics on some heavy-duty statistical and probability theorising along with tried and well-tested ways of working with statistical data. That is, they are not one-off, casual methods, but fully informed and tested ways of judging empirical adequacy. And just like the examples in the above section using arithmetic or algebra – what makes sense in technical reasoning may not make sense in economic terms, either as an account of the pathway or the outcome. Statistical adequacy does not necessarily align with plausibility in the economic sense. Thus, an element initially considered an important cause by the economist could pass the statistical test but be at such a low parameter value to imply it was economically insignificant. In addition, there are the various *ceteris paribus* tests of economic relevance: to judge variables as insignificant because of their marginal relevance, to judge others as irrelevant enough to be omitted, and to locate those variables that remain stable in value in the period whose relevance cannot be judged. Judging the economic qualities of the econometric model requires attention to signals like the size and sign of coefficients in causal relationship models, the patterns in time-series data, the time dynamics of relationships, the internal cross-relationships between different equations of the model, etc. There are also standard recipes for the process of model building: for example, start the modelling simple and make it more expansive to cover omitted variables, or v.v., start with a big encompassing model and slimming it down to get rid of irrelevant factors; testing it out on different groups of agents, or time periods, to assess its range and scope. The framing here is to find plausible models, models that seem to fit the data from the actual world well enough, or satisfactorily enough, to be accepted as working objects for the economist[9].

Another, but less common mode of exploring plausible worlds is by the construction of counterfactual worlds[10]. The most well-known counterfactual for economists is that by Robert Fogel (1964), investigating the importance of the railways in the development of the American economy in the 19th century. This was a tour de force of American economic history, asking the question: What would have happened to American economic growth and development if there had been no railways? It was widely assumed in the 1960s (when he asked this question) that railways had been essential to the growth and the development of the economy – both geographically, and in terms of the growth and distribution of the sectors of the economy. His counterfactual world was designed to 'test' this widely held belief, not by trying to prove it true or false, but to see how far a plausible counterfactual analysis would challenge those beliefs both in the alternative path of history and in its final outcome.

Assuming no railroads could be an extremely far-reaching claim, depending on how far back and sideways this question implied: for example, the counterfactual of assuming no invention of steam technology is not very helpful to Fogel's question for it rewrites so much of the past as to lose contact with the problem he explored. Fogel narrowed his counterfactual question to judge only the effect between the beginning and end of the main growth of the railway system in the USA up till 1890, and by figuring out the alternative costs of transporting the main four agricultural crops of the period by water and by road. But having narrowed down the scope in these directions, he immediately had to open it up to ask what would have happened in this alternative world: where would people have lived (the alternative settlement pattern, given this period was when 'the West' was opened up); what economic crops would they have grown; and how would they have gotten them to market? With no railways, water and roadways would have been used, but more radically, less of the geographical scope of the USA would have been 'settled' and opened up to intensive economic engagement of the kind that happened during that century. The counterfactual question required the economist to imagine and fill in an alternative world account of what would have happened in those 60 years, so Fogel drew out maps of his imagined new waterway transport routes: designing canals and river-widening for an imagined alternative economy that enabled access to the maximum possible settlement and agriculture.

In other words, even to answer his limited counterfactual question about alternative transport costs, he had to imagine an entirely different history of alternative growth, settlement and distribution patterns within the USA. There was no formal model at work until his final calculations, but rather the application of basic economic assumptions and standard economic calculations about individuals' efficient reactions to circumstances to optimise settlement and output and so the overall growth of the economy – all in very different ways in the counterfactual world from that actual world. That is, his counterfactual world was assumed to work according to the same economic reasoning as the world with railways did, just its development and its outcomes would be very different.

The recipe for using counterfactuals to create 'plausible worlds' is best laid out by Geoffrey Hawthorn (1991), who does so primarily for social scientists and historians, but perhaps it could equally well apply in other sciences. His 3-rule recipe tells the scientist/historian: to choose the starting point of the counterfactual on the basis of facts (not theories) that do not require a rewrite of the past; to discern and make use of the agency of relevant materials in the world that will drive the counterfactual direction from that starting point; and to minimise the use of theories or models unless they fit the question and circumstances very closely. Fogel seems to have followed just such ground rules. In contrast, perhaps as a light-hearted critique of Fogel, McAfee (1983) used a mathematically model-driven starting point to ask what would have happened by 2000 if Columbus had not discovered America but fell off the edge of the flat earth? A non-sensical question (given the knowledge set of 1983!) breaks all of Hawthorn's injunctions for plausible world counterfactuals. In Hawthorn's recipe, it is the strict limitations or constraints on which to base the counterfactual drivers to fill in the counterfactual path that depend on actual world knowledge (not on any theoretical assumptions) to provide plausibility for the alternative world sketched out. In both McAfee and Fogel, the alternative world used for comparison is clear, it is the actual world in its historical development, but with very different recipes, constraints, and so outcomes.

The reaction to Fogel's work, in the context of this discussion about the ability to distinguish between possible and impossible worlds was quite revealing. Economists of that moment (the 1960s) found the final social savings figure that Fogel delivered (the effect of not having railways) was the equivalent of only 1 year's growth, an implausible number to them, because it was so tiny compared to their widely shared belief that the railways had been essential to 19th century American economic growth and development. Most did not so much decry the counterfactual itself, or the methods Fogel used, but rather the extent of the limitations he introduced – tracing out effects for only four crops, no effects on manufacturers, on passenger transport/business travel, on retail distribution, on big finance – that is no exploration of all the backward and forward linkages that these other economists had been busy writing about for many years. For those economists, Fogel's alternative world picture was only a small part of the story, perhaps it set a lower-bound to calculations of the effect of the railways, but it missed out too much for their comfort. (There were also arguments about the precise nature of the calculations, but these are not especially relevant for this argument on possible vs impossible worlds.)

This alternative counterfactual world he had created was not considered 'impossible' in any technical sense for it was made up of plausible 'factual' alternatives that his contemporary economists recognised as such. Although, taken together, they created a quite startling alternative world that was considered implausible because radically incomplete. But neither did it seem entirely 'plausible' as an alternative imagined world. Here, the actual world and its factual characteristics play a stronger role than in the calculus of Malthus's alternative world. In Fogel's case, the factual characteristics both constrain and shape the alternative world, and it is the measured aggregate outcome that was regarded as implausible, not the counterfactual world analysis as it goes along. It was not nonsensical – in the sense of negative-lived lives – nor impossible in the sense of unimaginable in the actual world as in the implications from the calculations of Malthus's simple 'model'. Rather, it was simply the incomplete answer to the counterfactual question: plausible as an account of some of the pathways, implausible (because incomplete) in its final full outcome.

An alternative counterfactual practice of creating alternative worlds provides a contrast. These are ones that explicitly set out to explore alternative factual worlds in as broad a way as possible to reflect both more deeply and more broadly on the actual world. This is the agenda for a group of papers that provide a strong contrast to Fogel's work, that appeared under the title *Unmaking the West* (Tetlock et al. 2008). The basic question is to rethink the rise of the West, politically and economically, by exploring if that account can be undermined in some critical ways or at some critical points by tracing out ways in which it might have happened differently. How could it have been that the West did not become the dominant economic and political power over the last 300 years or so, but that China and Asia did? There were good reasons for this framing, as recent literature on Chinese economic history, and as the current position of China and India in the world's industrial economies reflect.

This is not about exploring how else the West might have been made, but how the West might never have been made. So, the book opens with an imagined group of Chinese scholars, working in an imagined world in which the Chinese had been the dominant culture and power over the past centuries, being asked to imagine some small key or 'hinge' events where their own (fictional) ascendency might have been stymied. The contributing authors to this volume (grounded in politics, economics, and history) were asked to trace the economic, social or political historical outcome that might follow from such a very small, but significant, hinge event in Western history that might have led to that imagined Asian ascendency. For example, what if England had remained Catholic in the 16th century? What if Britain had no coal, no colonies, and no modern science? Would any of these hinge events have created a path in which there was no industrial revolution in Britain? Tracing alternative worlds from these 'turning points', leads each writer to 'create' a set of changes in the world, and then trace out second and third-round counterfactual changes. Maybe these further changes are reversionary – taking the economy or society or polity back onto its original path, or maybe they are expansionary – taking the alternative world ever further away from the actual world.

The group of writers was asked to abide by "three exacting quality-control questions" (p. 9), but in comparison with Hawthorn's recipe, they are not so 'exacting'. Their first 'minimal rewrite rule' requires the counterfactual to start with a very small change in the historical record; this hinge event is rather like Hawthorn's first rule. But thereafter, the constraints are not so evident; rather, the basis for creating these expansionary alternative worlds is to follow the causal implications of such hinge events being different – what leads on to what and how. Rather than to constrain or limit the counterfactual, the recipe here gives full rein to the use of reasoning in tracing through implications far and wide. These are not just exercises in description, for there would be no alternative world spun out without framing the considerable differences that occur in the narrative accounts, i.e., in loosely causal or associational accounts of where and how those differences would create that alternative path. In other words, narrative reasoning about the path was needed to justify the alternative world outcome as in Fogel's case, but now it is an account much more loosely constrained by history, whereas by contrast, McAfee's highly imagined counterfactual history is led by his mathematical model world.

We are back in a sense to the 'anything can happen', or rather 'anything can have happened'. However, the constraints here are not model-based constraints or even strong factual constraints, but alternative fact construction using social science and historical knowledge and methods. How should we label these alternative worlds found in *Unmaking the West*? Given the first careful choice of hinge events as starting points for their counterfactuals, we might label them possible worlds, but their multiple further rounds of counterfactual possibilities, spinning out further paths that take them ever further from actual events, might suggest we should label them implausible (not impossible, because they are well traced out alternatives). By contrast, the tight factual constraints in Fogel's counterfactual world enabled him to judge, even measure, the difference from that single first counterfactual choice, and so judge the path taken by his alternative factual world as plausible, even while his contemporaries found the counterfactual world he created incomplete, and so the outcome implausible.

It has become apparent from these counterfactual cases how important the contrast or comparison element is in judging the validity of alternative world accounts, both in reasoning and in plausibility terms[11]. In some of these cases, the comparison is provided by history as the alternative to the actual world, yet the modes of doing so vary from tracing out one very 'factual' alternative world (Fogel's work) to numerous imagined alternatives (in imagining the economic development of the West compared to Asia) to opening up a completely different past alternative that has fragile connections and so comparisons with actual history (McAfee's case). In all these cases, history is the actual world, and the counterfactual alternatives seemed to be judged mainly on plausibility grounds, but the reasoning is not absent – those counterfactuals have to be driven by some economic agency that needs to be filled in and some reasoned pathways that may not be so evidenced. That economic reasoning side is more obvious in Malthus modelling case: we are led to imagine (with difficulty) a very different alternative future history, driven by the model assumptions and reasoning to an initial implausible outcome that is then saved by further economic reasoning. But history is not the only contrasting world that is useful in making these judgements. The little theoretical mathematical model worlds that generate variations in pathways and outcomes (the examples of model simulations, negative lives, etc.) tell us that much less specific background knowledge of the world is equally valuable in providing contrasts and comparisons that inform judgements about the alternative world built in such models.

In these economic cases, historical actuals have played an important role, but the actual worlds of other sciences may not be historical ones. Many sciences are involved in making their own alternative actual worlds: synthetic biology, chemical synthesis, drug 'discovery', etc. Here the key task may be to recreate or synthesise something that already exists in the world, or it might be to create something new that has certain qualities, i.e., to turn a possible object into an actual world one. Here, the task is to design a possible object and then to 'engineer' that something into an actual world object. Perhaps the possible object is the same as one that already exists, but is not yet something that can be made; perhaps it is a slightly different alternative actual, or perhaps the problem it is to find an alternative pathway or recipe to making an already recognised actual. This is a minor part of economists' work – but one such example is the case of auctions, where the use of theories and close study of how different kinds of auctions work were used in order to design auctions to attain certain aims. These fit the 'how possibly' creative work of synthetic biology discussed by Koskinen (2017), who also points out the importance of information that comes from learning how things can not be done (p. 504). Here, the contrasting/comparative actuals are not historical contrasts, yet if we take the range of approaches in chemical or biological synthesis as indicative, it seems that the range of both possible and actual alternative worlds could be pretty unconstrained. This could well extend the analysis of the modes of creating and judging possible alternative worlds as used by scientists.

12.5 Triangulation: Where Well-Reasoned and Plausible Worlds Come Together

The first recipe discussed above was to use the reasoning facility that comes with a model to see if it created reasonable and well-reasoned worlds, alternative worlds that were possible in those senses, but that reasonableness may also have included some fit to the actual world in order to have the judgement of being a plausible world. The second recipe was to use the factual aspects that come with 'applying' models to see if the evidence-based reasoning with models looked plausible and produced plausible outcomes. Of course, a model usually embeds some theoretical claims or concepts, so that reasoning with them is not 'theory free'. In other words, we can see two elements of fit being used here in both approaches. Yet the theory-side is not, as Mari and Giordani argue (above), only an exploration of the concepts in the theory side but might rather be an exploration of the reasoning the theory entails or allows, and judgements of reasonableness that are not ones of logic but of fit with some knowledge of the actual world as in Malthus. It is important that both the pathways found in using the models, and the model outcomes (predictions, final points, etc.) need to be considered in these judgements of reasonableness and plausibility; and I repeat, these are judgements of the community of scientists involved based on their wider field knowledge. This third recipe is perhaps not so much a different recipe as just more explicit in its use of both resources – a triangulation process between the model, its reasoning resources, and the evidence base relevant to the model – to judge the quality of the model according to the alternative worlds its usage suggests.

'Analytical narratives' offer one such version for generating alternative worlds in the social sciences, ones in which there is a back-and-forth pattern of analysis between model reasoning and qualitative evidence, using the models to analyse the factual and using the factual to constrain the particular choices within versions of a model and between different models. The model here effectively stands in as a shorthand to explore alternative accounts of 'reality' against the 'narrative', which is offered by qualitative evidence of what actually happened. The introduction and usage of this mode of working in political science has been primarily engaged in using game-theory models to explore the roles of institutions in political economy. They largely address questions about why political institutions emerge and work differently in different contexts, different times and places, and in solving different problems. This is not a way of testing any general theory, or making big wide claims, but a means, again, of exploring how the world works by triangulating between evidence of the actual worlds and reasoning about alternative theoretical models as possible designs of institutions and their impacts[12].

Another version of this 'analytic narratives' recipe (with a slightly different name) is found in economic history, which takes a sequential view of the role of models and narrative evidence (Bates et al., 1998). That is, a question is raised about why and how some economic institutions formed, disappeared, or perhaps collapsed. For example, Why did the possibility of buying out from military conscription disappear? Why did absolute monarchies disappear? What accounts for the formation and subsequent collapse of an international coffee cartel? The latter account (Bates, 1998) begins with a narrative evidence account of that cartel formation and then applies a game theory model to explain that formation. Bates then proceeds through an alternating sequence of narratives, statistical evidence, and different models to capture the formation, history, and then collapse of the cartel. Why is the sequence needed? Because one model is not sufficient to explain the whole historical evidence sequence, rather, different models provide analysis for different aspects at different periods. This sequence of model application to evidence, followed by a new model applied to the remaining evidence, etc., continues until the scholar is happy that the narrative of evidence is covered by the sequence of models, and so the whole 'explanation' of events is achieved.

These model-narrative-evidence recipes are best understood as exploring the actual world by systematic use of alternative worlds embedded in models – either in statistically evidenced models or in theory-based models – that can be applied in formal and informal ways to explore alternative explanations for the specific phenomena at issue.

Another set of examples shows this triangulation at work in economics research, which aims to deal with different sets of model paths and outcomes within the same domain that ought to align, but initially fail to do so. Here models are used to explore the misalignment of model-evidence directly by counterfactual methods, leading both to developments in the model to adapt to the problem set, and enquiries into the data to choose (or develop) data that is more closely relevant to the problem question in hand. The salient point here, and the difference with the counterfactuals discussed earlier (such as Fogel's case), is that here, both sides of the comparison are model worlds (that is, the alternative to one model account is another model account of the same or a parallel world), not the development of a fictional alternative world to a factual one that existed.

Ramey's work (2019a,b; Orchard et al. 2023) offers a variety of examples using counterfactuals in this way that she calls 'plausibility tests'. The context for her work (often with collaborators) was provided by US government interventions over the past two decades designed to support the economy in times of crisis. This was done by paying out either tax rebates (or direct grants) to households to support them, or massive government spending programmes; both kinds of actions were taken in order to avoid a major economic downturn. The problem she investigates is where different models produce inconsistent results for the same event, or produce extremely unlikely results when comparing model outcomes from similar actions at different times. Her basic approach is to see these comparative model sites as implying or raising counterfactual questions and then to pursue the calculations to investigate the lack of fit between closely comparable model results and so 'test' the plausibility of one model's result with another model's results. These investigations created a positive programme of how to search for reconciliation that Ramey has labelled a "Macro Counterfactual Plausibility Analysis".

One of these cases examined is the effect of the 2008 tax rebate in the US,[13] in which a key number that characterises the individual household effect (i.e., at the 'micro' level) was found to disagree very considerably with the parallel number that is taken to characterise the aggregate economy effect (i.e., at the 'macro' level). It is not entirely clear that they should be exactly aligned, but the wide differences in statistical work produced considerable surprise amongst economists. The team developed counterfactual questions to explore the two alternative model-world accounts at work here, i.e., as portrayed in the micro and macro models. So neither was taken as the fixed-base case (i.e., equivalent of the actual world history as in Fogel's counterfactual that explored an alternative fictional world) – rather interpretation and numbers in both accounts were potentially revisable since both were model accounts.

A simple analysis of the macro data for this 2008 event suggested that despite this sudden increase in income for households, there was no sudden equivalent spike in aggregate spending. This was not initially seen as problematic, the inference being that households saved rather than spent their sudden windfall. But research soon showed that if one studied the micro-household data, it seemed that, indeed, households did respond with a spike increase in their spending, and the majority of the spike went on buying automobiles rather than normal household goods. On their own, the household findings also made sense to economists: households use the rebate to buy a durable good, not more everyday consumption. But in putting these two findings together (each of which made separate sense for economists focussing on micro and those concentrated on macro problems and their data), they were not just surprising, but suspect: why don't they agree? Were these alternative model-world pictures consistent, or impossible? Suspicious of the results, these economists calculated a simple counterfactual on the micro-model, ignoring the macro: 'imagine no tax rebates' (here using a fictional counterfactual), and found that without the tax rebates, there would have been a very large fall of 87% spending on automobiles at the micro-level according to the model and data. The equivalent macro counterfactual calculations assuming no rebates suggest that the household expenditure on autos would have dropped from $17bn in March 2008 to $3bn in June, whereas the actual lowest level throughout the period was $12bn in April 2009. As Ramey's group expressed it: "This counterfactual [outcome] strains credulity" (Orchard et al. 2023, p. 2). Clearly, that set of results does not make sense when taken together. A further counterfactual exercise on the same topic generated an outcome labelled "preposterous"(Ramey 2019a, 125)!

To make better sense of this 2008 episode, the counterfactual question then changed to ask: 'what if the micro number were used in recalibrating a standard macro model?' What kind of counterfactual path would this counterfactual macro model reveal by plotting the difference in spending paths between the two models – the standard one and the recalibrated one? The results suggested that the counterfactual micro-model path and outcomes are implausible because various historical events that, evaluated post hoc (10+ years later), would have made the micro estimates implausibly high, and so brought down the difference between the counterfactual micro-model path and outcomes compared to the actual macro-model path and outcomes. At the same time, looking at the data with 'improved econometric techniques' meant some adjustments that lowered the micro parameter better to fit the macro measurement.

In this kind of counterfactual work using models on both sides of the comparisons, both the data and the model are potentially adaptable. This search for plausibility and reason in modelling and results comes from a search for alignment in both the theoretical and empirical domains. Both data and models need to be 'the right tools for the job', that is the right tools for answering some particular question put to a model about a specific problem or situation that has arisen in the actual world. The use of counterfactual questions, and exploring their answering paths and outcomes, provide information on what is not plausible and so prod triangulation efforts to align the right data and right model for the specific question to be answered, which may itself only be revealed during the alignment process as the scientists learn more and more about the situation. Such alignment offers another criterion for the validity of the model reasoning and the plausible explanation of the world using model alternatives. In general, these triangulations involved searching out data more relevant to the problem, and the use of models compatible with the counterfactual questions and the specifics of the data. Plausibility then comes not only in the agreement of the numbers or paths with fictional or real historical events treated as comparable counterfactuals, but confidence in both data and model being more carefully aligned to the question. Here we see very practical ways in which models are used as a thinking or mediating tool in exploring how the world works by triangulating between alternative theoretical worlds and alternative actual world materials.

12.6 Conclusion

This account began with the problem of how to distinguish possible world accounts in sciences from impossible ones in such a way to validate the former as useful accounts. Using a more neutral terminology of 'alternative' worlds, it quickly became clear from the economic examples that there are some worlds outlined using modelling or counterfactual approaches that are possible but implausible or incomplete in some way, and some that are technically possible but non-sensical or unbelievable in subject matters for various reasons. Developing the terminology of alternative worlds, and their validation, prompted judgements based on the reasoned implications and the plausibility of those alternative worlds with respect to field knowledge. Alternative worlds could be validated as plausible worlds and well-reasoned worlds with respect to theoretical and empirical knowledge in the field, and preferably both.

It is perhaps worth noting that economists following philosophy of economics injunctions, have traditionally worried about the assumptions that go into a model or the accuracy of predictions from a model, and validated them on those grounds. This is beginning to change, as we can see from the triangulation work of the final section (for example, in the work of Ramey and crew), but new criteria appropriate for modelling and counterfactual work have not been worked through. As suggested here, the appropriate criteria for models and counterfactuals are much broader than those earlier recipes: the focus is not just on the beginning or end, but on the value of the account all the way through the middle, i.e., through the reasoning middle. This does not mean assumptions and predictions do not matter, but they are not the only things that matter, rather that exploring alternative worlds with models and counterfactuals may be a much more informative process about both theories and our world than only one or the other. The counter-factualist spinning alternatives to the actual worlds might well change their earlier interpretation of the events of the actual world, just as the modellers might well revise their theories from such alternative-world explorations. Maybe the railroads did not really matter that much to US economic development; maybe highly idealised models or well-trained econometric models don't tell you much about processes or outcomes. But exploring the possibilities of alternative worlds with models and counterfactuals may well reveal much about both the actual world, and about the kind of theory that offers a reasonable and plausible account of that world.

How should this exploratory process be characterised? What kind of ingredients are involved? Sjölin Wirling and Grüne-Yanoff's survey of the literature (see earlier and in this volume) suggested imagination and background knowledge were involved. The argument made here is that spinning alternative worlds requires *both* imagination *and* background knowledge *and that they must work together to be effective.* The background knowledge evident in our cases encompasses lots of different kinds of things, from very specific subject matter to quite general matters, from laws and theories to formal and informal empirical knowledge, and comparable case knowledge. Imagination is also clearly a part of the recipe of generating alternative world accounts, perhaps more so with counterfactuals where there is potentially greater freedom to spin alternative worlds than in working with models where the assumptions and structure of relations fix certain boundaries.

Spinning alternative worlds in models and counterfactuals with lots of imagination but no background knowledge is a recipe for creating impossible and implausible outcomes: plausible and possible worlds may come from luck, but not from design. From the model side: internal coherence in a model is clearly a plus, but not enough on its own to generate plausible and possible worlds rather than falling over into impossible ones. From the background knowledge side, factual or theoretical knowledge from the scientist's background, either in modelling or in counterfactuals, is not enough to generate the additional knowledge that reasoning through alternative worlds in modelling or with counterfactuals can create.

What we find in the cases discussed is not only the combination of imagination and background knowledge, but also the importance of limitations, either framed by the dimensions and design of the model itself, or the particular decisions in counterfactuals. Without limits on the imagination, anything can happen. Background knowledge, theoretical and factual, does not only provide those constraints. Rather, it is what enables scientists to use models as exploration tools, tools to explore the possibilities of different values of parameters, of different sets of elements, of different design decisions, etc. Using models in this way enables scientists to test out the implications of their imagined world. But it is the limitations given by the model or the counterfactual question or the background knowledge that make the feedback from alternative imagined worlds useful in figuring out the validity of the model (what range of phenomena are covered, etc.; what happens when you change something in the model, etc.). Learning that something that is impossible in the model world is often as useful as learning that something is possible, just as in learning about the im/possibilities in the actual world. Imagination is needed not just to construct models and counterfactuals, but to work through their implications, so explorations with models and counterfactuals are indirect tests of our imaginations. Is that world really credible, really possible, really plausible? How do we judge – by using the imagination in various ways to test the construction, but also to test the model's usefulness in relation to how the world works? We use our models to explore our imagination of alternative worlds that might be possible, and we use our imagination to test out the boundaries and possibilities of our model or counterfactual world.

Notes

1 Thanks go the participants at the 'Modelling the Possible' workshop hosted by Vienna in 2021, especially to Tarja Knuuttila and Rami Koskinen, Till Grüne-Yanoff, Andrea Loettgers, Alfred Nordmann, Mauricio Suarez, Paul Teller, and others. Thanks also go to two anonymous referees whose thoughtful comments helped me to revise the paper.
2 See references therein for the most salient philosophy of science literature on possible worlds; further references are also found in two accounts that have parallels with mine here but are about designing and engineering alternative worlds in science: namely Nordmann (this volume) and Koskinen (2017).
3 The problem is framed here as an account of the practises of using models, and how scientists learn from them, not about a particular kind of model, nor about conceptualising and/or defining the philosophical qualities of possible world models, or subclasses of such possible-world models (as, for example, in Grüne-Yanoff and Verreault-Julien, 2021).
4 Where these are cases I have worked on before, full accounts, and references to relevant philosophical literature, can be found in the references given.
5 Discussed fully in Morgan (2012, Chapter 3).
6 For a full account, see Morgan (2021).
7 See Morgan (2022), the introductory chapter of Morgan, Hajek, and Berry (2022) on 'narrative science'. Several other chapters in that book are particularly concerned with the narrative reasoning that goes on with models: John Beatty on reasoning backwards in evolutionary theory; Paula Olmos on the narratives of 'just-so reasoning'; John Huss on competing model-based explanations of extinctions; and Teru Miyake on earthquake narratives. For a 'handbook' account of the relationships of narratives to models, see Morgan (2024) and for economics examples, see Morgan and Stapleford (2023) and particularly Biddle (2023).
8 This case, and some comparable ones, are discussed in Morgan (2012, Chapter 6).
9 There is a vast literature about all this inside econometrics; for two papers which provide a small entry into some of the issues from a viewpoint compatible with this paper, see Morgan (1988) and Boumans and Morgan (2001).
10 Of course there is a huge literature on counterfactuals in philosophy of science; I restrict coverage here to their usage in fairly specific traditions in social science. The case of Fogel's counterfactual is analysed in detail in Morgan (2014).
11 I thank one of the referees for pushing me to think more clearly about the 'alternative' worlds and their role in comparisons that drive the judgements of both counterfactual and model worlds.
12 See for example papers by Quack and Herfeld (2023) and Skarbeck and Skarbeck (2023), and references therein.
13 Another site for these plausibility tests was judging the economic outcomes of the American Recovery and Reinvestment Act (ARRA) of 2010, see Ramey (2019b).

References

Bates, Robert. 1998. "The International Coffee Organisation: An International Institution." In Analytic Narratives, edited by Robert Bates et al., 194–230. Princeton: Princeton University Press.

Bates, Robert et al. 1998. *Analytic Narratives*. Princeton: Princeton University Press.

Beatty, John. 2022. "Narrative Solutions to a Common Evolutionary Problem." In *Narrative Science*, edited by Mary S. Morgan, Kim Hajek, and Dominic Berry, 405–23. Cambridge: Cambridge University Press.

Biddle, Jeff. 2023. "Narratives and Empirical Strategies in Zvi Griliches' Early Research." *Narrative in Economics: Historical Experiences*, edited by Mary S. Morgan and Thomas Stapleford, Special Issue of *History of Political Economy*, 55(3): 447–70.

Boumans, M.J. 1999. "Built-In Justification." In *Models as Mediators*, edited by Mary S. Morgan, and Margaret Morrison, 66–96. Cambridge: Cambridge University Press.

Boumans, Marcel, and Mary S Morgan. 2001. "*Ceteris Paribus* Conditions: Materiality and the Application of Economic Theories."*Journal of Economic Methodology* 8(1): 11–26.

Fogel, R.W. 1964. *Railroads and American Economic Growth: Essays in Econometric History*. Baltimore: Johns Hopkins Press.

Grüne-Yanoff, Till, and Philippe Verreault-Julien. 2021. "How-Possibly Explanations in Economics: Anything Goes?."*Journal of Economic Methodology* 28(1): 114–23.

Hawthorn, Geoffrey. 1991. *Plausible Worlds*. Cambridge: Cambridge University Press.

Hughes, R.I.G. 1999. "The Ising Model, Computer Simulation, and Universal Physics." In *Models as Mediators*, edited by Mary S. Morgan, and Margaret Morrison, 97–145. Cambridge: Cambridge University Press.

Humphrey, Thomas. 1996. "The Early History of the Box Diagram." *Federal Reserve Bank of Richmond Economic Quarterly* 82(1): 37–75.

Huss, John E. 2022. "Mass Extinctions and Narratives of Recurrence." In *Narrative Science*, edited by Mary S. Morgan, Kim Hajek, and Dominic Berry, 61–81. Cambridge: Cambridge University Press.

Klein, Ursula. 1999. "Techniques of Modelling and Paper-Tools in Classical Chemistry." In *Models as Mediators*, edited by Mary S. Morgan, and Margaret Morrison, 146–67. Cambridge: Cambridge University Press.

Knuuttila, Tarja. 2011. "Modelling and Representing: An Artefactual Approach to Model-Based Representation." *Studies in History and Philosophy of Science* 42: 262–71.

Koskinen, Rami. 2017. "Synthetic Biology and the Search for Alternative Genetic Systems: Taking How-Possibly Models Seriously." *European Journal of Philosophy of Science* 7: 493–506.

Malthus, Thomas Robert. 1798. *An Essay on the Principle of Population*, edited by Philip Appleman. Critical Edition. New York: Norton, 1976.

Mari, Luca, and Alessandro Giordani. 2014. "Modelling Measurement: Error and Uncertainty." In *Error and Uncertainty in Scientific Practice*, edited by Marcel Boumans, Giora Hon, and Arthur C. Petersen, 79–96. London: Pickering & Chatto.

McAfee, R. Preston. 1983. "American Economic Growth and the Voyage of Columbus." *American Economic Review* 73: 735–40.

Miyake, Teru. 2022. "Reasoning from Narratives and Models: Reconstructing the Tohoku Earthquake." In *Narrative Science: Reasoning, Representing and Knowing Since 1800*, edited by Mary S. Morgan, Kim M. Hajek, and Dominic J. Berry, 104–121. Cambridge: Cambridge University Press.

Morgan, Mary S. 1988. "Finding a Satisfactory Empirical Model." In *The Popperian Legacy in Economics*, edited by Neil. de Marchi, 199–211. Cambridge: Cambridge University Press. (Reprinted in *The Philosophy and Methodology of Economics* Ed: B. Caldwell (Edward Elgar, 1993).)

Morgan, Mary S. 2012. *The World in the Model: How Economists Work and Think*. Cambridge: Cambridge University Press.

Morgan, Mary S. 2014. "What If? Models, Fact and Fiction in Economics." *Journal of the British Academy* 2: 231–68.

Morgan, Mary S. 2021. "Narrative Inference with and without Statistics: Making Sense of Economic Cycles with Malthus and Kondratiev." *History of Political Economy* 53(SI): 113–38.

Morgan, Mary S. 2022. "Narrative: A General Purpose Technology for Science." In *Narrative Science: Reasoning, Representing and Knowing Since 1800*, edited by. Morgan, Mary. S, Kim M. Hajek and, Dominic J., Berry, 3–30. Cambridge: Cambridge University Press.

Morgan, Mary S., and Margaret Morrison, eds. 1999. *Models as Mediators*. Cambridge University Press., *Ideas in Context* Series.

Morgan, Mary S., Kim M. Hajek, and Dominic J. Berry, eds. 2022. *Narrative Science: Reasoning, Representing and Knowing Since 1800*. Cambridge: Cambridge University Press. Open Access.

Morgan, Mary S, and Tom Stapleford, eds. 2023. *Narrative in Economics: Historical Experiences* Special Issue of *History of Political Economy*, 55(3).

Morgan, Mary S, and Thomas A Stapleford. 2023. "Narrative in Economics: A New Turn on the Past." *Narrative in Economics: Historical Experiences*, edited by Mary S. Morgan and Tom Stapleford, 395–422. Special issue of *History of Political Economy*, 55(3).

Morrison, Margaret, and Mary S Morgan. 1999. "Models as Mediating Instruments." In *Models as Mediators*, edited by Mary S. Morgan, and Margaret Morrison, 10–37. Cambridge: Cambridge University Press.

Nordmann, Alfred. 2023. "Prospective Modeling" (this volume).

Olmos, Paula 2022. "Just-So What?". *Narrative Science: Reasoning, Representing and Knowing Since 1800*, edited by Mary S. Morgan, Kim M. Hajek, and Dominic J. Berry, 424–44. Cambridge: Cambridge University Press.

Orchard, Jacob, Valerie A Ramey, and Johannes F Wieland. 2023. "Micro-MPCs and Macro-Counterfactuals: The Case of the 2008 Rebates" Technical Report, Working Paper, June 27, 2023.

Quack, Alexandra, and Catherine Herfeld. 2023. "The Role of Narratives in Transferring Rational Choice Models." *History of Political Economy* 55(3): 549–76.

Ramey, Valerie. 2019a. "Fiscal Policy: Tax and Spending Multipliers in the United States." In *Evolution or Revolution: Rethinking Macroeconomic Policy After the Great Recession*, edited by Olivier Blanchard, and Lawrence H. Summers, MIT Press and the Petersen Institute for International Economics, 225–254. Cambridge, MA: MIT Press.

Ramey, Valerie. 2019b. "Ten Years After the Financial Crisis: What Have We Learned from the Renaissance in Fiscal Research." *Journal of Economic Perspectives* 33(2): 89–114.

Samuelson, P.A. 1939. "Interactions between the Multiplier Analysis and the Principle of Acceleration." *Review of Economics and Statistics* 21: 75–8.

Sjölin Wirling, Ylwa, and Til Grüne-Yanoff. 2024. "Through the Prism of Modal Epistemology: Perspective on Modal Modelling" (this volume)

Skarbeck, David, and Emily Skarbeck. 2023. "Analytic Narratives in Political Economy." *History of Political Economy* 55(4): 609–38.

Tetlock, Philip E., R. N. Lebow, and G. Parker, eds. 2008. *Unmaking the West*. Ann Arbor: University of Michigan Press.

13 Are Pandemics a Necessary Evil? The Role of Epidemiological and Immunological Models in Understanding and Preventing Diseases

Gregor P. Greslehner

13.1 Introduction

The COVID-19 pandemic challenged many things we (as a society, scientists, philosophers, or just ordinary humans) had taken for granted and forced us to deal with events and hardships we might have deemed somewhat theoretically possible, yet did not expect to experience ourselves directly. Perhaps one could find some comfort in thinking that something like this *had* to happen—out of natural necessity beyond our control—or on the contrary, be scared by the prospect that other pandemics will only be a matter of time. More interesting than personal reassurance, however, are the theoretical issues tied to the notions of the necessity and possibility of pandemics (both "intra- and inter-pandemic," i.e., during pandemics and when comparing different pandemic events). The questions I want to address here concern the modal nature of pandemics and endemics (as the difference lies mainly in how geographically widespread these events are, they will be treated in parallel here), and in particular pathogen-host coevolution. The aim is to develop a deeper understanding of how pandemics play out, as well as to question and examine some common beliefs about what is necessary and possible with respect to pandemics, virus evolution, host-pathogen coevolution, and evolutionary trade-offs.

Philosophy of science has not been deeply invested in topics about pandemics and epidemiology, especially before COVID-19. This is perhaps due to the fact that these branches of science differ quite a bit from the usual objects being studied by philosophers, without overarching theories and often with little experimental support for their claims (Broadbent 2013). Even its scientific status is sometimes called into question, despite its historical importance (Nature Communications Editorial 2018). However, the philosophy of epidemiology is gaining ground and attention (Kaplan and Valles 2021), a welcome and much-needed trend. Perhaps this is also why many of the evolutionary "just-so stories" about the trajectories of pandemics have been perpetuated relatively unchecked for so long.

DOI: 10.4324/9781003342816-17

However, this is also a good opportunity for thinking about what is possible in their trajectories, what constraints exist to make some events more likely than others, and whether some things might happen necessarily. Besides spinning more or less convincing narratives and the practice of epidemiological modeling, there is some unexplored territory here, which would not only allow us to see more clearly through the recent events of the COVID-19 pandemic but also learn more about evolutionary and coevolutionary trajectories in general. While there are historical contingencies to individual pandemics, making them somewhat random events, there are several trends and tendencies that can be extracted and discussed, which also apply outside of any pathological context. These can be used to understand, explain, and predict what is possible and necessary about pandemics and other evolutionary trends more generally. They also provide an interesting case of applying how-possibly models, as well as questioning the aims, limits, and scope of these kinds of models. In this chapter, I show that pandemics, with their unusual explanatory scope, provide an excellent case to argue for the epistemic value of how-possibly models from a different angle than the existing philosophical debates focused on other scientific fields.

13.2   Patterns of Pandemics

A lot of contingent factors contribute to how a particular pandemic plays out. However, certain trends, constraints, the law of large numbers, large populations, and long periods of time make some events almost certain. While politicians have repeatedly claimed to be surprised by new waves or the emergence of new virus variants throughout the COVID-19 pandemic, scientists were also surprised by the unpredictable steps the SARS-CoV-2 virus (Kupferschmidt 2022) took. Epidemiological models have often indeed predicted these events (in addition to somewhat accurate predictions about the number of cases in the near future to inform decision-making). Looking at the history of infectious diseases, one is inclined to see patterns of such diseases that render the surprise largely unwarranted, given the accuracy and predictive power of epidemiological models. In the midst of an ongoing pandemic, the lack of data is often the biggest reason for uncertainties and gaps in our knowledge, whereas general regularities can be modeled and predicted. Useful information can also be obtained from ancient pandemics, reaching back thousands of years, even sharing genetic information with the current culprit SARS-CoV-2 (Souilmi et al. 2021). This also shows the shared modeling structure between history and biology in epidemiological modeling.

Even before the COVID-19 pandemic started, many voices had warned that the next pandemic was not a question of "if" but "when." The World Health Organization entertained early surveillance systems for pathogens of concern and urged preparing for a hypothetical "disease X."[1] Once the difficult challenges of dealing with the catastrophic effects and aftermath of the most recent pandemic are over, the next one is inevitable. New zoonotic spillovers, like the latest emergence of monkeypox (Tomori and Ogoina 2022), appear to be frequent events. Hopefully next time, we will be better prepared, have improved our understanding of pandemic models—and will have learned some lessons in how to best mitigate the damage of a pandemic. Crucial lessons will be derived from the practice and reconsideration of how-possibly models. The lessons learned are also interesting from a philosophy of science perspective, not to mention their importance to making well-informed policy decisions (Grüne-Yanoff 2021; Van Basshuysen et al. 2021).

Pandemics have plagued humanity for thousands of years, as recorded by historical examples in ancient texts (Høiby 2021). History books are full of pandemic events and diseases that often heavily impacted entire populations (just think of the bubonic plague in Europe, smallpox and other diseases brought by colonizers to the Americas, the 1918 Spanish flu, AIDS, etc.). Understandably, dealing with these most unpleasant and harmful events has not been among the most popular to inspire philosophical reflections. While philosophical reflections on COVID-19 have recently been collected (Leonelli 2022), many open questions remain, in particular with respect to its modal aspects.

There seems to be a certain element of regularity, inevitability, and necessity to pandemics or endemics. Even though the COVID-19 pandemic had not been predicted as such (meaning the exact time of occurrence, strain of virus, health impact, duration, etc.), the certainty of one like it—and our lack of preparedness—had been voiced by many, including public health agencies, epidemiologists, and even individuals like Bill Gates. Of course, this has also fueled speculations and conspiracy theories (another necessary evil, it seems).[2]

What can we learn from observed patterns from the history of past pandemics? Analogical reasoning, a powerful inference tool in many respects, can be misleading when jumping to conclusions about one pandemic to another (Peckham 2020). We should not just wait and observe how the virus evolves, either. Besides close monitoring, there are some theoretical trends and limitations, as well as constraints on virus evolution that allow us to anticipate possible trajectories and how to respond to them (Starr et al. 2022), which can be captured in scientific models.

There are also evolutionary "countermeasures" for how organisms have found solutions to these reoccurring threats: adaptive immune systems (Sette and Crotty 2021). In ingenious ways, immune systems have evolved the capacity to deal with almost any possible structural and functional features of pathogens—at least in some individuals within a population and given enough time to react. Both on the population and the organism level, the coevolution with microorganisms and their potential pathogenic threat has allowed for the possibility of responding and adapting to them. Epidemiological and immunological models, in turn, allow us to understand and predict these patterns of possibility—and ideally, to react accordingly.

Advances in immunology and vaccine development reflect our advanced understanding, and sometimes our lack thereof. For instance, while COVID-19 vaccines have been successfully developed at record speed, vaccines against HIV have proven notoriously difficult. To some extent, this has been due to a reductionist perspective and lock-and-key models of molecular binding between antibodies and antigens (Greslehner 2021; Van Regenmortel 2019). The new mRNA technology promises to side-step some of these issues and allow the development of vaccines that were previously unsuccessful.

Evolution appears to have equipped us with the proper means to deal with pathogens and pandemics, but on a conceptual level, we still have much to learn. The way the COVID-19 pandemic went down does not inspire much hope that we might be better prepared next time. Both on a scholarly and public health level, efforts were quite chaotic, with an increasing part of the population not being well-informed and not supporting different measures. Perhaps a better perspective on what is possible and necessary with the COVID-19 pandemic—and pandemics in general—would also be helpful in improving this situation. Epidemiology is full of black boxes, but it is not always useful to try to fill in all the gaps (DiMarco 2021). Thus, a philosophy of science perspective on the role of epidemiological and immunological models is a promising start to better understanding pandemics and what is possible and necessary about them. In addition to the complex biology underlying pandemics, we must not overlook how pandemic models can impact human behavior—greatly affecting the actual outcomes of the modeled process in question, thus creating feedback loops. This lesson is particularly important when pandemic models are being used to advise policy decisions and serve society (Atkeson 2021; Saltelli et al. 2020; Van Basshuysen et al. 2021). However, the scope of this paper is on the biology side of things and the modal aspects of pandemic models to inform the aims and impact of applying these models further down the line. In any event, the nature and scope of how-possibly models can be advanced in this hitherto relatively untouched area of science.

13.3   What Is Necessary and Possible about Pandemics and Pandemic Trajectories?

As we have seen in the previous section, there are a number of trends in how pandemics unfold, but it is difficult to pin them down with certainty. Individual events and even regular occurrences defy any result strictly tying them to necessity and beyond just possibility in a meaningful way. Perhaps we might fare better not by looking at pandemic events as a whole, but by looking at some particular aspects that have repeatedly been claimed to involve some kind of necessity. First, the popular and prima facie plausible idea that pathogens necessarily evolve towards being less virulent; and second, as another potential silver lining—the necessity of pandemics to have an end. Even if there are several possibilities of *how* pandemics end, the fact that they ultimately do so necessarily, might give us some hope for current and future crises. But again, wishful thinking is not a proper guide in dealing with possibility and necessity here. The notions of possibility and necessity have been used in many different contexts, meaning a variety of different things. Let us narrow down the debate to more specific aspects in the following two subsections.

13.3.1 *Do Pathogens Necessarily Become Less Virulent?*

A long-lasting debate in epidemiology concerns host-parasite coevolution and the question of whether pathogens necessarily evolve towards lower virulence (Anderson and May 1982). These "models of parasite virulence" (Frank 1996) are still relevant for the virulence trajectory of SARS-CoV-2 (Alizon and Sofonea 2021). While the empirical evidence for a general trend towards less virulence is rather mixed (Acevedo et al. 2019), some general lessons are insightful for modeling epidemiological and immunological trade-offs like these. Viruses can indeed evolve towards being *more* virulent (Wertheim 2022), when natural selection favors a higher level of virulence for its transmission, as has for example been observed with some variants of HIV and SARS-CoV-2. The once popular transmission-virulence trade-off hypothesis or "law of declining virulence" (Méthot 2012) is severely limited (Alizon and Michalakis 2015) and definitely does not allow us to assume that viruses were to evolve necessarily towards being less virulent or pathogenic, as the interplay between epidemiology and evolution allows for a wider range of different outcomes, especially during the early phases of a pandemic. The theory behind this hypothesis does work quite well in some cases (Berngruber et al. 2013), but many questions remain open and a general, vague trend should not be considered to be necessary or a law. There are more interesting dynamics to extract and explain behind the underlying evolutionary processes, especially their different possible trajectories, which can be mapped out with predictions and models.

What makes some microbes pathogens in the first place? Being a pathogen is not an intrinsic property (Méthot and Alizon 2014) and most viruses do not cause harm (Casanova and Abel 2021). What lies at the root of most pathogenicity—thus making pandemic events possible in the first place—is that the virus and host have not had time to adapt to each other to find a balance that is acceptable for both parties (Alizon and Sofonea 2021; Méthot 2012). This is also why most pandemics happen after zoonotic spillover events, i.e., when a virus moves from one host to another, without having had time to adapt. There are many ways in which this can go wrong, and it is not clear in which directions things adapt until a viable trade-off is found. What makes hosts sick in these cases is often not so much the result of the activities of the virus itself, but the immune system's reaction to it. In the case of severe COVID-19, where likely the "cytokine storm," i.e., an overreaction by the immune system, causes much of the damage and symptoms (Moore and June 2020). The systemic and long-term properties are also a different and complicated story, which is why we still have much to learn about so-called "long COVID." The same applies to further virus variants, which we cannot expect to become less virulent by necessity. By considering different how-possibly models of virus adaptation and evolutionary trajectories, we can develop a much better grasp of the space of possibilities and range of different outcomes and scenarios of pandemics. An important distinction to be made here concerns the aims of modeling possibilities: are they supposed to predict certain outcomes that are considered to be the most likely—or, are they hypothetical projections of what would happen under certain (often deliberately unrealistic) assumptions (Fuller 2021; Schroeder 2021)? Conflating these two practices lies at the bottom of much criticism and disappointment in pandemic models. Unlike weather forecasts with the explicit aim to predict conditions (with a certain degree of uncertainty) that are regularly tested against data from reality, pandemics are much rarer events with incomplete data. Often, the best way to advance scientific knowledge is by exploring the range of possibilities without any regard to attempting to predict likely outcomes, but by exploring "what if" scenarios (Pastore y Piontti et al. 2019, 53).

13.3.2 *How Do Pandemics End?*

Taken together, these prospects about what is possible and necessary also allow us to end on a positive note: pandemic models tell us that and how pandemic events end. Also, SARS-CoV-2 eventually turned into an endemic virus that does not hold the human population of the entire globe hostage indefinitely. Pandemics might be a necessary evil, but they also necessarily end at some point. What is crucial is both the accuracy of epidemiological and immunological modeling and—perhaps even more importantly—the efficient communication of model-based inferences to decision-makers and the general public. Philosophy of epidemiology (Broadbent 2013) and philosophy of immunology (Pradeu 2020) offer important contributions to better understanding these issues as well as their modal aspects and their modeling practices.

How epidemics end is not just a biological issue, but also heavily influenced by other social and political criteria (Charters and Heitman 2021). In fact, there are likely multiple "endings" of a pandemic. Epidemiology and its philosophy are also well-advised to take these issues seriously and develop further definitions and criteria for discussing these issues. From a practical and individual point of view, this means again that we will not wake up one day and discover that an ongoing pandemic is finally over. It is a gradual and contextual process, a "transition to endemicity" (Antia and Halloran 2021), in which a stable equilibrium between pathogen and host is reached. Oftentimes, these are slow processes leading to an almost unnoticed level of the pathogenic virus circulating in the population, continuing to adapt and transform. Important lessons here can be drawn from how the seasonal H1N1 influenza might be traced back to the 1918 "Spanish flu" pandemic (Patrono et al. 2022). Perhaps framing these adaptive processes in the usual war metaphors of an "arms race" between the pathogen and the host's immune system is not a useful way to look at these things (Zach and Greslehner 2023).

The idea that we could achieve "victory" over COVID-19 with herd immunity or elimination seems to be impossible (Aschwanden 2021; Morens, Folkers, and Fauci 2022), given the unattainably high immunity threshold through vaccination or infection it would take to do so. Naturally, pandemics end when the pathogen causing it ceases to spread. It is unlikely—or even impossible—that SARS-CoV-2 can be completely eradicated; we can expect the virus to stick around and continue to evolve (Day et al. 2020). However, it is unclear to what extent it will remain a major public health issue. From experience with past pandemics, vaccination has often proved to be the most important tool to either eradicate pathogens completely, or at least reduce the levels of spreading within a population to small enough numbers, making it an endemic disease. This way, and considering how things have actually turned out, COVID-19 might eventually be comparable to the flu, but it is way too soon to make that call. With more advanced vaccinations and immunity in the population, endemicity appears to be the most likely outcome for COVID-19. However, there is again no necessity, new variants might force us back to square one. Taken together, the evolutionary trajectory for SARS-CoV-2 is not over at this point, and the best we have are estimates and models indicating that the path toward endemicity is to be expected. In addition to modeling how things could have possibly turned out, they can also be informative beyond the mere possibilities by assessing the likelihood of different outcomes within particular time frames and in the long run.

One way or another, pandemics do end necessarily: either through eradication of the pathogen, extinction of hosts, or by becoming endemic. Somewhat fortunately, extreme cases are rather the exception here, and the adaptive potential of natural selection clearly leans towards reaching a trade-off level of endemicity that we can also expect for COVID-19. The good news, then, is that also this pandemic will be over eventually. The question remains, however, what path has been taken to get there and how to proceed from there in order to be better prepared for future pandemics. A proper understanding of the evolutionary and immunological mechanisms and models underlying the pathogen and disease is crucial for mitigating the damage that might occur. How-possibly models do not just serve to illustrate various possibilities here in an abstract way, as they allow for the assessment and preparation of various different outcomes and scenarios. The value of these models often does not lie in their predictive success, but in considering rare "what if scenarios" that could act like "bottlenecks," i.e., low probability events that one can hypothetically let play out and prepare for the devastating effects such events might have despite their low probability. During the COVID-19 pandemic, it was often practiced to model what would happen if no protective measures were taken at all, as compared to if everyone perfectly adhered to safety measures. These were deliberately unrealistic assumptions from extreme scenarios to mark out the range of likely scenarios in between. Overlooking these different aims of pandemic models—unconditional predictions versus conditional projections—often gave rise to misunderstandings and misinterpretations (Schroeder 2021).

This also entails realistic expectations about what vaccines can and cannot do: "We're Asking the Impossible of Vaccines: complete protection against infection has long been hailed as the holy grail of vaccination. It might simply be unachievable" (Wu 2021). Relying exclusively on the success of vaccines appears therefore to be insufficient. Other public health measures have been and will be important. The fact that preventive measures are effective is also shown by the fact that the attempts to mitigate COVID-19 have also worked towards eliminating some lineages of influenza (Dhanasekaran et al. 2022). It is clear that "the covid-19 pandemic will end with public health tools, not clinical ones" (Karan 2022).

13.4 What the Pandemic and Immunological Models (Do Not) Tell Us

Since there are no general heuristics for inferring causality or predicting in epidemiology, a good approach might be to ask ourselves, "what could possibly go wrong?" (Broadbent 2011) and prepare accordingly (also for the modeling practice, for which the phrase is intended as a shorthand). In addition to just explaining how some specific pandemic phenomenon can possibly be explained, the predictive character of how things might turn out in the future, as well as projecting how things might play out under specific hypothetical circumstances, is crucial. Epidemiological models inform decisions, with different merits and flaws (Iranzo and Pérez-González 2021). The fact that future pandemics are not only possible, but a "necessary evil," is definitely one that we cannot justify or afford to ignore.

The Lancet has put forward an extensive list of "lessons for the future from the COVID-19 pandemic" (Sachs et al. 2022). There are indeed many things to learn and consider to be better prepared for the future (Krammer 2020). From a philosophical point of view, many aspects of the pandemic and its ramifications have already been put forward (Leonelli 2022), and many more can be expected. For the philosophical interest of this chapter, considering past and future possibilities about pandemics and their impact on how to think about evolution has left us with a couple of general insights, as well as some lessons about modeling practices in different fields. In epidemiology and other theoretical work on pandemics, the practices and aims of how-possibly models stand out as a means to provide general lessons for other fields in which these models play a role as well. Among the issues discussed here, are the differences between, e.g., compartmental models and agent-based models, predictions, explanations, and interventions (Iranzo and Pérez-González 2021).

Philosophers of science might also consider working more on the issue of prediction, not merely as an afterthought to explanation (Douglas 2009), but to put more emphasis on description for understanding (Findl and Suárez 2021). With more data, COVID-19 might become more predictable (Callaway 2022), but the conceptual tools to think about pandemics and understand what is necessary and possible is a theoretical problem, where interdisciplinary collaboration—including philosophy—is essential (Laplane et al. 2019).

For instance, immune evasion appears to be a necessary path for the virus to stay with a host. This insight also allows us to identify targets for vaccines and therapies, like the spike protein for the COVID-19 vaccines. Evolution can explore several evolutionary trajectories in parallel, thus providing multiple moving targets. It is never clear *a priori* that a virus being more virulent is more or less advantageous, or equally, that a host having a less tolerant immune system is a disadvantage (Weber et al. 2022). Importantly, there are always coevolutionary selection processes at work. Evolutionary trade-offs will be found along these fluid evolutionary landscapes. These landscapes are not set in stone, we can influence them with our behavior and public health policies. To be able to do so requires some understanding of the constraints, likely outcomes, possibilities, and necessities of the different outcomes at play. It is not sufficient to merely think about and work with known facts that characterize actualized systems and processes—modeling the possible is an invaluable tool when it comes to pandemics.

Rather than a "new normal," we might instead think about the "new necessary" (Dace et al. 2021), because we definitely have to prepare ourselves for future pandemics—our collective responsibility. A first step is to treat pandemics "as a natural evolutionary phenomenon" (Lederberg 1988) and not simply believe hypotheses about trajectories that sound plausible (or something we wish for). We should also acknowledge what we do not know and cannot know, but we can be aware of a high number of possibilities and prepare for the most likely and most impactful ones. Doing so also means to include the evolutionary, ecological, social, and economic aspects of how and why pandemics occur (Boem 2021). Put concisely, "epidemiology is ecosystem science" (Lee 2021). Since pandemics are arguably "necessary evils" after all, we can at least limit the chances for the virus to evolve. As we have seen above, there is absolutely no necessity for the virus to become "milder" and evolve towards being less harmful. While such dangerous "bottleneck" events have low probabilities, considering their possibility is crucial for taking preventive measures. The best way to prepare for these possibilities is still to reduce transmission and the likelihood of more pathogenic variants emerging. At the same time, spillover events can also be limited, e.g., by how we treat livestock and other situations where humans and animals come into close contact. On the other hand, let's not forget Orgel's second rule: "evolution is cleverer than you are" (Dunitz and Joyce 2013). Taken together, we might not be able to outsmart the virus, but we can limit the number of chances there are for these events to take place. They are ultimately not preventable altogether, but being aware of their possibilities, and modeling them to understand and prepare for them, puts us in a good position to deal with pandemics in the future. Hopefully, we will learn important lessons from the history of pandemics.

13.5   Advancing the Debate on How-Possibly Models

In several recent philosophical accounts, how-possibly models have been discussed in opposition to how-actually models. The epistemic value and aims of the former have frequently been a matter of debate (Grüne-Yanoff and Verreault-Julien 2021). Different kinds of how-possibly models and their explanatory and epistemic aims have been discussed (Reydon 2012; Verreault-Julien 2019). Pandemic modeling can be an unusual but informative setting for this kind of modeling practice, allowing us to reassess the epistemic merits of how-possibly models. Unlike in many other cases where how-possibly models might have a provisional character, until a full explanation in the form of so-called how-actually models based on known causes can be found (some ideally in the form of identifying the underlying mechanism, see e.g., Craver and Darden 2013), the modeling practice and aims in epidemiology seem to be different. Here, the goal is not to provide and explain the exact and actual mechanism of disease transmission in a population (the cruise ship *Diamond Princess* provided such a small and relatively controlled environment in which the mechanisms of disease transmission of COVID-19 could be modeled (Azimi et al. 2021)). Sometimes, that might be the aim in retrospect, or to test safety measures in different settings. Most pandemic models and predictions, however, do not target how things actually transpire or what necessarily had to happen. Rather, it is about outlining what is possible, and what is likely, predicting different outcomes, and suggesting measures that should be taken accordingly—while being fully aware of the modal aspects. Despite targeting possibilities rather than the facts of the matter, these pandemic models do have epistemic value and use beyond the 'usual suspects' of scientific explanation. Thus, epidemiological models provide a prime object for philosophers studying modeling practices about possibilities, refreshing the debate with new application cases.

Forecasting the weather, modeling the climate, historical events, or evolutionary trajectories in biology—all these scenarios are about estimating and predicting what is possible. Capturing the current temperature and temperature distribution at any given point in time does not require weather models, but assessing likely scenarios and trends from what is known at the time does. The accuracy and success of these how-possibly models prove they are not to be regarded as something preliminary that can and should ultimately be replaced with a how-actually model; some epistemic gaps are inevitable and taken into account by design. In a way, these are higher-level explanatory and modeling aims in comparison to explaining a specific mechanism or phenomenon. But that does not undermine the epistemic justification for this kind of modeling. On the contrary, how-possibly models are *the* explanatorily relevant tool for this kind of scenario. Epidemiology is less concerned with understanding and explaining actual phenomena, than it is with modeling possible scenarios as its part and parcel.

My modest aim in this contribution is to point out that modeling practices like those in epidemiology will be informative in advancing the philosophical debate. Some features of pandemic models are unlike those usually discussed in other fields of biology, economics, or other fields of science. Similar to Rami Koskinen's discussion of synthetic biology (Koskinen 2017), I believe that considering possibilities has genuine epistemic merit in biology, including other areas of the life sciences that are considering future trends in evolution in general, or what is possible and necessary about pandemics in particular. Considering evolutionary possibilities, trends, and constraints will be much more informative than ruminating about the necessity of pandemics—which is a somewhat grim truth, but by itself not that interesting a fact requiring further investigation. Beyond being a "necessary evil" to be reckoned with and prepared for, there are more insightful modeling practices at play concerning pandemics, shedding new light on how-possibly models in epidemiology and their epistemic role in general.

13.6   Conclusion

Single events can hardly ever be claimed to have occurred necessarily—at least not with a detailed account of where, when, how, and who. So, it cannot be said that the COVID-19 pandemic was a necessary event in how exactly things went down. However, some tendencies made the occurrence of a pandemic event like this (and more in the near future) an almost certainty—as has been predicted by many researchers and epidemiologists. There were, of course, some uncertainties as to which pathogen would be responsible, its mode and rate of transmission, etc. But the fact that a pandemic was to be expected, and that our preparations left much to be desired, made the predicament worse than it needed to be when the pandemic did in fact occur. Thus, while the occurrence of the event itself might be considered a necessary evil, the details of how things happened, and how our preparation and response looked were not. The best way to learn from past pandemics is to take the lessons of what did and could have happened seriously, not just hope for the best. Despite their wide range of possibilities, pandemics are events that follow certain patterns and have dynamics that can be modeled and predicted. Accurate assessment of different outcomes is difficult but possible with sufficient data. Due to the high-level nature of pandemic events and the scarcity of data—especially in the middle of such an event taking place—the aim of modeling here might not be to explain how things happened (or had to happen), but to provide a spectrum of possibilities one should consider and prepare for.

These trends, tendencies, how-possibly models, as well as constraints on the evolutionary trajectories that pathogens and hosts can take allow us to make predictions about and understand the dynamics of pandemics (Tenthorey, Emerman and Malik 2022). This is good news, as the current practices in agriculture, animal treatment, and climate catastrophe increase the likelihood of more pandemics in the future (Carlson et al. 2022).

Within the evolutionary dynamics of pathogen-host adaptation, a common assumption used to be that the pathogen will necessarily evolve towards being less pathogenic, i.e., becoming "mild." However, as the explanation of the underlying evolutionary mechanisms and trade-offs shows, this can go either way. While there are often good reasons for the pathogen to reduce the severity of disease within a host, this is by no means a necessity. Several diseases show that sometimes the host can and will be killed in extreme cases.

While it remains unpredictable what exactly comes next, both for COVID-19 and unknown future pandemics, there are no laws, no certainties—we do have models that can successfully predict that some things are more likely than others. Epidemiological and immunological models, while far from being perfect, are useful tools (Holmdahl and Buckee 2020) and sometimes do allow accurate predictions. While single events that depend on a lot of contingent factors can rarely be said to be necessary in a particular instance, some evolutionary constraints and dynamics permit theorizing about what is possible and necessary—all of which can be modeled in very informative ways.

Thus, to answer the title question: the fact that pandemics occur might be considered a necessary evil we have to deal with. Unlike some opponents of the measures against COVID-19, however, "learning to live" with a pandemic does not mean ignoring it and prematurely behaving as if it were over when it very much is not. As we have discussed at length above, there is no guarantee that pathogenic viruses will always evolve towards being less harmful. The good news is that there is no selective pressure for becoming more harmful either; natural selection is open to both possibilities in the long run, as long as neither the pathogen's nor host's fitness is affected to the degree of extinction. However, some conditions make outbreaks possible and more likely. Knowing what is possible, impossible, or necessary about virus evolution is an essential piece of the puzzle of understanding and modeling pandemics. This endeavor will require multiple disciplines working together. As in many other areas of science, there are good reasons to expect philosophy to be able to make important contributions (Laplane et al. 2019).

Acknowledgments

Many thanks to Rami Koskinen and Till Grüne-Yanoff for their helpful comments and suggestions.

This project has received funding from the European Research Council (ERC) under the European Union's Horizon 2020 research and innovation program (grant agreement no. 818772).

This research was funded in whole or in part by the Austrian Science Fund (FWF) [10.55776/COE3]. For open access purposes, the author has applied a CC BY public copyright license to any author-accepted manuscript version arising from this submission.

Notes

1 https://www.who.int/activities/prioritizing-diseases-for-research-and-development-in-emergency-contexts [access: 30.09.2022]
2 The spread of conspiracy theories has itself been modelled with epidemiological models (Kauk, Kreysa, and Schweinberger 2021). Thanks to Till Grüne-Yanoff for pointing this out to me.

References

Acevedo, Miguel A, Forrest P Dillemuth, Andrew J Flick, Matthew J Faldyn, and Bret D Elderd. 2019. "Virulence-driven Trade-offs in Disease Transmission: A Meta-analysis." *Evolution* 73(4): 636–47. https://doi.org/10.1111/evo.13692.

Alizon, Samuel, and Yannis Michalakis. 2015. "Adaptive Virulence Evolution: The Good Old Fitness-Based Approach." *Trends in Ecology & Evolution* 30(5): 248–54. https://doi.org/10.1016/j.tree.2015.02.009.

Alizon, Samuel, and Mircea T Sofonea. 2021. "SARS-CoV-2 Virulence Evolution: Avirulence Theory, Immunity and Trade-offs." *Journal of Evolutionary Biology* 34(12): 1867–77. https://doi.org/10.1111/jeb.13896.

Anderson, R. M, and R. M May. 1982. "Coevolution of Hosts and Parasites." *Parasitology* 85(2): 411–26. https://doi.org/10.1017/S0031182000055360.

Antia, Rustom, and M. Elizabeth Halloran. 2021. "Transition to Endemicity: Understanding COVID-19."*Immunity* 54(10): 2172–76. https://doi.org/10.1016/j.immuni.2021.09.019.

Aschwanden, Christie. 2021. "Five Reasons Why COVID Herd Immunity Is Probably Impossible." *Nature* 591(7851): 520–22. https://doi.org/10.1038/d41586-021-00728-2.

Atkeson, Andrew. 2021. "COVID's Lessons for Future Modeling of Pandemics." *NBER Reporter*, no. 1, 10–15.

Azimi, Parham, Zahra Keshavarz, Jose Guillermo Cedeno Laurent, Brent Stephens, and Joseph G Allen. 2021. "Mechanistic Transmission Modeling of COVID-19 on the *Diamond Princess* Cruise Ship Demonstrates the Importance of Aerosol Transmission." *Proceedings of the National Academy of Sciences* 118(8): e2015482118. https://doi.org/10.1073/pnas.2015482118.

Berngruber, Thomas W, Rémy Froissart, Marc Choisy, and Sylvain Gandon. 2013. "Evolution of Virulence in Emerging Epidemics." Edited by François Balloux." *PLoS Pathogens* 9(3): e1003209. https://doi.org/10.1371/journal.ppat.1003209.

Boem, Federico. 2021. "Modeling Pandemic: Proximate and Ultimate Causes." *Argumenta* 6(2): 57–77. https://doi.org/10.14275/2465-2334/202113.BOE.

Broadbent, Alex. 2011. "What Could Possibly Go Wrong? — A Heuristic for Predicting Population Health Outcomes of Interventions." *Preventive Medicine* 53(4–5): 256–59. https://doi.org/10.1016/j.ypmed.2011.09.009.

Broadbent, Alex. 2013. *Philosophy of Epidemiology*. London: Palgrave Macmillan.

Callaway, Ewen. 2022. "Are COVID Surges Becoming More Predictable? New Omicron Variants Offer a Hint." *Nature* 605(7909): 204–6. https://doi.org/10.1038/d41586-022-01240-x.

Carlson, Colin J, Gregory F Albery, Cory Merow, Christopher H Trisos, Casey M Zipfel, Evan A Eskew, Kevin J Olival, Noam Ross, and Shweta Bansal. 2022. "Climate Change Increases Cross-Species Viral Transmission Risk." *Nature* 607(7919): 555–62. https://doi.org/10.1038/s41586-022-04788-w.

Casanova, Jean-Laurent, and Laurent Abel. 2021. "Lethal Infectious Diseases as Inborn Errors of Immunity: Toward a Synthesis of the Germ and Genetic Theories." *Annual Review of Pathology: Mechanisms of Disease* 16(1): 23–50. https://doi.org/10.1146/annurev-pathol-031920-101429.

Charters, Erica, and Kristin Heitman. 2021. "How Epidemics End." *Centaurus* 63(1): 210–24. https://doi.org/10.1111/1600-0498.12370.

Craver, Carl F, and Lindley Darden. 2013. *In Search of Mechanisms: Discoveries Across the Life Sciences*. Chicago: University of Chicago Press.

Dace, Hermione, Brianna Miller, Rania Ramli, Daniel Sleat, Eva Thorne, and Ryan Wain. 2021. *The New Necessary: How We Future-Proof for the Next Pandemic*. London: Tony Blair Institute for Global Change. https://institute.global/policy/new-necessary-how-we-future-proof-next-pandemic

Day, Troy, Sylvain Gandon, Sébastien Lion, and Sarah P Otto. 2020. "On the Evolutionary Epidemiology of SARS-CoV-2." *Current Biology* 30(15): R849–57. https://doi.org/10.1016/j.cub.2020.06.031.

Dhanasekaran, Vijaykrishna, Sheena Sullivan, Kimberly M Edwards, Ruopeng Xie, Arseniy Khvorov, Sophie A Valkenburg, Benjamin J Cowling, and Ian G Barr. 2022. "Human Seasonal Influenza under COVID-19 and the Potential Consequences of Influenza Lineage Elimination." *Nature Communications* 13(1): 1721. https://doi.org/10.1038/s41467-022-29402-5.

DiMarco, Marina. 2021. "Wishful Intelligibility, Black Boxes, and Epidemiological Explanation." *Philosophy of Science* 88(5): 824–34. https://doi.org/10.1086/715222.

Douglas, Heather E. 2009. "Reintroducing Prediction to Explanation." *Philosophy of Science* 76(4): 444–63.

Dunitz, Jack D., and Gerald F. Joyce. 2013. "Leslie Eleazer Orgel. 12 January 1927— 27 October 2007." *Biographical Memoirs of Fellows of the Royal Society* 59: 277–89. http://doi.org/10.1098/rsbm.2013.0002.

Findl, Johannes, and Javier Suárez. 2021. "Descriptive Understanding and Prediction in COVID-19 Modelling." *History and Philosophy of the Life Sciences* 43(4): 107. https://doi.org/10.1007/s40656-021-00461-z.

Frank, Steven A. 1996. "Models of Parasite Virulence." *The Quarterly Review of Biology* 71(1): 37–78.

Fuller, Jonathan. 2021. "What Are the COVID-19 Models Modeling (Philosophically Speaking)?." *History and Philosophy of the Life Sciences* 43(2): 47. https://doi.org/10.1007/s40656-021-00407-5.

Greslehner, Gregor P. 2021. "The Vaccinologist's 'Dirty Little Secret': A Better Understanding of Structure-Function Relationships of Viral Immunogens Might Advance Rational HIV Vaccine Design." *Archives of Virology* 166(5): 1297–303. https://doi.org/10.1007/s00705-021-04982-7.

Grüne-Yanoff, Till. 2021. "Choosing the Right Model for Policy Decision-Making: The Case of Smallpox Epidemiology." *Synthese* 198(S10): 2463–84. https://doi.org/10.1007/s11229-018-1827-3.

Grüne-Yanoff, Till, and Philippe Verreault-Julien. 2021. "How-Possibly Explanations in Economics: Anything Goes?." *Journal of Economic Methodology* 28(1): 114–23. https://doi.org/10.1080/1350178X.2020.1868779.

Høiby, Niels. 2021. "Pandemics: Past, Present, Future: That Is Like Choosing between Cholera and Plague." *APMIS* 129(7): 352–71. https://doi.org/10.1111/apm.13098.

Holmdahl, Inga, and Caroline Buckee. 2020. "Wrong but Useful — What Covid-19 Epidemiologic Models Can and Cannot Tell Us." *New England Journal of Medicine* 383(4): 303–5. https://doi.org/10.1056/NEJMp2016822.

Iranzo, Valeriano, and Saúl Pérez-González. 2021. "Epidemiological Models and COVID-19: A Comparative View." *History and Philosophy of the Life Sciences* 43(3): 104. https://doi.org/10.1007/s40656-021-00457-9.

Kaplan, Jonathan Michael, and Sean A Valles. 2021. "Reflecting on What Philosophy of Epidemiology Is and Does, as the Field Comes into Its Own: Introduction to the Special Issue on Philosophy of Epidemiology." *Synthese* 198(S10): 2383–92. https://doi.org/10.1007/s11229-019-02252-3.

Karan, Abraar. 2022. "The Covid-19 Pandemic Will End with Public Health Tools, Not Clinical Ones." *BMJ*. https://doi.org/10.1136/bmj.o1561.

Kauk, Julian, Helene Kreysa, and Stefan R Schweinberger. 2021. "Understanding and Countering the Spread of Conspiracy Theories in Social Networks: Evidence from Epidemiological Models of Twitter Data." *PLOS ONE* 16(8): e0256179. https://doi.org/10.1371/journal.pone.0256179.

Koskinen, Rami. 2017. "Synthetic Biology and the Search for Alternative Genetic Systems: Taking How-Possibly Models Seriously." *European Journal for Philosophy of Science* 7(3): 493–506. https://doi.org/10.1007/s13194-017-0176-2.

Krammer, Florian. 2020. "Pandemic Vaccines: How Are We Going to Be Better Prepared Next Time?." *Med* 1(1): 28–32. https://doi.org/10.1016/j.medj.2020.11.004.

Kupferschmidt, Kai. 2022. "As Omicron Rages on, Virus' Path Remains Unpredictable." *Science* 377(6604): 352–53. https://doi.org/10.1126/science.ade0166.

Laplane, Lucie, Paolo Mantovani, Ralph Adolphs, Hasok Chang, Alberto Mantovani, Margaret McFall-Ngai, Carlo Rovelli, Elliott Sober, and Thomas Pradeu. 2019. "Opinion: Why Science Needs Philosophy." *Proceedings of the National Academy of Sciences* 116(10): 3948–52.

Lederberg, Joshua. 1988. "Pandemic as a Natural Evolutionary Phenomenon." *Social Research* 55(3): 343–59.

Lee, Keekok. 2021. "Epidemiology Is Ecosystem Science." *Synthese* 198(S10): 2539–67. https://doi.org/10.1007/s11229-019-02129-5.

Leonelli, Sabina. 2022. "Introduction: Biomedical Knowledge in a Time of COVID-19." *History and Philosophy of the Life Sciences* 44(3): 40. https://doi.org/10.1007/s40656-022-00520-z.

Méthot, Pierre-Olivier. 2012. "Why Do Parasites Harm Their Host? On the Origin and Legacy of Theobald Smith's 'Law of Declining Virulence'—1900-1980." *History and Philosophy of the Life Sciences* 34(4): 561–601.

Méthot, Pierre-Olivier, and Samuel Alizon. 2014. "What Is a Pathogen? Toward a Process View of Host-Parasite Interactions." *Virulence* 5(8): 775–85. https://doi.org/10.4161/21505594.2014.960726.

Moore, John B, and Carl H June. 2020. "Cytokine Release Syndrome in Severe COVID-19." *Science* 368(6490): 473–74. https://doi.org/10.1126/science.abb8925.

Morens, David M, Gregory K Folkers, and Anthony S Fauci. 2022. "The Concept of Classical Herd Immunity May Not Apply to COVID-19." *The Journal of Infectious Diseases* 226(2): 195–98. https://doi.org/10.1093/infdis/jiac109.

Nature Communications Editorial. 2018. "Epidemiology Is a Science of High Importance." *Nature Communications* 9(1): 1703. https://doi.org/10.1038/s41467-018-04243-3.

Pastore y Piontti, Ana, Nicola Perra, Luca Rossi, Nicole Samay, and Alessandro Vespignani. 2019. *Charting the Next Pandemic: Modeling Infectious Disease Spreading in the Data Science Age*. Cham: Springer International Publishing. https://doi.org/10.1007/978-3-319-93290-3.

Patrono, Livia V, Bram Vrancken, Matthias Budt, Ariane Düx, Sebastian Lequime, Sengül Boral, and M. Thomas P Gilbert, et al. 2022. "Archival Influenza Virus Genomes from Europe Reveal Genomic Variability During the 1918 Pandemic." *Nature Communications* 13(1): 2314. https://doi.org/10.1038/s41467-022-29614-9.

Peckham, Robert. 2020. "COVID-19 and the Anti-Lessons of History." *The Lancet* 395(10227): 850–52. https://doi.org/10.1016/S0140-6736(20)30468-2.

Pradeu, Thomas. 2020. *Philosophy of Immunology*. Cambridge: Cambridge University Press.

Reydon, Thomas A.C. 2012. "How-Possibly Explanations as Genuine Explanations and Helpful Heuristics: A Comment on Forber." *Studies in History and Philosophy of Science Part C: Studies in History and Philosophy of Biological and Biomedical Sciences* 43(1): 302–10. https://doi.org/10.1016/j.shpsc.2011.10.015.

Sachs, Jeffrey D, Salim S Abdool Karim, Lara Aknin, Joseph Allen, Kirsten Brosbøl, Francesca Colombo, and Gabriela Cuevas Barron, et al. 2022. "The *Lancet* Commission on Lessons for the Future from the COVID-19 Pandemic." *The Lancet.* 400(10359): 1224–80. https://doi.org/10.1016/S0140-6736(22)01585-9.

Saltelli, Andrea, Gabriele Bammer, Isabelle Bruno, Erica Charters, Monica Di Fiore, Emmanuel Didier, and Wendy Nelson Espeland, et al. 2020. "Five Ways to Ensure That Models Serve Society: A Manifesto." *Nature* 582 (7813): 482–84. https://doi.org/10.1038/d41586-020-01812-9.

Schroeder, S. Andrew. 2021. "How to Interpret Covid-19 Predictions: Reassessing the IHME's Model." *Philosophy of Medicine* 2(1): 1–7. https://doi.org/10.5195/pom.2021.43.

Sette, Alessandro, and Shane Crotty. 2021. "Adaptive Immunity to SARS-CoV-2 and COVID-19." *Cell* 184(4): 861–80. https://doi.org/10.1016/j.cell.2021.01.007.

Souilmi, Yassine, M. Elise Lauterbur, Ray Tobler, Christian D Huber, Angad S Johar, Shayli Varasteh Moradi, Wayne A Johnston, Nevan J Krogan, Kirill Alexandrov, and David Enard. 2021. "An Ancient Viral Epidemic Involving Host Coronavirus Interacting Genes More than 20,000 Years Ago in East Asia." *Current Biology* 31(16): 3504–3514.e9. https://doi.org/10.1016/j.cub.2021.05.067.

Starr, Tyler N, Allison J Greaney, William W Hannon, Andrea N Loes, Kevin Hauser, Josh R Dillen, and Elena Ferri, et al. 2022. "Shifting Mutational Constraints in the SARS-CoV-2 Receptor-Binding Domain During Viral Evolution." *Science* 377 (6604): 420–24. https://doi.org/10.1126/science.abo7896.

Tenthorey, Jeannette L, Michael Emerman, and Harmit S Malik. 2022. "Evolutionary Landscapes of Host-Virus Arms Races." *Annual Review of Immunology* 40(1): 271–94. https://doi.org/10.1146/annurev-immunol-072621-084422.

Tomori, Oyewale, and Dimie Ogoina. 2022. "Monkeypox: The Consequences of Neglecting a Disease, Anywhere." *Science* 377(6612): 1261–63. https://doi.org/10.1126/science.add3668.

Van Basshuysen, Philippe, Lucie White, Donal Khosrowi, and Mathias Frisch. 2021. "Three Ways in Which Pandemic Models May Perform a Pandemic." *Erasmus Journal for Philosophy and Economics* 14(1): 110–27. https://doi.org/10.23941/ejpe.v14i1.582.

Van Regenmortel, Marc H. V. 2019. *HIV/AIDS: Immunochemistry, Reductionism and Vaccine Design*. Cham: Springer.

Verreault-Julien, Philippe. 2019. "How Could Models Possibly Provide How-Possibly Explanations?." *Studies in History and Philosophy of Science Part A* 73(February): 22–33. https://doi.org/10.1016/j.shpsa.2018.06.008.

Weber, Jesse N, Natalie C Steinel, Foen Peng, Kum Chuan Shim, Brian K Lohman, Lauren E Fuess, Swapna Subramanian, Stephen P De Lisle, and Daniel I Bolnick. 2022. "Evolutionary Gain and Loss of a Pathological Immune Response to Parasitism." *Science* 377(6611): 1206–11. https://doi.org/10.1126/science.abo3411.
Wertheim, Joel O. 2022. "When Viruses Become More Virulent." *Science* 375(6580): 493–94. https://doi.org/10.1126/science.abn4887.
Wu, Katherine J. 2021. "We're Asking the Impossible of Vaccines." The Atlantic. September 9, 2021. https://www.theatlantic.com/science/archive/2021/09/sterilizing-immunity-myth-covid-19-vaccines/620023/.
Zach, Martin, and Gregor P Greslehner. 2023. "Understanding Immunity: An Alternative Framework beyond Defense and Strength." *Biology & Philosophy* 38(1): 7. https://doi.org/10.1007/s10539-023-09893-2.

Index

For Product Safety Concerns and Information please contact our EU
representative GPSR@taylorandfrancis.com
Taylor & Francis Verlag GmbH, Kaufingerstraße 24, 80331 München, Germany